T0281068

To my wife, Jan

David M. Bressoud
Mathematics and Computer Science Department
Macalester College
1600 Grand Avenue
Saint Paul, MN 55105
USA
bressoud@macalester.edu

Mathematics Subject Classification (2000): 26B12, 00A35

Figure 1.12 is reprinted from Kepler's *Mysterium Cosmographicum* as reprinted by C.H. Beck Verlag in Volume 8 of Johannes Kepler: *Gesammelte Werke*, Munich, 1963.

Library of Congress Cataloging-in-Publication Data
Bressoud, David M., 1950–
    Second year calculus: from celestial mechanics to special
relativity / David M. Bressoud.
        p.   cm. — (Undergraduate texts in mathematics. Readings in
    mathematics)
    Includes bibliographical references and index.
    ISBN 0-387-97606-X. – ISBN 3-540-97606-X
    1. Calculus   I. Title.   II. Series.
    QA303.B88257 : 1991   91-3846

Printed on acid-free paper.

Photocomposed copy prepared from the author's LaTeX files.
Printed and bound by R.R. Donnelley and Sons, Harrisonburg, VA.
Printed in the United States of America.

9 8 7 6 5 4 (Corrected fourth printing, 2001)

ISBN 0-387-97606-X
ISBN 3-540-97606-X   SPIN 10794520

Springer-Verlag   New York Berlin Heidelberg
A member of BertelsmannSpringer Science+Business Media GmbH

David M. Bressoud

Second Year Calculus
From Celestial Mechanics to Special Relativity

With 98 Illustrations

 Springer

Graduate Texts in Mathematics

Readings in Mathematics

Ebbinghaus/Hermes/Hirzebruch/Koecher/Mainzer/Neukirch/Prestel/Remmert: *Numbers*
Fulton/Harris: *Representation Theory: A First Course*
Remmert: *Theory of Complex Functions*

Undergraduate Texts in Mathematics

Readings in Mathematics

Anglin: *Mathematics: A Concise History and Philosophy*
Anglin/Lambek: *The Heritage of Thales*
Bressoud: *Second Year Calculus*
Hairer/Wanner: *Analysis by Its History*
Hämmerlin/Hoffmann: *Numerical Mathematics*
Isaac: *The Pleasures of Probability*
Samuel: *Projective Geometry*

Undergraduate Texts in Mathematics

Readings in Mathematics

Editors

S. Axler
F.W. Gehring
P.R. Halmos

Springer
New York
Berlin
Heidelberg
Barcelona
Hong Kong
London
Milan
Paris
Singapore
Tokyo

Preface

This is a textbook on multivariate and vector calculus, but it is also a story. It is a story that begins and ends with revolutions in our understanding of the physical world in which we live. It carries us from the birth of the mechanized view of the world in Isaac Newton's *Mathematical Principles of Natural Philosophy*, in which mathematics becomes the ultimate tool for modeling physical reality, to the dawn of a radically new and often counterintuitive age in Albert Einstein's Special Theory of Relativity, in which it is the mathematical model that suggests new aspects of that reality.

This is also a chance to have some fun with mathematics. Here is the promised reward of being able to do something interesting and useful with the calculus that you have mastered in the past year. A colleague once told me of a high school experience in which he had the opportunity to participate in a special class that promised to reveal the basic tools of mathematics. Visions of orbit calculations and other mathematical applications danced through his head in anticipation. He was disappointed when the course turned out to be set theory. Presented here is the course for which he was hoping. We shall compute orbits and rocket trajectories, see how to model flows and force fields, derive the laws of electricity and magnetism, and show how observations of mathematical symmetry lead to the conclusion that matter and energy are interchangeable.

If I stand accused of blurring the line between mathematics and physics, I enthusiastically plead "Guilty!" Mathematics is often viewed as all technique, the foundation for interesting studies of the world but dull and tedious in and of itself. I hope that this book will reveal to you some of the intimate interplay between mathematics and our understanding of the physical universe and, in the process, illuminate some of the intrinsic beauty of mathematics itself.

I have also tried to emphasize the mathematical structure underlying this subject. For this, I have taken my inspiration from two of the great texts on several variable calculus: Tom Apostol's *Calculus* and Harold Edwards' *Advanced Calculus*. The former was my own textbook as an undergraduate, and I admire its clarity and precision and, above all, its treatment of the derivative of a vector field as a linear transformation. Edwards opened the world of differential forms to me, and I am especially indebted to him for revealing the natural progression from the fundamental theorem of calculus to Maxwell's equations to special relativity. I have taken what I found most imitation as sincere flattery.

The physicist and occasional mathematician Freeman Dyson, in an address to the American Mathematical Society in 1972, spoke of the passage from the equations of electricity and magnetism to the insights of special relativity as one of the opportunities missed by mathematicians as they divorced themselves from the problems of physics. Even today, most mathematicians have a poor appreciation of Maxwell's contributions. An excerpt from Dyson's comments is included as Appendix A. I hope that many of you will be motivated to read all of his address and to participate in the reintegration of physics and mathematics that I feel has begun.

This book grew out of an honors calculus class at Penn State. It is intended to be covered in one 15-week semester of four classroom hours per week. That is a fast pace, and yet it is one that dedicated students can maintain. There are no interchangeable parts. This text was designed to be used as a whole. That does not mean that each chapter need be given the same emphasis. Chapters 1, 3, and 11, as well as Sections 8.2, 8.5, 9.5, and 10.9, can be touched lightly and rapidly with the student responsible for reading and assimilating much of the material. But, I hope that the instructor will not give these portions too scant attention for they provide the flesh of a subject that too often is reduced to the dry bones of technique.

There are many people to whom I owe a debt of gratitude for help with this book, among them Don Albers and Freeman Dyson for their early encouragement; Ray Ayoub, Allan Krall, Steve Maurer, and David Rosen, who went through much of the manuscript and suggested improvements; the National Security Agency for its support; and my editor at Springer-Verlag, who expressed consistent confidence in this project and made a number of valuable suggestions. But, most especially, I wish to thank the students who struggled through the first draft of the text in the fall of 1990 and pointed out many of my misprints, as well as places where the explanations were obscure, the examples inadequate, and the exercises impossible. I do not claim that they or I have now found all such faults, but this is a better book for their efforts. They are Jeffrey Caruso, Robert Colbert, David Druist, Mark Flood, Kathleen Galvin, Darren Gibula, Steven Gradess, Stanley Hsu, Steven Jackson, John Johnson, Timothy Keane, Brian Ledell, Kurt Ludwick, Lara Palmer, Brian Pavlakovic, Christine Penney, Julie Richards, Alexander Richman, Nicola Schussler, Andrew Shropshire, Michael Smith, Peter Stone, Xiong Sun, Christopher Tatnall, Melissa Wallner, Marc Weinstein, and Jill Wyant.

To anyone who requests it, I will send a current list of corrections for this book, and I ask your help in finding misprints and errors. Regular mail should be sent to Macalester College, St. Paul, MN 55105, USA; e-mail to bressoud@macalstr.edu.

David M. Bressoud   March 6, 1991

Contents

Preface   **vii**

1 $F = ma$ **1**
   1.1 Prelude to Newton's *Principia*   1
   1.2 Equal Area in Equal Time   5
   1.3 The Law of Gravity   9
   1.4 Exercises 16
   1.5 Reprise with Calculus 18
   1.6 Exercises 26

2 Vector Algebra   **29**
   2.1 Basic Notions 29
   2.2 The Dot Product 34
   2.3 The Cross Product 39
   2.4 Using Vector Algebra 46
   2.5 Exercises 50

3 Celestial Mechanics   **53**
   3.1 The Calculus of Curves 53
   3.2 Exercises 65
   3.3 Orbital Mechanics 66
   3.4 Exercises 75

4 Differential Forms   **77**
   4.1 Some History 77
   4.2 Differential 1-Forms 79
   4.3 Exercises 86
   4.4 Constant Differential 2-Forms 89
   4.5 Exercises 96
   4.6 Constant Differential k-Forms 99
   4.7 Prospects 105
   4.8 Exercises 107

5 Line Integrals, Multiple Integrals   111
    5.1 The Riemann Integral 111
    5.2 Line Integrals 113
    5.3 Exercises 119
    5.4 Multiple Integrals 120
    5.5 Using Multiple Integrals 131
    5.6 Exercises 134

6 Linear Transformations   139
    6.1 Basic Notions 139
    6.2 Determinants 146
    6.3 History and Comments 157
    6.4 Exercises 158
    6.5 Invertibility 163
    6.6 Exercises 169

7 Differential Calculus   171
    7.1 Limits 171
    7.2 Exercises 178
    7.3 Directional Derivatives 181
    7.4 The Derivative 187
    7.5 Exercises 197
    7.6 The Chain Rule 201
    7.7 Using the Gradient 205
    7.8 Exercises 207

8 Integration by Pullback   211
    8.1 Change of Variables 211
    8.2 Interlude with Lagrange 213
    8.3 Exercises 216
    8.4 The Surface Integral 221
    8.5 Heat Flow 228
    8.6 Exercises 230

9 Techniques of Differential Calculus   233
    9.1 Implicit Differentiation 233
    9.2 Invertibility 238
    9.3 Exercises 244
    9.4 Locating Extrema 248
    9.5 Taylor's Formula in Several Variables 254
    9.6 Exercises 262
    9.7 Lagrange Multipliers 266
    9.8 Exercises 277

10 The Fundamental Theorem of Calculus   **279**
  10.1 Overview 279
  10.2 Independence of Path 286
  10.3 Exercises 294
  10.4 The Divergence Theorems 297
  10.5 Exercises 310
  10.6 Stokes' Theorem 314
  10.7 Summary for $\mathbf{R}^3$ 321
  10.8 Exercises 323
  10.9 Potential Theory 326

11 $E = mc^2$ **333**
  11.1 Prelude to Maxwell's *Dynamical Theory* 333
  11.2 Flow in Space–Time 338
  11.3 Electromagnetic Potential 345
  11.4 Exercises 349
  11.5 Special Relativity 352
  11.6 Exercises 360

Appendices

A An Opportunity Missed   **361**

B Bibliography   **365**

C Clues and Solutions   **367**

  Index   **382**

1

$$F = ma$$

The heavens declare the glory of God,
and the firmament shows his handiwork.
   —*Psalm 19*

Had I been present at the creation, I would have given some
useful hints for the better ordering of the universe.
   —attributed to King *Alfonso X* of Castile (1221–1284)

[Newton] has so clearly laid open and set before our eyes the
most beautiful frame of the System of the World, that if King
Alfonso were now alive, he would not complain for want of the
graces either of simplicity or of harmony in it.
   —*Roger Cotes*, from the Preface to the second edition of
Philosophiæ Naturalis Principia Mathematica (1713)

1.1 Prelude to Newton's *Principia*

Popular mathematical history attributes to Isaac Newton (1642–1727) and
Gottfried Wilhelm Leibniz (1646–1716) the distinction of having invented
calculus. Of course, it is not nearly so simple as that. Techniques for evaluating areas and volumes as limits of computable quantities go back to the
Greeks of the classical era. The rules for differentiating polynomials and the
uses of these derivatives were current before Newton or Leibniz were born.
Even the fundamental theorem of calculus, relating integral and differential
calculus, was known to Isaac Barrow (1630–1677), Newton's teacher. Yet
it is not inappropriate to date calculus from these two men for they were
the first to grasp the power and universal applicability of the fundamental
theorem of calculus. They were the first to see an inchoate collection of
results as the body of a single unified theory.

   Newton's preeminent application of calculus is his account of celestial
mechanics in *Philosophiæ Naturalis Principia Mathematica* or *Mathematical Principles of Natural Philosophy*. Ironically, he makes very little specific
mention of calculus in it. This may, in part, be due to the fact that calculus
was still sufficiently new that he felt it would be suspect. In part, it is a
reflection of an earlier age in which mathematicians jealously guarded powerful new techniques and only revealed the fruits of their labors. Newton's

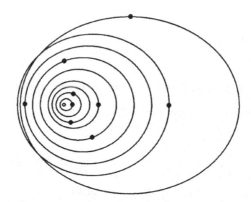

FIGURE 1.1. The solar system as we understand it.

Principia is couched in the language of Euclidean geometry. It is worthwhile for us to look at some of his arguments in that language, if only to convince ourselves that there must be a better way of presenting it. But, before we begin the exploration of Newton's work, I think it is important to understand why this is seen as the most important book in science ever written.

We are all familiar with the modern understanding of our solar system. The sun sits at the center of a ring of essentially concentric orbits, each planet following an elliptical path around the sun, which sits not at the center but at one focus of the ellipse. It is the picture given in Figure 1.1. No one has ever seen this picture. Even today there are no satellites sitting at that vantage point tracking the orbits and plotting their paths, much less in the seventeenth century when this picture was first proposed by Johann Kepler (1571–1630). Kepler and the astronomers before and since him have constructed their understanding of the solar system by watching the sun and planets, observable as points of light, move across the sky. We only see this two-dimensional projection of their paths. We often speak of the ignorance of medieval astronomers who placed the earth immovable at the center with the sun and planets revolving in circular orbits about it, but that is a very reasonable model consistent with the two great precepts of the scientific method: observe carefully and explain simply.

One clear observable is the immovability of the earth on which we stand. There is no sensation of movement, of hurtling 'round the sun. In fact, when the earth does dislodge, as in an earthquake, it is terrifying not just because of the physical danger, but there is also an innate sense that something that should not move is shifting. Well into the seventeenth century this remained

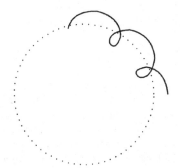

FIGURE 1.2. Part of the curve traced by a point on an epicycle.

an argument of some of the most learned scientists that the sun could not be at the center.

A second clear observable is that most stars hold fixed positions in the canopy of the night sky as it rotates, once every 24 hours, and moves through its yearlong progression. But, there are certain stars, known as the planets (from the greek πλανήτηϛ meaning "wanderer"), that move against this fixed backdrop. They stay in a narrow band about the *ecliptic* or path of the sun during the year. To assume that the sun and planets revolve in circular orbits of constant speed with the earth at the center is a simple explanation of what has been observed.

The problems arose as the observations became better. One of the first inconsistencies to be noticed was that the four seasons are not of equal length. It is 92.5 days from the vernal equinox to the summer solstice, 94 days from then until the autumnal equinox. The summer half of the year is therefore 186.5 days, well over half of the 365.25 days that constitute a full year. Hipparchus of Rhodes and Alexandria (d. *circa* 125 B.C.), considered the father of trigonometry, explained this phenomenon by moving the earth off the center of the circle.

There were more inconsistencies. The speed of a planet as it travels along the ecliptic is not constant and in fact it will sometimes reverse direction and back up. This was eventually explained by Ptolemy of Alexandria (d. 168 A.D.) through the use of epicycles and equants. An *epicycle* is a small circle whose center moves along a larger circle (Figure 1.2). If the planet moves uniformly around the epicycle while the center of the epicycle circles the earth, then seen from the earth the planet will seem to speed up, slow down, briefly reverse direction, and then return to its original direction. This explains the type of motion observed, but it is not enough to clarify all of the variations in speed. The *equant* is a point located off the center of the circle such that the movement of the center of the epicycle appears uniform, not from the center of the large circle, but from the equant. Together these gave a model that was not simple but at least agreed with the observations to within the possible accuracy of that time.

Kepler's Revelation

The system proposed by Nicolaus Copernicus (1473–1543) with the sun at the center and the earth and planets following circular orbits around it had the great advantage of simplicity. The problem was that astronomical measurements had become far more precise. The Copernican system was inaccurate, off by as much as four to five degrees at a time when astronomical measurements were becoming accurate to within a few minutes or even fractions of minutes (60 minutes = 1 degree). Tycho Brahe (1546–1601) spent the last 20 years of his life making such detailed observations of planetary positions. His data were bequeathed to his assistant, Johann Kepler, who faced the daunting task of finding a model that was both simple and agreed with his mentor's calculations. In this, he succeeded brilliantly. As he himself wrote in the preface to his chapter on the planets in *Harmonice Mundi*:

> It is not eighteen months since I caught the first glimpse of light, three months since the dawn, very few days since the unveiled sun, most admirable to gaze upon, burst upon me. Nothing restrains me; I shall indulge my sacred fury; I shall triumph over mankind by the honest confession that I have stolen the golden vases of the Egyptians to build up a tabernacle for my God far from the confines of Egypt. If you forgive me, I rejoice; if you are angry, I can bear it; the die is cast, the book is written, to be read either now or by posterity, I care not which; it may well wait a century for a reader, as God himself has waited six thousand years for someone to behold his work.

Kepler's "unveiled sun" was the realization that Tycho Brahe's measurements could be explained by the following three laws.

1. A planetary orbit sweeps out equal area in equal time.

2. A planetary orbit is an ellipse with the sun at one focus.

3. The square of the period of the orbit is directly proportional to the cube of the mean distance (the average of the closest and farthest distances from the sun).

Few scientists had problems with the first or third law, but the second proved a considerable stumbling block to those who had been schooled to think in terms of straight lines and circles. Why an ellipse? Why should the heavens, a pure manifestation of God's handiwork, be ruled by a degenerate form of the circle?

It was Newton's *Principia*, published in 1687, that ultimately confirmed Kepler's model and restored this lost purity by showing that elliptical orbits as described by Kepler are equivalent to the statement that gravitational acceleration is inversely proportional to the square of the distance to the orbiting body. Simplicity of geometric form was replaced by simplicity of underlying forces.

1.2   Equal Area in Equal Time

This is not the place for a detailed analysis of the *Philosophiæ Naturalis Principia Mathematica*. All I seek is to convey some of the flavor of what Newton accomplished. He begins with definitions of mass, momentum (what Newton calls *motion*, mass times velocity), various forces, and acceleration, and then states what he calls his three "Axioms, or Laws of Motion".

1. Every body continues in its state of rest, or of uniform motion in a right line, unless it is compelled to change that state by forces impressed upon it.

2. The change of motion is proportional to the motive force impressed, and is made in the direction of the right line in which that force is impressed.

3. To every action there is always opposed an equal reaction; or, the mutual actions of two bodies upon each other are always equal, and directed to contrary parts.

It is the second axiom that gives us our equation

$$F = ma,$$

or force equals mass times acceleration. If the mass is constant, then ma is the derivative of mv; it is the rate at which the motion or momentum is changing. Equality, not just proportionality, is achieved by a judicious choice of units. It is important to realize that here and in what follows Newton works with small instantaneous forces that create discrete changes in the velocity of the moving body. It is the genius of calculus that continuous changes can be approximated by, and are in fact the limit of, such discrete changes. Without the tools of calculus at hand, Newton works with instantaneous forces and discrete changes and then argues that in the limit these give a valid picture of the continuous phenomena of our world.

In elaborating on his second axiom, Newton specifies that this new component to the velocity that has been created by the instantaneous force

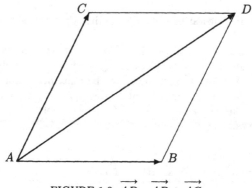

FIGURE 1.3. $\overrightarrow{AD} = \overrightarrow{AB} + \overrightarrow{AC}$.

should be added to the previous velocity and that this addition is accomplished by what is today known as the *parallelogram rule*. We represent velocity by a *vector* specifying direction and magnitude. Vectors are usually drawn as arrows or directed line segments whose lengths are proportional to the respective magnitudes. It is important to remember, however, that the vector is *only* the magnitude and direction of the directed line segment and that the same vector can have different endpoints. We shall denote the line segment from point A to point B by AB. If we are talking about the line segment, then it has no direction; AB is the same as BA. To denote the vector represented by the directed line segment from A to B we shall use $\overrightarrow{AB}$.

The vector from B to A has the opposite direction and so

$$\overrightarrow{BA} = - \overrightarrow{AB}.$$

But the vector $\overrightarrow{AB}$ is not wedded to the points A and B. In Figure 1.3, the vector from A to B and the vector from C to D are *the same vector* since they represent the same magnitude and direction. If we do not change magnitude or direction, then the vector is the same no matter where we put it.

If $\overrightarrow{AB}$ represents the initial velocity and $\overrightarrow{AC}$ the change in velocity created by the instantaneous force, then we locate the point D so that $ABDC$ is a parallelogram. The vector $\overrightarrow{AD}$ is the resulting velocity vector (Figure 1.3). It will be convenient to denote the length of a vector $\overrightarrow{AD}$ by $|AD|$. Note that the length of the vector $\overrightarrow{AD}$ is the same as the length of the line segment AD.

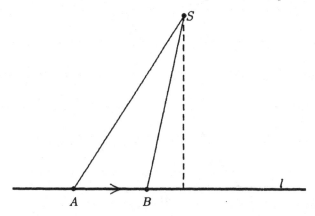

FIGURE 1.4. Area swept out when there are no forces.

Newton's Proof of Kepler's First Law

Newton begins his study of Kepler's laws by showing that the first law is a consequence of the force of gravity acting along the straight line connecting the sun to the orbiting planet.

Theorem 1.1 (Newton's Proposition I) *Let S be a fixed point and let P be a moving particle such that the only forces acting on P at any given time lie in the direction of the line connecting S to P at that moment. Then, the path followed by P will lie within a single plane, and the area swept out by the line connecting S to P will be the same for any equal length of time.*

It is worth noting how much more this says than just Kepler's first law. Equal area in equal time is a consequence of any type of force that is purely radial, no matter how wildly it may fluctuate. A precise definition of a *plane* will have to await Chapter 2. For now, it is enough to use your intuitive understanding of a plane as a flat surface.

Proof: We first observe that the theorem holds when there are no forces acting on P. By Newton's first law of motion, P travels in a straight line at a constant velocity. This means that it travels equal distances in equal time. If we call this line l, then we may assume that S does not lie on l for otherwise the area swept out is always zero. Our point P stays in the plane defined by the point S and the line l. We fix a specified length of time and choose any two points A and B on l such that P travels from A to B in this specified time (see Figure 1.4).

The area swept out by P as it travels from A to B is the area of the triangle SAB, which is one half $|AB|$ times the perpendicular distance from

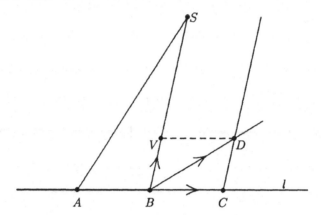

FIGURE 1.5. Area swept out under radial force.

S to l. Both the distance from A to B and the perpendicular distance from S to l are independent of the choice of starting point, A, and so the area is also independent of the choice of A.

We now assume that when the particle reaches B it is acted on by an instantaneous force along the straight line SB. By Newton's second law of motion, this adds a new component parallel to BS to the existing velocity of the particle P. We denote it by the vector $\overrightarrow{BV}$. Note that V does not have to lie between B and S. The force could equally well be pushing P away from S, in which case V would lie below the line l. The rest of this argument would remain valid.

We now mark a point C on l (Figure 1.5) such that $|AB| = |BC|$. The vector $\overrightarrow{BC}$ represents the original velocity vector at the point B before the instantaneous force was applied. The resulting velocity is thus given by $\overrightarrow{BD}$, where D is found by using the parallelogram rule. Since D lies in the plane of S and l, our particle stays in this same plane.

The area swept out in our second specified length of time is the area of the triangle SBD. Since the line through D and C is parallel to the line through S and B, the area of this triangle is one half $|SB|$ times the perpendicular distance between BS and CD. But, this is also the area of the triangle SBC, which is equal to the area of the triangle SAB. Thus equal area has been swept out.

Q.E.D.

The initials *Q.E.D.* stand for *Quod Erat Demonstrandum*, a Latin translation of the phrase ὅπερ ἔδει δεῖξαι with which Euclid concluded most of his proofs. It means "what was to be proved" and signifies that the proof has been concluded.

Newton's Proposition II observes that if the area of triangle SBD is equal to the area of SAB (and thus of SBC), then the line through C and D must be parallel to BS, and therefore the force acting on the particle when it was at B was in the direction of BS. We thus see that the property of sweeping out equal area in equal time is more than a consequence of the force acting radially; it is actually equivalent to it. We have proved the following theorem.

Theorem 1.2 (Newton's Proposition II) *Let S be a fixed point and let P be a moving particle that stays in a fixed plane containing S and sweeps out equal area in equal time; then, the only forces acting on P are radial forces from S.*

1.3   The Law of Gravity

The story that Newton discovered gravity as the result of a falling apple is apocryphal. It is also misleading because Newton did not discover gravity. People had long been aware that there is something that causes apples to fall off trees and invariably head toward the ground. But, Newton did discover two important aspects of gravity that revolutionized our understanding of the mechanics of the universe. The first was that the same force that pulls apples to the ground is responsible for keeping planets in their orbits. The second was that the strength of this force is inversely proportional to the square of the distance between the two bodies involved.

It is common today to begin with the assertion that gravitational attraction is inversely proportional to the square of the distance and then from that to derive that planetary orbits must be ellipses with the sun at one focus. We shall eventually do this. If one is pressed to explain why this particular law of gravitational attraction holds, recourse to explanation by intimidation is too often used. There is, apparently, no other conceivable law of gravity. Newton, in his *Principia*, begins by exploring other possible laws. He then assumes the validity of Kepler's first two laws and shows that they imply his law of gravitational attraction. It is a rather appealing argument and utilizes a great deal of the geometry of conic sections, most of which Newton gleaned from *Conics* by Appolonius of Perga (262–170 B.C.).

I shall present Newton's argument very much the way he did, but there are some preliminary definitions and results on ellipses that we shall need, and here I intend to cheat and use analytic geometry, the algebraic description of geometric curves, to speed us through this basic material.

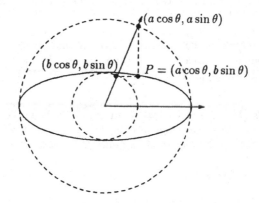

FIGURE 1.6. The point P determined by the angle θ.

Properties of Ellipses

The general equation of an ellipse with its major axis on the x axis is

$$\frac{x^2}{a^2} + \frac{y^2}{b^2} = 1,$$

where we shall take $a \geq b > 0$. This ellipse crosses the x axis at $(\pm a, 0)$ and the y axis at $(0, \pm b)$. The quantity a is called the *semimajor axis*, b is called the *semiminor axis*. The center of the ellipse is at the origin O. An arbitrary point P on the ellipse can be described or *parametrized* by the angle θ by

$$P = (a \cos \theta,\ b \sin \theta).$$

Note that θ is usually *not* the angle between OP and the positive x axis (Figure 1.6). If we set

$$P' = -P = (-a \cos \theta, -b \sin \theta),$$

then the line PP' passes through the origin. We call it a *diameter* of the ellipse (Figure 1.7).

We define points Q and Q' by changing the parameter θ by $\pm \pi/2$:

$$\begin{aligned} Q &= (a \cos(\theta + \pi/2), b \sin(\theta + \pi/2)) \\ &= (-a \sin \theta, b \cos \theta), \\ Q' &= (a \cos(\theta - \pi/2), b \sin(\theta - \pi/2)) \\ &= (a \sin \theta, -b \cos \theta). \end{aligned}$$

The diameter QQ' is the *conjugate diameter* to PP'. Note that QQ' usually does not meet PP' at a right angle. However, pairs of conjugate diameters

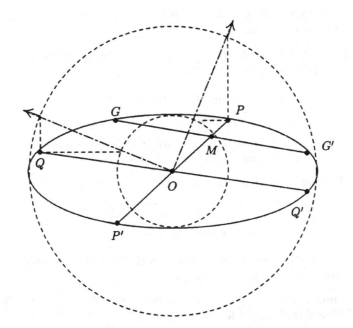

FIGURE 1.7. Conjugate diameters.

do possess a number of very nice properties given in the following lemmas, whose proofs are left as exercises.

Lemma 1.1 *The tangent to the ellipse at P is parallel to the conjugate diameter to PP'.*

Lemma 1.2 *Any chord of the ellipse that is parallel to the tangent at P is bisected by the diameter PP' (Figure 1.7).*

Lemma 1.3 *If the chord GG' is parallel to the tangent at P and intersects the diameter $\overrightarrow{PP'}$ at M, then (Figure 1.7)*

$$\frac{|GM|^2}{|OQ|^2} = \frac{|PM||P'M|}{|OP|^2}.$$

Lemma 1.4 *The area of the parallelogram whose vertices lie at the endpoints of two conjugate diameters is always $2ab$.*

The *foci* of our ellipse are the points

$$\begin{aligned} F_1 &= (-\sqrt{a^2 - b^2}, 0), \\ F_2 &= (\sqrt{a^2 - b^2}, 0). \end{aligned}$$

The reason for their name comes from the following two lemmas, whose proofs are also left to the exercises.

Lemma 1.5 *Let P be any point on the ellipse. The sum of the distances $|F_1P|$ and $|F_2P|$ is the constant $2a$.*

Lemma 1.6 *Let P be any point on the ellipse. The line that bisects the angle $\angle F_1PF_2$ is perpendicular to the tangent to the ellipse at P.*

These two lemmas imply that if the inside of our ellipse is mirrored and we place a light source that flashes for a single instant at the focus F_1, then all of the reflected light rays will meet at the same moment at the other focus, F_2 (Figure 1.8).

The final lemma we shall need is not a standard result on ellipses and is proved by Newton in *Principia*.

Lemma 1.7 *Let P be any point on the ellipse, QQ' the conjugate diameter to PP', and let E be the point of intersection of F_1P and QQ' (Figure 1.9). Then, the distance from E to P is the length of the semimajor axis:*

$$|EP| = a.$$

Proof: Let O be the point at the origin. We draw the line through F_2 parallel to QQ' and let I denote its point of intersection with F_1P. We

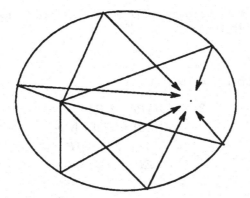

FIGURE 1.8. A mirrored ellipse.

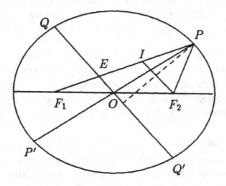

FIGURE 1.9. Proof of Lemma 1.7.

draw the line bisecting the angle $\angle F_1PF_2$. By Lemmas 1.1 and 1.6, this angle bisector is perpendicular to the line segment F_2I. We thus see that

$$|IP| = |PF_2|.$$

By the similarity of the triangles F_1IF_2 and F_1EO and the fact that $|F_1F_2|$ is twice $|F_1O|$, $|F_1I|$ must be twice $|F_1E|$ or

$$|F_1E| = |EI|.$$

Now, by Lemma 1.5, we see that

$$\begin{aligned}
2a &= |F_1P| + |PF_2| \\
&= |F_1E| + |EP| + |PF_2| \\
&= |EI| + |EP| + |IP| \\
&= 2|EP|.
\end{aligned}$$

Q.E.D.

Newton's Proof of the Law of Gravity

We are now equipped to follow Newton's derivation of the law of gravitational attraction.

Theorem 1.3 (Newton's Proposition XI) *If a particle P moves along an ellipse in such manner that its only acceleration or change in velocity is always directed along the line from P to the focus F_1, then the magnitude of that acceleration, and thus the magnitude of the attracting force, is inversely proportional to the square of the distance between P and F_1.*

Proof: As before, we let O be the center of the ellipse, PP' the diameter with one endpoint at P, QQ' its conjugate diameter, and a and b the semimajor and semiminor axes, respectively (Figure 1.10). We draw the line bisecting the angle $\angle F_1PF_2$ and let K be its point of intersection with the conjugate diameter QQ'. By Lemmas 1.1 and 1.6, PK is perpendicular to QQ'.

We draw the tangent to the ellipse at P and mark a point R on it so that the vector $\overrightarrow{PR}$ represents the velocity of our particle as it reaches the point P. We draw the line through R parallel to PF_1 and let S denote its first point of intersection with the ellipse. If our particle is subject to an instantaneous change in velocity at P, which will then carry it to S, then the change in velocity must be represented by the vector $\overrightarrow{PX}$, where X lies on the line PF_1 and $|PX| = |RS|$.

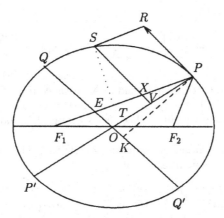

FIGURE 1.10. Proof of Theorem 1.3.

Let A denote the area swept out as our particle moves from P to S and let L denote the constant $2b^2/a$ (called the *principal latus rectum*). What Newton actually proves is that

$$\frac{|RS|}{A^2} \quad \text{approaches} \quad \frac{4}{L|PF_1|^2}$$

in the limit as R approaches P. Since the area swept out, A, is constant by Theorem 1.1, the magnitude of the change in velocity is inversely proportional to $|PF_1|^2$, the square of the distance between P and the focus. In the remainder of the argument, equalities that are only asymptotically true will be identified by using $\cong$.

There are three more points we need to construct: E, the intersection of PF_1 and QQ'; V, the intersection of OP and the extension of the line SX; and T, obtained by dropping the perpendicular from S to the line PF_1. The area A, swept out as the particle moves from P to S, is approximated by $|PF_1||ST|/2$.

We observe three equalities and one asymptotic equality

$$\frac{L|RS|}{L|PV|} = \frac{|RS|}{|PV|} = \frac{|PX|}{|PV|} = \frac{|PE|}{|OP|} = \frac{a}{|OP|}, \qquad (1.1)$$

where the third equality follows from the fact that XV is parallel to EO and the last equality arises from Lemma 1.7,

$$\frac{L|PV|}{|PV||VP'|} = \frac{L}{|VP'|}, \qquad (1.2)$$

$$\frac{|PV||VP'|}{|SV|^2} = \frac{|OP|^2}{|OQ|^2}, \qquad (1.3)$$

by Lemma 1.3, and

$$\frac{|SV|^2}{|ST|^2} \cong \frac{|SX|^2}{|ST|^2} = \frac{|PE|^2}{|PK|^2} = \frac{a^2}{|PK|^2} = \frac{|OQ|^2}{b^2}, \qquad (1.4)$$

where the approximate equality follows because both $|SV|$ and $|SX|$ are asymptotically equal to the perpendicular distance from S to the line PK (see Exercise 13, Section 1.4), the first equality arises through use of similar triangles, the second equality follows from Lemma 1.7, and the last follows from Lemma 1.4.

Multiplying the left sides of Equations (1.1) through (1.4) and then the right sides of these equalities, we obtain

$$\frac{L|RS|}{L|PV|} \frac{L|PV|}{|PV||VP'|} \frac{|PV||VP'|}{|SV|^2} \frac{|SV|^2}{|ST|^2} \cong \frac{a}{|OP|} \frac{L}{|VP'|} \frac{|OP|^2}{|OQ|^2} \frac{|OQ|^2}{b^2}.$$

We simplify both sides and observe that $|VP'|$ is very close to $2|OP|$ when $|PR|$ is small:

$$\frac{L|RS|}{|ST|^2} \cong \frac{La|OP|}{|VP'|b^2} = \frac{2|OP|}{|VP'|} \cong 1.$$

Dividing both sides by $L|PF_1|^2$ and using our approximation for the area, $A \cong |ST||PF_1|/2$, we finally get

$$\frac{|RS|}{(|ST||PF_1|)^2} \cong \frac{1}{L|PF_1|^2},$$

$$\frac{|RS|}{A^2} \cong \frac{4}{L|PF_1|^2}. \qquad (1.5)$$

Q.E.D.

1.4 Exercises

1. For the ellipse $x^2/9 + y^2/4 = 1$ and each of the following points P, find the endpoints of the conjugate diameter QQ'.

   (a) $P = (3, 0)$.
   (b) $P = (3\sqrt{2}/2, \sqrt{2})$.
   (c) $P = (-2\sqrt{2}, 2/3)$.

2. For the ellipse $x^2/25 + y^2/9 = 1$, let $P = (3, 12/5)$ and $G = (0, 3)$. Find the point H on the ellipse so that the chord GH is parallel to the conjugate diameter QQ'.

In Exercises 3–12, we are working with the general ellipse $x^2/a^2 + y^2/b^2 = 1$ and a point $P = (a \cos\theta, b\sin\theta)$ on the ellipse.

3. Show that the slope of the conjugate diameter QQ' is $-(b/a)\cot\theta$.

4. Prove Lemma 1.1 by using Exercise 3 and showing that the slope of the tangent to the ellipse at P is also $-(b/a)\cot\theta$.

In Exercises 5–9, let ϕ be an arbitrary angle and define

$$G = G(\phi) = (a\cos(\theta + \phi), b\sin(\theta + \phi)),$$

$$G' = G'(\phi) = (a\cos(\theta - \phi), b\sin(\theta - \phi)).$$

Note that $Q = G(\pi/2), Q' = G'(\pi/2)$.

5. Show that the slope of GG' is $-(b/a)\cot\theta$ no matter which value of ϕ we choose, and thus GG' is always parallel to QQ'.

6. Show that a straight line intersects an ellipse at a maximum of two points. Thus, if we select a point H on the ellipse so that GH is parallel to QQ', then $H = G'$.

7. Prove Lemma 1.2 by using Exercise 6 and showing that the midpoint of GG' is at
$$M = \cos\phi(a\cos\theta, b\sin\theta).$$

8. Prove Lemma 1.3 by showing that
$$
\begin{aligned}
|PM||P'M| &= \sin^2\phi\, |OP|^2, \\
|GM|^2 &= \sin^2\phi\, |OQ|^2.
\end{aligned}
$$

9. Show that the area of triangle QOP is $ab/2$. Use this fact to prove Lemma 1.4.

In Exercises 10–12, let F_1 and F_2 be the foci of the ellipse and let $c = |F_1O| = |F_2O|$, $c^2 = a^2 - b^2$.

10. Show that $|F_1P| = a + c\cos\theta$ and $|PF_2| = a - c\cos\theta$, thus proving Lemma 1.5.

11. Draw the line l bisecting the angle $\angle F_1PF_2$, and then draw the line m through F_2 perpendicular to l (Figure1.11). Let $R = (s, t)$ be the point of intersection of m with F_1P. Prove that $|PR| = a - c\cos\theta$. Using similar triangles, prove that

$$s + c = 2c\cos\theta\frac{c + a\cos\theta}{a + c\cos\theta},$$

$$t = \frac{2bc\sin\theta\cos\theta}{a + c\cos\theta}.$$

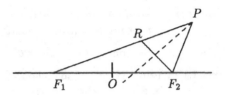

FIGURE 1.11. Exercise 11.

12. Using the results of Exercise 11, prove that the slope of RF_2 is

$$\frac{t}{s-c} = \frac{-b}{a} \cot \theta,$$

thus proving Lemma 1.6.

13. In Figure 1.10, extend the line through S, X, and V so that it intersects the line PK at W. Let $\theta = \angle RPS$ and $\phi = \angle F_1PK$. Show that $|SX|/|SW| = 1 - \tan \theta \tan \phi$. As S approaches P, ϕ stays fixed but θ approaches 0. Use this to conclude that $|SX|/|SV|$ approaches 1.

14. The quote from *Harmonice Mundi* is something of a cheat. It only refers to the third law, the first two having been published ten years earlier in *Astronomia Nova*. Kepler also had a fourth law governing the relative distances of the planets from the sun. It is now conveniently forgotten by most scientists since it was wrong. What was Kepler's fourth law? Hint: see Figure 1.12, taken from his book *Mysterium Cosmographicum*, published in 1596.

1.5   Reprise with Calculus

While I find Newton's proof of Theorem 1.1 very appealing in its simplicity and the clarity with which it illuminates the connection between radial force and equal areas, his proof of Theorem 1.3 is not as transparent as one would wish. There is much more to Newton's *Principia*. He goes on to derive Kepler's laws from the law of gravity and then to explore the consequences of his insights. The entirety of *Principia* consists of three volumes. But, at this point, I want to leave Newton and find a simpler language for explaining celestial mechanics.

The search for a better idiom in which to understand our mathematics is going to be a recurrent theme throughout this book. It is not always easy to make the transition; new concepts are often at a high level of

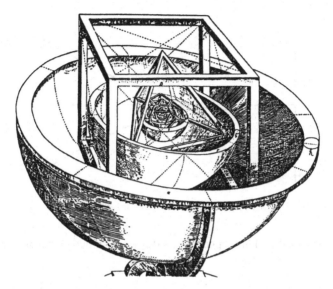

FIGURE 1.12. Kepler's fourth law?

abstraction and difficult to grasp. But, once you are comfortable with the new terminology, it can greatly clarify relationships and proofs. I feel that the effort expended is more than repaid in a better understanding of the material at hand and an enhanced ability to build on it. It is worth keeping in mind, however, that we always have choices and that future generations may look upon our expressions and proofs as unnecessarily convoluted.

Trajectories as Functions

In moving into the language of calculus, we first need to describe the trajectory of our moving particle as a function. We can think of its position $\vec{r}(t)$ as a function of time. For this chapter, we shall stay in the x, y plane. The x and y coordinates are each functions of time:

$$\vec{r}(t) = (x(t), y(t)).$$

For any specific value of t, we can think of $\vec{r}(t)$ as *either a point in the plane or as the vector from the origin to this point*. While initially somewhat confusing, it is very convenient to be able to move freely between these two interpretations.

We speak of $\vec{r}(t)$ as a function from one real variable, t, to two real

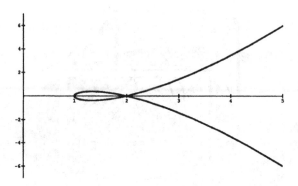

FIGURE 1.13. $\vec{r}(t) = (t^2 + 1, t^3 - t)$.

variables, x and y. This aspect of $\vec{r}$ is described in notational shorthand:

$$\vec{r} : \mathbf{R} \longrightarrow \mathbf{R}^2.$$

A function from one variable to one variable such as $f(t) = t^3 - t$ is called a *scalar function*. A function from one variable to more than one variable such as $\vec{r}(t)$ is called a *vector function*. We shall restrict our attention to functions for which the derivative is defined at all or almost all values of t.

A nice example is the path (Figure 1.13) for which the position at time t is given by

$$\begin{aligned} x(t) &= t^2 + 1, \\ y(t) &= t^3 - t. \end{aligned}$$

The velocity vector at time t, $\vec{v}(t)$, is the rate at which the position is changing:

$$\vec{v} = \frac{d\vec{r}}{dt},$$

and is uniquely determined by the rate at which the x coordinate of our particle is changing:

$$\frac{dx}{dt} = 2t$$

and the rate at which the y coordinate is changing

$$\frac{dy}{dt} = 3t^2 - 1,$$

so that we have for our example

$$\vec{v}(t) = \left(\frac{dx}{dt}, \frac{dy}{dt} \right) = (2t, 3t^2 - 1).$$

Here is where we can exploit the ambiguity between points and vectors. It is natural to think of the velocity as a vector. At time $t = 2$, our particle is at the point $(5, 6)$ moving with a velocity $(4, 11)$, that is, the velocity has the same magnitude and direction as the vector from $(0, 0)$ to $(4, 11)$. But, we can also view the velocity as a vector function describing its own path in the x, y plane and ask how the velocity is changing. The rate of change of the velocity is the acceleration $\vec{a}(t)$, which is the derivative of the velocity:

$$\vec{a} = \frac{d\vec{v}}{dt} = \left(\frac{d^2 x}{dt^2}, \frac{d^2 y}{dt^2} \right) = (2, 6t).$$

For the problem at hand, that of understanding celestial mechanics, it is easiest to use the polar coordinates $r(t)$, the distance to the origin at time t, and $\theta(t)$, the angle between $\vec{r}(t)$ and the positive x axis. We shall need to assume that we stay away from the origin so that $r(t)$ is never zero. Note that $r(t)$ is the magnitude of the vector $\vec{r}(t)$:

$$r(t) = |\vec{r}(t)|. \tag{1.6}$$

The relationships between rectangular and polar coordinates are given by

$$x = r \cos \theta, \quad y = r \sin \theta, \tag{1.7}$$

$$r = \sqrt{x^2 + y^2}, \quad \tan \theta = y/x. \tag{1.8}$$

Local Coordinates

It is also convenient to compute in terms of *local coordinates* that change as our particle moves. In particular, we shall want to decompose the acceleration into a component parallel to the vector $\vec{r}(t)$ and a second component perpendicular to it. This can be done by defining a *unit vector* or vector of length 1 in the direction of $\vec{r}$ by

$$\vec{u}_r(t) = \frac{\vec{r}(t)}{r(t)} = \frac{(r(t) \cos \theta(t), r(t) \sin \theta(t))}{r(t)} = (\cos \theta(t), \sin \theta(t)) \tag{1.9}$$

and a perpendicular unit vector (Figure 1.14)

$$\vec{u}_\theta(t) = (- \sin \theta(t), \cos \theta(t)). \tag{1.10}$$

It is important to keep in mind that $\vec{u}_r$ and $\vec{u}_\theta$ are functions of t. In particular, they have derivatives that are related:

$$\frac{d\vec{u}_r}{dt} = \left(-\sin \theta \, \frac{d\theta}{dt}, \cos \theta \, \frac{d\theta}{dt} \right) = \frac{d\theta}{dt} \, \vec{u}_\theta, \tag{1.11}$$

$$\frac{d\vec{u}_\theta}{dt} = \left(-\cos \theta \, \frac{d\theta}{dt}, -\sin \theta \, \frac{d\theta}{dt} \right) = -\frac{d\theta}{dt} \, \vec{u}_r. \tag{1.12}$$

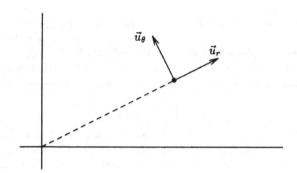

FIGURE 1.14. The local coordinates $\vec{u}_r$ and $\vec{u}_\theta$.

Since the product rule for derivatives works on each coordinate, it also holds for the product of a scalar function times a vector function:

$$\frac{d}{dt}(f(t)\vec{r}(t)) = \frac{df}{dt}\vec{r}(t) + f(t)\frac{d\vec{r}(t)}{dt}. \tag{1.13}$$

Combining these results with the fact that

$$\vec{r} = r\vec{u}_r, \tag{1.14}$$

we see that

$$\vec{v} = \frac{d\vec{r}}{dt} = \frac{d}{dt}(r\,\vec{u}_r) = \frac{dr}{dt}\vec{u}_r + r\frac{d\vec{u}_r}{dt} = \frac{dr}{dt}\vec{u}_r + r\frac{d\theta}{dt}\vec{u}_\theta, \tag{1.15}$$

$$
\begin{aligned}
\vec{a} &= \frac{d\vec{v}}{dt} = \frac{d}{dt}\left(\frac{dr}{dt}\,\vec{u}_r\right) + \frac{d}{dt}\left(r\frac{d\theta}{dt}\,\vec{u}_\theta\right) \\
&= \frac{d^2r}{dt^2}\vec{u}_r + \frac{dr}{dt}\frac{d\theta}{dt}\vec{u}_\theta + \frac{dr}{dt}\frac{d\theta}{dt}\vec{u}_\theta + r\frac{d^2\theta}{dt^2}\vec{u}_\theta + r\frac{d\theta}{dt}\left(-\frac{d\theta}{dt}\,\vec{u}_r\right) \\
&= \left(\frac{d^2r}{dt^2} - r\left(\frac{d\theta}{dt}\right)^2\right)\vec{u}_r + \left(r\frac{d^2\theta}{dt^2} + 2\frac{dr}{dt}\frac{d\theta}{dt}\right)\vec{u}_\theta \\
&= \left(\frac{d^2r}{dt^2} - r\left(\frac{d\theta}{dt}\right)^2\right)\vec{u}_r + \frac{1}{r}\frac{d}{dt}\left(r^2\frac{d\theta}{dt}\right)\vec{u}_\theta. \tag{1.16}
\end{aligned}
$$

Modern Proof of Kepler's First Law

Equation (1.16) tells us that acceleration is entirely radial, that is, parallel to $\vec{r}$, if and only if

$$\frac{1}{r}\frac{d}{dt}\left(r^2\frac{d\theta}{dt}\right) = 0,$$

which is equivalent to saying that $r(t)^2\, d\theta/dt$ is a constant independent of t. We have proven the following result.

Lemma 1.8 *If the position of a particle over time is described by the vector function $\vec{r}(t)$, then the acceleration is purely radial if and only if $r(t)^2\, d\theta/dt$ is a constant independent of t.*

Combining this with the next lemma gives us another proof of Theorems 1.1 and 1.2.

Lemma 1.9 *If the position of a particle over time is described by the vector function $\vec{r}(t)$, then the rate at which the radial vector sweeps out area is given by*

$$\frac{dA}{dt} = \frac{1}{2}r(t)^2\frac{d\theta}{dt}. \tag{1.17}$$

Proof: Given a circle with center at the origin and radius r, the area of the sector swept out by the radius as it moves through an angle of $\Delta\theta$ is given by $(r^2/2)\Delta\theta$. It follows that if ΔA is the area swept out by the radial vector from time s to time t and if the distance from the origin stays constant during this time interval, then

$$\Delta A = \frac{r^2}{2}\,\Delta\theta,$$

where $\Delta\theta = \theta(t) - \theta(s)$.

If r does not stay constant, then we can find two points in the interval $[s,t]$, call them t_1 and t_2, where r takes on its minimum and maximum values, respectively, over this interval:

$$r(t_1) \leq r \leq r(t_2).$$

It follows that

$$\frac{r(t_1)^2}{2}\Delta\theta \leq \Delta A \leq \frac{r(t_2)^2}{2}\Delta\theta.$$

We now divide through by $\Delta t = t - s$:

$$\frac{r(t_1)^2}{2}\frac{\Delta\theta}{\Delta t} \leq \frac{\Delta A}{\Delta t} \leq \frac{r(t_2)^2}{2}\frac{\Delta\theta}{\Delta t}$$

and take the limit as s approaches t. This forces t_1 and t_2 to also approach t and yields

$$\frac{r(t)^2}{2}\frac{d\theta}{dt} \le \frac{dA}{dt} \le \frac{r(t)^2}{2}\frac{d\theta}{dt}.$$

Q.E.D.

Modern Proof of the Law of Gravity

We shall now use calculus to demonstrate that if the acceleration is purely radial and the path is an ellipse with one focus at the origin, then the acceleration is inversely proportional to the square of the distance from the origin.

Lemma 1.10 *The general equation of an ellipse with one focus at the origin and major axis on the x axis is given in polar coordinates by*

$$r(1 + \varepsilon \cos \theta) = c, \tag{1.18}$$

where ε and c are real constants, $|\varepsilon| < 1$, $c > 0$. The semimajor axis is $c/(1 - \varepsilon^2)$ and the semiminor axis is $c/\sqrt{1 - \varepsilon^2}$.

Proof: Equation (1.18) can be rewritten as

$$r = c - \varepsilon r \cos \theta.$$

We use Equations (1.7) and (1.8) to convert this to rectangular coordinates and then square each side to obtain

$$x^2 + y^2 = c^2 - 2c\varepsilon x + \varepsilon^2 x^2,$$

which can be rewritten as

$$(1 - \varepsilon^2)\left(x^2 + \frac{2c\varepsilon x}{1 - \varepsilon^2} + \frac{c^2\varepsilon^2}{(1 - \varepsilon^2)^2}\right) + y^2 = c^2 + \frac{c^2\varepsilon^2}{1 - \varepsilon^2}$$

$$= \frac{c^2}{1 - \varepsilon^2},$$

$$\frac{(x + c\varepsilon/(1 - \varepsilon^2))^2}{a^2} + \frac{y^2}{b^2} = 1, \tag{1.19}$$

where $a = c/(1 - \varepsilon^2)$, $b = c/\sqrt{1 - \varepsilon^2}$.

The center of this ellipse is at $(-c\varepsilon/(1 - \varepsilon^2), 0)$ which is $c|\varepsilon|/(1 - \varepsilon^2)$ from the origin. This is precisely the distance of the focus from the center of the ellipse:

$$\sqrt{a^2 - b^2} = \sqrt{\frac{c^2}{(1-\varepsilon^2)^2} - \frac{c^2}{1-\varepsilon^2}} = \sqrt{\frac{c^2\varepsilon^2}{(1-\varepsilon^2)^2}} = \frac{c|\varepsilon|}{1-\varepsilon^2}.$$

Q.E.D.

Theorem 1.4 *If a particle moves so that its acceleration is always radial and if the particle follows the curve given by Equation (1.18), then the acceleration is*

$$\vec{a}(t) = \frac{-k^2}{cr^2}\vec{u}_r, \tag{1.20}$$

where $k = 2\,(dA/dt)$ is a constant.

Proof: The fact that dA/dt is constant follows from Theorem 1.2. From Lemma 1.9 we know that

$$r^2\frac{d\theta}{dt} = k. \tag{1.21}$$

If we solve for $d\theta/dt$:

$$\frac{d\theta}{dt} = kr^{-2}, \tag{1.22}$$

then we can substitute into our expression for the acceleration [Equation (1.16)]:

$$\vec{a} = \left(\frac{d^2r}{dt^2} - \frac{k^2}{r^3}\right)\vec{u}_r. \tag{1.23}$$

We solve Equation (1.18) for r and then differentiate with respect to t:

$$r = \frac{c}{1 + \varepsilon\cos\theta},$$

$$\frac{dr}{dt} = \frac{-c\varepsilon(-\sin\theta)}{(1 + \varepsilon\cos\theta)^2}\frac{d\theta}{dt} = \frac{c\varepsilon\sin\theta}{c^2r^{-2}}\frac{k}{r^2},$$

where the last equality uses Equations (1.18) and (1.22). Simplifying this expression, we obtain

$$\frac{dr}{dt} = \frac{k\varepsilon}{c}\sin\theta.$$

We differentiate a second time and again use Equations (1.18) [in the form $\cos\theta = (c - r)/r\varepsilon$] and (1.22) to simplify:

$$\frac{d^2r}{dt^2} = \frac{k\varepsilon}{c}\cos\theta\frac{d\theta}{dt} = \frac{k\varepsilon}{c}\frac{(c-r)}{r\varepsilon}\frac{k}{r^2} = \frac{k^2}{r^3} - \frac{k^2}{cr^2},$$

and therefore

$$\vec{a} = \left(\frac{d^2r}{dt^2} - \frac{k^2}{r^3}\right)\vec{u}_r = -\frac{k^2}{cr^2}\vec{u}_r. \tag{1.24}$$

Q.E.D.

1.6 Exercises

1. For each of the following paths given by $\vec{r}(t)$, find the position of the particle at the specified times and sketch the path.

    (a) $\vec{r}(t) = (3\cos t, \sin t), t = 0, \pi/3, 3\pi/4, 3\pi/2$.
    (b) $\vec{r}(t) = (\sin t, t^2 - 1), t = -\pi/2, 0, \pi/2, \pi$.
    (c) $\vec{r}(t) = (\sqrt{t}, \sqrt{t}/(t+1)), t = 1/4, 1, 2, 4$.
    (d) $\vec{r}(t) = (t - 2\sin t, 1 - 2\cos t), t = -\pi/3, 0, \pi, 2\pi$.

2. For each of the vector functions in Exercise 1, find the velocity $\vec{v}$ at the indicated times.

3. For each of the vector functions in Exercise 1, find the acceleration $\vec{a}$ at the indicated times.

4. Using the relationships of Equation (1.7), prove that

$$\frac{d\theta}{dt} = \frac{x\,(dy/dt) - y\,(dx/dt)}{x^2 + y^2}. \qquad (1.25)$$

    It follows that acceleration is radial if and only if $x\,(dy/dt)$ and $y\,(dx/dt)$ differ by a constant.

    In Exercises 5–10, let $\vec{r}(t) = (t^2 - t, t\sqrt{2t - t^2}), 0 \le t \le 2$.

5. Sketch the curve described by $\vec{r}(t)$.

6. Find $r(t)$.

7. Find $\vec{u}_r$ and $\vec{u}_\theta$ as functions of t.

8. Find dr/dt and $d\theta/dt$. (Hint: use Exercise 4.)

9. Express the velocity in terms of the local coordinates $\vec{u}_r$ and $\vec{u}_\theta$.

10. Express the acceleration in terms of the local coordinates $\vec{u}_r$ and $\vec{u}_\theta$.

11. Compare and contrast the proof of Theorems 1.1 and 1.2 given in Section 1.5 with Newton's original proof. Which proof do you like better? Why?

12. The constant ε in Equation (1.18) is called the *eccentricity*. What happens if ε is larger than 1? equal to 1? equal to 0? less than 0?

13. If $\varepsilon = 1$ in Equation (1.18), then there is a value of θ, $\theta = \pi$, for which r is not defined. If $\varepsilon = -1$, then r is not defined when $\theta = 0$. If $|\varepsilon|$ is larger that one, then there is an interval of values for θ over which r is not defined. Find this interval in terms of ε and explain its relationship to the path $r(1 + \varepsilon \cos\theta) = c$ when $|\varepsilon| > 1$.

14. Prove that the constant c in Equation (1.18) is half of the *latus rectum* L of the ellipse.

15. Comparing Equations (1.5) and (1.20) we see that $|RS|$ is the change in velocity per unit time, which corresponds to $|\vec{a}(t)|$; $2A$ is twice the area swept out per unit time, which corresponds to k; $|PF_1|$ is the distance from the sun, which corresponds to r; but L is twice c. Explain this discrepancy.

16. Find the acceleration in terms of distance from the origin of a particle moving along an ellipse with its center at the origin (instead of having one focus at the origin) and sweeping out equal area in equal time.

17. Find the acceleration in terms of distance from the origin of a particle moving along the logarithmic spiral

$$r = e^{-c\theta},$$

where c is an arbitrary constant, given that the particle sweeps out equal area in equal time.

18. Given a particle that sweeps out an equal area in equal time and whose path is given by $r = f(\theta)$, show that the acceleration is given by

$$\vec{a} = \frac{k^2}{r^3} \left[\frac{f''(\theta)}{f(\theta)} - 2 \left(\frac{f'(\theta)}{f(\theta)} \right)^2 - 1 \right] \vec{u}_r.$$

19. Find the acceleration in terms of distance from the origin of a particle with constant angular velocity,

$$\frac{d\theta}{dt} = k,$$

which follows an elliptical orbit with the sun at one focus,

$$r(1 + \varepsilon \cos \theta) = c.$$

20. If we use complex coordinates to represent the points in the plane of the orbit, then we have the correspondence

$$\vec{u}_r = (\cos \theta, \sin \theta) \longleftrightarrow e^{i\theta} = \cos \theta + i \sin \theta.$$

Show that $\vec{u}_\theta$ corresponds to $ie^{i\theta}$.

21. Show that if

$$\vec{r} = re^{i\theta}, \tag{1.26}$$

where r and θ are functions of time t, then

$$\vec{v} = \frac{d\vec{r}}{dt} = \frac{dr}{dt} e^{i\theta} + r\frac{d\theta}{dt} ie^{i\theta} \qquad (1.27)$$

and

$$\vec{a} = \frac{d^2\vec{r}}{dt^2} = \left(\frac{d^2r}{dt^2} - r\left(\frac{d\theta}{dt} \right)^2 \right) e^{i\theta} + \frac{1}{r}\frac{d}{dt}\left(r^2 \frac{d\theta}{dt} \right) ie^{i\theta}. \qquad (1.28)$$

2

Vector Algebra

2.1 Basic Notions

While the terminology of calculus that we have at hand is certainly sufficient to prove the converse of Theorem 1.3, namely, that Newton's law of gravity implies that planets must move in elliptical orbits with the sun at one focus, this and other arguments we are to make will be greatly simplified if we adopt a language developed in the late nineteenth century, that of vector algebra.

Vector algebra came into its own not because it helped in understanding celestial mechanics, but because it clarified the then emerging explanations of electricity and magnetism. With roots in the work on quaternions by William Hamilton (1805–1865) and the calculus of extension by Hermann Günther Grassman (1809–1877), vector algebra began to gain acceptance when it was employed by James Clerk Maxwell (1831–1879) in his explanations of electricity and magnetism in the 1870s. It received its first full published exposition in 1893 in the first volume of Oliver Heaviside's (1850–1925) *Electromagnetic Theory*. With the publication of *Vector Analysis* in 1901 by Edwin B. Wilson (1879–1964), based on lectures by J. Willard Gibbs (1839–1903), the language of vector algebra became entrenched in mathematical physics.

Yet, as we shall see, electricity and magnetism are much more succinctly explained in the language of differential forms. Everything we do in this chapter can be and eventually will be restated in terms of differential forms. There are problems, however, in leaping directly into them. They represent a much higher level of abstraction than we need at present. Vectors are more concrete and recognizable objects, and vector algebra is still an important tool that you are likely to run across elsewhere.

Addition and Scalar Multiplication

We shall restrict our attention in this chapter to vectors in three dimensions, although most of what will be said carries over into an arbitrary number of dimensions. We already have representations of vectors as points: to say that $\vec{r} = (2, 4, 3)$ means that the magnitude of $\vec{r}$ is the distance from the origin to $(2, 4, 3)$, and the direction of $\vec{r}$ is the direction from the origin

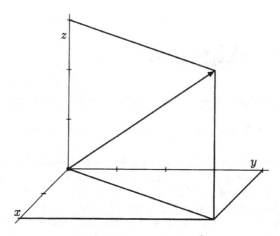

FIGURE 2.1. $2\vec{\imath} + 4\vec{\jmath} + 3\vec{k}$.

to $(2, 4, 3)$. This representation of $\vec{r}$ as a point in x, y, z space in effect decomposes $\vec{r}$ into a sum of three vectors: a vector of length 2 in the x direction, a vector of length 4 in the y direction, and a vector of length 3 in the z direction (Figure 2.1). To make this decomposition slightly more explicit, we set $\vec{\imath}$, $\vec{\jmath}$, and $\vec{k}$ to be the unit vectors in the x, y, and z directions, respectively. We can then write

$$\vec{r} = 2\vec{\imath} + 4\vec{\jmath} + 3\vec{k}.$$

We also know how to add vectors by using the parallelogram rule described in Section 1.2. From the parallelogram rule, we see that vector addition is commutative:

$$\vec{r} + \vec{s} = \vec{s} + \vec{r}, \tag{2.1}$$

because it does not matter which side of the parallelogram we lay down first. Vector addition is associative:

$$(\vec{r} + \vec{s}) + \vec{t} = \vec{r} + (\vec{s} + \vec{t}). \tag{2.2}$$

The sum of $\vec{r}$, $\vec{s}$, and $\vec{t}$ is the vector to the far corner of the parallelepiped formed by these three vectors (Figure 2.2). We also have two distributive laws:

$$(a + b)\vec{r} = a\vec{r} + b\vec{r}, \tag{2.3}$$

$$a(\vec{r} + \vec{s}) = a\vec{r} + a\vec{s}. \tag{2.4}$$

Equation (2.4) is valid because magnification of $\vec{r} + \vec{s}$ by a factor a corresponds to magnifying the entire parallelogram, which means magnifying

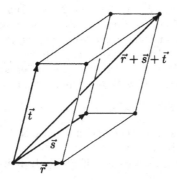

FIGURE 2.2. $\vec{r} + \vec{s} + \vec{t}$

each side by a factor of a. Reversing the direction of $\vec{r} + \vec{s}$ (that is, multiplication by -1) corresponds to reversing the directions of both $\vec{r}$ and $\vec{s}$.

Together these properties imply that we can add any two vectors by adding the respective coordinates of their representations:

$$
\begin{aligned}
(2,4,3) + (5,-1,2) &= 2\vec{\imath} + 4\vec{\jmath} + 3\vec{k} + 5\vec{\imath} - \vec{\jmath} + 2\vec{k} \\
&= 7\vec{\imath} + 3\vec{\jmath} + 5\vec{k} \\
&= (7,3,5).
\end{aligned}
$$

We multiply a vector by a real number by multiplying each of its coordinates:

$$
\begin{aligned}
3(5,-1,2) &= 3(5\vec{\imath} - \vec{\jmath} + 2\vec{k}) \\
&= 15\vec{\imath} - 3\vec{\jmath} + 6\vec{k} \\
&= (15,-3,6).
\end{aligned}
$$

Given two points, say $\vec{r}$ and $\vec{s}$, the vector from $\vec{r}$ to $\vec{s}$ is $\vec{s} - \vec{r}$. Here again we are exploiting the duality between points and vectors and using the fact that

$$
\vec{r} + (\vec{s} - \vec{r}) = \vec{s}.
$$

The vector from $(2,4,3)$ to $(5,-1,2)$ is $(3,-5,-1)$. Remember that this says the distance and direction from $(2,4,3)$ to $(5,-1,2)$ is the same as the distance and direction from the origin to $(3,-5,-1)$.

Parallel Vectors and Lines

We define vectors $\vec{r}$ and $\vec{s}$ to be *parallel* if and only if there is a real number a such that $\vec{r} = a\vec{s}$ or $\vec{s} = a\vec{r}$. The zero vector, $\vec{0}$, is thus parallel to every other vector. As examples, $(2, 4, 3)$, $(4, 8, 6) = 2(2, 4, 3)$, $(10/3, 20/3, 5) = (5/3)(2, 4, 3)$, and $(-1, -2, -3/2) = -(1/2)(2, 4, 3)$ are all parallel vectors.

A nonzero vector $\vec{r}$ defines a unique *line* through the origin, namely, the line passing through $\vec{r}$ and the origin. We can parametrize this line by using the real variable t and define it as the set of points: $\{t\vec{r} \mid t \in \mathbf{R}\}$, where $\mathbf{R}$ denotes the set of real numbers. Note that any other nonzero vector parallel to $\vec{r}$ gives the same line, but with a different parametrization:

$$\{(2t, 4t, 3t) \mid t \in \mathbf{R}\} = \{(4s, 8s, 6s) \mid s \in \mathbf{R}\}.$$

These two sets are the same. The only difference is the value of the parameter at a given point in the set.

A point (x, y, z) lies on this line if and only if it satisfies the vector equation

$$(x, y, z) = (2t, 4t, 3t)$$

for some t. This can also be written as a system of three simultaneous equations:

$$\begin{aligned} x &= 2t, \\ y &= 4t, \\ z &= 3t. \end{aligned}$$

If desired, we can eliminate t from these equations and reduce it to a system of two equations in x, y, and z:

$$\begin{aligned} 2x &= y, \\ 3x &= 2z. \end{aligned}$$

Better than this we cannot do. We always need two equations to describe a line in three dimensions.

If we want a line that passes through some point other than the origin, for example through $\vec{a} = (3, 5, -1)$, but is still parallel to the vector $\vec{r} = (2, 4, 3)$, then we add $\vec{a}$ to each of the points in the original set. This new line is described as the set

$$\{\vec{a} + t\vec{r} \mid t \in \mathbf{R}\}.$$

Our system of equations becomes

$$\begin{aligned} x &= 3 + 2t, \\ y &= 5 + 4t, \\ z &= -1 + 3t, \end{aligned}$$

or equivalently,

$$\begin{aligned} 2x &= y + 1, \\ 3x &= 2z + 11. \end{aligned}$$

Planes

A *plane* containing the origin is uniquely determined by two nonparallel vectors, call them $\vec{r}$ and $\vec{s}$. Specifically, it is the set of all *linear combinations* of $\vec{r}$ and $\vec{s}$:

$$\{u\vec{r} + v\vec{s} \mid u,\, v \in \mathbf{R}\}.$$

We say that $\vec{r}$ and $\vec{s}$ *span* this plane. The significant difference between a line and a plane is that a line has one free parameter while a plane has two free parameters. As with the line, there is more than one choice of parametrization for the same plane. We can start with any two points in this plane as long as the corresponding vectors are not parallel.

Each point in our plane corresponds to a *unique* pair of real numbers, (u, v). To see this, suppose that

$$u_1\vec{r} + v_1\vec{s} = u_2\vec{r} + v_2\vec{s}.$$

If $u_1 = u_2$, then $v_1\vec{s} = v_2\vec{s}$, and so $v_1 = v_2$. In this case, the two representations of the point are identical. If $u_1 \neq u_2$, then we can solve for $\vec{r}$ in terms of $\vec{s}$:

$$\begin{aligned}
u_1\vec{r} - u_2\vec{r} &= v_2\vec{s} - v_1\vec{s}, \\
(u_1 - u_2)\vec{r} &= (v_2 - v_1)\vec{s}, \\
\vec{r} &= \frac{v_2 - v_1}{u_1 - u_2}\vec{s}.
\end{aligned}$$

This case contradicts our assumption that $\vec{r}$ and $\vec{s}$ are not parallel.

If our plane is defined by the vectors $\vec{r} = (2, 4, 3)$ and $\vec{s} = (5, -1, 2)$, then any point (x, y, z) in our plane must satisfy the vector equation

$$(x, y, z) = (2u, 4u, 3u) + (5v, -v, 2v),$$

which we can write as a system of three equations:

$$\begin{aligned}
x &= 2u + 5v, \\
y &= 4u - v, \\
z &= 3u + 2v.
\end{aligned}$$

If $\vec{r}$ and $\vec{s}$ are not parallel, then it is always possible to reduce these three equations to a single equation in x, y, and z:

$$x + y - 2z = 0.$$

To work backward from the equation of a plane to a parametrization, we need only find two nonparallel vectors satisfying this equation. For example, $(1,1,1)$ and $(2,0,1)$ will do. Notice that this does not give us the same parametrization with which we started, but rather a new and equally valid parametrization describing the same plane:

$$\{u(1,\ 1,\ 1) + v(2,\ 0,\ 1)\ |\ u, v \in \mathbf{R}\}.$$

That our original vectors (2,4,3) and (5, −1, 2) lie in this plane is verified by observing that

$$
\begin{aligned}
(2,\ 4,\ 3) &= 4(1,\ 1,\ 1) - (2,\ 0,\ 1), \\
(5,\ -1,\ 2) &= -(1,\ 1,\ 1) + 3(2,\ 0,\ 1).
\end{aligned}
$$

If we want a plane that is parallel to the original plane but passes through $\vec{a} = (3, 5, -1)$ instead of the origin, then we add this vector to each of the points in our original plane. Our system of three equations becomes

$$
\begin{aligned}
x &= 3 + 2u + 5v, \\
y &= 5 + 4u - v, \\
z &= -1 + 3u + 2v,
\end{aligned}
$$

or equivalently,

$$x + y - 2z = 10.$$

If we are given this equation and want to find a parametrization, we need to find a point in this plane, for example (5,5,0), and then find two vectors from that point to two other points in this plane: the vector from (5,5,0) to (6,6,1) is (1,1,1); the vector from (5,5,0) to (7,5,1) is (2,0,1). We say that (1,1,1) and (2,0,1) also *span* this plane. One of the possible parametrizations is thus given by

$$\{(5,\ 5,\ 0) + u(1,\ 1,\ 1) + v(2,\ 0,\ 1)\ |\ u, v \in \mathbf{R}\}.$$

2.2 The Dot Product

Work is force times distance. If the force acts in the same direction as the moving object, then computing the work is simply a matter of multiplying these quantities. But, if the force is at an angle to the motion (Figure 2.3), then only the component of the force in the direction of motion contributes to the work.

We let the vector $\vec{F}$ represent the direction and magnitude of the force, $\vec{d}$ the direction and magnitude of the distance moved, and θ the angle between them. The magnitude of the effective force is $|\vec{F}| \cos \theta$, and so the work is

$$\text{Work} = |\vec{F}||\vec{d}| \cos \theta.$$

The *scalar* or *dot product* is a notation for representing this particular product of vectors:

$$\vec{F} \cdot \vec{d} = |\vec{F}||\vec{d}| \cos \theta. \tag{2.5}$$

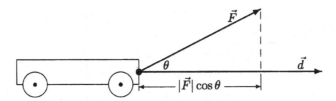

FIGURE 2.3. The component of the force in the direction of the distance.

The particular case where the vectors are equal, and thus $\theta = 0$, gives us

$$\vec{r} \cdot \vec{r} = |\vec{r}|^2, \tag{2.6}$$

so that

$$|\vec{r}| = \sqrt{\vec{r} \cdot \vec{r}}. \tag{2.7}$$

As a consequence, we see that

$$\vec{\imath} \cdot \vec{\imath} = \vec{\jmath} \cdot \vec{\jmath} = \vec{k} \cdot \vec{k} = 1. \tag{2.8}$$

Since $\cos \theta = \cos(-\theta)$, the dot product is commutative:

$$\vec{r} \cdot \vec{s} = \vec{s} \cdot \vec{r}. \tag{2.9}$$

It also follows from the definition that for any real number a,

$$a(\vec{r} \cdot \vec{s}) = a\vec{r} \cdot \vec{s} = \vec{r} \cdot a\vec{s}. \tag{2.10}$$

Because $\cos \pi/2 = 0$, the dot product of perpendicular vectors is always 0 ,and therefore,

$$\vec{\imath} \cdot \vec{\jmath} = \vec{\imath} \cdot \vec{k} = \vec{\jmath} \cdot \vec{k} = 0. \tag{2.11}$$

Decomposition of a Vector

Given a unit vector, $\vec{u}$, this product is particularly useful when decomposing $\vec{r}$ into the sum of a component parallel to $\vec{u}$ and a component perpendicular to $\vec{u}$. This decomposition is unique (Figure 2.4), and we can represent it by

$$\vec{r} = \vec{r}_u + \vec{r}_{u\perp}. \tag{2.12}$$

The magnitude of $\vec{r}_u$ is $|\vec{r}|\,|\cos \theta|$, where θ is the angle between $\vec{r}$ and $\vec{u}$. More specifically, we have

$$\vec{r}_u = |\vec{r}| \cos \theta\, \vec{u} = (\vec{r} \cdot \vec{u})\, \vec{u}. \tag{2.13}$$

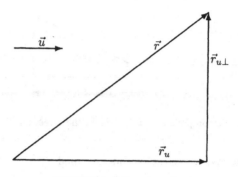

FIGURE 2.4. $\vec{r} = \vec{r}_u + \vec{r}_{u\perp}$.

If $\vec{s}$ is any nonzero vector, then $\vec{s}/|\vec{s}|$ is the unit vector in the direction of $\vec{s}$. If we define $\vec{r}_s$ to be the component of $\vec{r}$ in the direction of $\vec{s}$, then we see that

$$\vec{r}_s = |\vec{r}| \cos\theta \, \frac{\vec{s}}{|\vec{s}|} = \frac{\vec{r} \cdot \vec{s}}{|\vec{s}|} \frac{\vec{s}}{|\vec{s}|} = \frac{\vec{r} \cdot \vec{s}}{|\vec{s}|^2} \vec{s} = \left(\frac{\vec{r} \cdot \vec{s}}{\vec{s} \cdot \vec{s}} \right) \vec{s}. \qquad (2.14)$$

It follows that

$$\vec{r}_s \cdot \vec{s} = \left(\frac{\vec{r} \cdot \vec{s}}{\vec{s} \cdot \vec{s}} \right) \vec{s} \cdot \vec{s} = \vec{r} \cdot \vec{s}. \qquad (2.15)$$

The Distributive Law

The sum of any two vectors parallel to $\vec{t}$ is again a vector parallel to $\vec{t}$. The sum of two vectors perpendicular to $\vec{t}$ lies in the plane perpendicular to $\vec{t}$ and so is again a vector perpendicular to $\vec{t}$. This implies that if $\vec{v} = \vec{r} + \vec{s}$ then

$$\vec{v}_t = \vec{r}_t + \vec{s}_t, \quad \vec{v}_{t\perp} = \vec{r}_{t\perp} + \vec{s}_{t\perp}.$$

Equation (2.15) says that $\vec{v} \cdot \vec{t} = \vec{v}_t \cdot \vec{t}$. Combined with the observation we have just made, this implies that

$$(\vec{r} + \vec{s}) \cdot \vec{t} = \vec{v} \cdot \vec{t} = \vec{v}_t \cdot \vec{t} = (\vec{r}_t + \vec{s}_t) \cdot \vec{t}. \qquad (2.16)$$

We now use Equation (2.14) and our distributive law for the sum of two scalars times a vector [Equation (2.3)]:

$$(\vec{r} + \vec{s}) \cdot \vec{t} = \left(\frac{\vec{r} \cdot \vec{t}}{\vec{t} \cdot \vec{t}} \vec{t} + \frac{\vec{s} \cdot \vec{t}}{\vec{t} \cdot \vec{t}} \vec{t} \right) \cdot \vec{t} = \left(\frac{\vec{r} \cdot \vec{t}}{\vec{t} \cdot \vec{t}} + \frac{\vec{s} \cdot \vec{t}}{\vec{t} \cdot \vec{t}} \right) \vec{t} \cdot \vec{t} = (\vec{r} \cdot \vec{t}) + (\vec{s} \cdot \vec{t}). \qquad (2.17)$$

The distributive law now implies a simple rule for taking the dot product of two vectors written in coordinate form:

$$
\begin{aligned}
(a, b, c) \cdot (f, g, h) &= (a\vec{\imath} + b\vec{\jmath} + c\vec{k}) \cdot (f\vec{\imath} + g\vec{\jmath} + h\vec{k}) \\
&= af\vec{\imath} \cdot \vec{\imath} + ag\vec{\imath} \cdot \vec{\jmath} + ah\vec{\imath} \cdot \vec{k} \\
&\quad + bf\vec{\jmath} \cdot \vec{\imath} + bg\vec{\jmath} \cdot \vec{\jmath} + bh\vec{\jmath} \cdot \vec{k} \\
&\quad + cf\vec{k} \cdot \vec{\imath} + cg\vec{k} \cdot \vec{\jmath} + ch\vec{k} \cdot \vec{k} \\
&= af + bg + ch.
\end{aligned}
\tag{2.18}
$$

It follows from Equations (2.7) and (2.18) that

$$
|(a, b, c)| = \sqrt{(a, b, c) \cdot (a, b, c)} = \sqrt{a^2 + b^2 + c^2}.
\tag{2.19}
$$

The Triangle Inequality

We can also use the dot product to prove the *triangle inequality*,

$$
|\vec{r} + \vec{s}| \leq |\vec{r}| + |\vec{s}|,
\tag{2.20}
$$

as follows:

$$
\begin{aligned}
|\vec{r} + \vec{s}|^2 &= (\vec{r} + \vec{s}) \cdot (\vec{r} + \vec{s}) \\
&= |\vec{r}|^2 + 2\vec{r} \cdot \vec{s} + |\vec{s}|^2 \\
&= |\vec{r}|^2 + 2|\vec{r}||\vec{s}| \cos\theta + |\vec{s}|^2 \\
&\leq |\vec{r}|^2 + 2|\vec{r}||\vec{s}| + |\vec{s}|^2 \\
&= (|\vec{r}| + |\vec{s}|)^2.
\end{aligned}
$$

Furthermore, we see from this argument that equality occurs if and only if $\cos\theta = 1$, that is, if and only if $\vec{r}$ and $\vec{s}$ have the same direction.

Examples

To get some feeling for how the dot product works in practice, let us take our vectors $\vec{r} = (2, 4, 3)$ and $\vec{s} = (5, -1, 2)$. We see that

$$
\begin{aligned}
(2, 4, 3) \cdot (5, -1, 2) &= 2 \times 5 + 4 \times (-1) + 3 \times 2 \\
&= 12,
\end{aligned}
$$

$$
\begin{aligned}
|(2, 4, 3)| &= \sqrt{2^2 + 4^2 + 3^2} \\
&= \sqrt{29},
\end{aligned}
$$

$$|(5,-1,2)| = \sqrt{5^2 + (-1)^2 + 2^2}$$
$$= \sqrt{30},$$

$$|(2,4,3) + (5,-1,2)| = |(7,3,5)|$$
$$= \sqrt{49 + 9 + 25}$$
$$= \sqrt{73} \leq \sqrt{29} + \sqrt{30},$$

$$\vec{r}_s = \frac{(2,4,3)\cdot(5,-1,2)}{(5,-1,2)\cdot(5,-1,2)}(5,-1,2)$$
$$= \frac{12}{30}(5,-1,2)$$
$$= \left(2, \frac{-2}{5}, \frac{4}{5}\right),$$

$$\vec{r}_{s\perp} = \vec{r} - \vec{r}_s$$
$$= (2,4,3) - \left(2, \frac{-2}{5}, \frac{4}{5}\right)$$
$$= \left(0, \frac{22}{5}, \frac{11}{5}\right),$$

$$\vec{s}_r = \frac{(5,-1,2)\cdot(2,4,3)}{(2,4,3)\cdot(2,4,3)}(2,4,3)$$
$$= \frac{12}{29}(2,4,3)$$
$$= \left(\frac{24}{29}, \frac{48}{29}, \frac{36}{29}\right),$$

$$\vec{s}_{r\perp} = (5,-1,2) - \left(\frac{24}{29}, \frac{48}{29}, \frac{36}{29}\right)$$
$$= \left(\frac{121}{29}, \frac{-77}{29}, \frac{22}{29}\right).$$

Also, if θ is the angle between $(2,4,3)$ and $(5,-1,2)$, then we can solve for $\cos\theta$ in Equation (2.5),

$$\cos\theta = \frac{(2,4,3)\cdot(5,-1,2)}{|(2,4,3)||(5,-1,2)|} = \frac{12}{\sqrt{29 \times 30}} = 0.406838\ldots.$$

The dot product is also useful whenever we are defining objects in terms of perpendicularity since two vectors are perpendicular if and only if their dot product is 0. (The vector $\vec{0}$, in addition to being parallel to every vector, is also perpendicular to every vector.) For example, the plane through $\vec{a} = (-2,6,5)$, perpendicular to $\vec{r} = (1,5,-4)$, consists of those points (x,y,z)

for which $\vec{r}$ is perpendicular to the vector from $(-2, 6, 5)$ to (x, y, z):

$$
\begin{aligned}
0 &= (1, 5, -4) \cdot ((x, y, z) - (-2, 6, 5)) \\
&= (1, 5, -4) \cdot (x + 2, y - 6, z - 5) \\
&= x + 2 + 5y - 30 - 4z + 20 \\
&= x + 5y - 4z - 8.
\end{aligned}
$$

In general, the plane through $\vec{a} = (a_1, a_2, a_3)$, perpendicular to $\vec{r} = (r_1, r_2, r_3)$, satisfies the equation

$$
\begin{aligned}
0 &= (r_1, r_2, r_3) \cdot ((x, y, z) - (a_1, a_2, a_3)) \\
&= r_1 x + r_2 y + r_3 z - \vec{r} \cdot \vec{a}. \tag{2.21}
\end{aligned}
$$

Note that the coefficients of x, y, and z in the equation of a plane correspond to a vector perpendicular to that plane. It follows that parallel planes have equations that can be written so that they differ only in the constant term.

As another example, the minimum distance from the point $(7, -3, 4)$ to the plane $x + y - 2z = 10$ is the perpendicular distance. To find this distance, we first find a vector from any point on the plane, we can choose $(0, 0, -5)$, to $(7, -3, 4)$:

$$
\vec{t} = (7, -3, 4) - (0, 0, -5) = (7, -3, 9).
$$

The perpendicular distance is the magnitude of the component of $\vec{t}$ in the direction of the perpendicular to the plane, $\vec{r} = (1, 1, -2)$.

$$
|\vec{t}_r| = \frac{|(7, -3, 9) \cdot (1, 1, -2)|}{|(1, 1, -2)|} = \frac{|-14|}{\sqrt{6}} = \frac{7}{3}\sqrt{6}.
$$

2.3   The Cross Product

Consider a rigid spinning body with its center of mass at the origin. We can use a vector, $\vec{r}$, to represent its spin. The line from the origin to $\vec{r}$ is the axis of rotation, and we use the convention that if we look back toward the origin from $\vec{r}$ then the spin is counterclockwise. The magnitude, $|\vec{r}|$, represents the angular velocity which could be measured in radians per second. The faster the spin, the longer the vector. Let $\vec{s}$ be any point on the body and let θ be the angle from $\vec{r}$ to $\vec{s}$ (Figure 2.5). Since the point $\vec{s}$ is moving around the axis, it has a velocity, $\vec{v}$. The magnitude of this velocity, $|\vec{v}|$, is $|\vec{r}|$ times the distance of $\vec{s}$ from the axis of rotation, that is,

$$
|\vec{v}| = |\vec{r}||\vec{s}||\sin\theta|.
$$

The direction of $\vec{v}$ is perpendicular to the plane spanned by $\vec{r}$ and $\vec{s}$ and is completely determined by our spin convention. We can think of $\vec{v}$ as a peculiar product of $\vec{r}$ and $\vec{s}$ called the *vector* or *cross product*:

$$
\vec{v} = \vec{r} \times \vec{s}. \tag{2.22}
$$

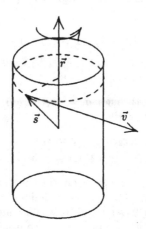

FIGURE 2.5. A rigid spinning body.

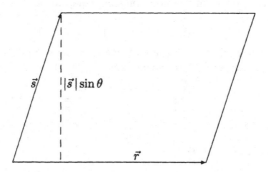

FIGURE 2.6. The magnitude of $\vec{r} \times \vec{s}$.

Note that the magnitude of $\vec{r} \times \vec{s}$ is the area of the parallelogram defined by $\vec{r}$ and $\vec{s}$ (Figure 2.6). If $\vec{r}$ and $\vec{s}$ are parallel, then $\theta = 0$ or π, and so $\sin \theta = 0$ and the cross product is $\vec{0}$. In particular, this tells us that

$$\vec{r} \times \vec{r} = \vec{0}. \tag{2.23}$$

The Right-Hand Rule

If $\vec{r}$ and $\vec{s}$ are not parallel, then $\vec{r}$ and $\vec{s}$ span a plane. The direction of $\vec{r} \times \vec{s}$ is taken to be perpendicular to this plane. That restricts us to two possible directions. The choice of which of these two we take is determined by the spin convention described previously which is referred to as the *right-hand*

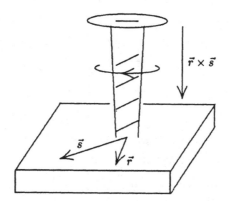

FIGURE 2.7. The right-hand rule: first version.

rule and can be described in several different but equivalent ways.

Imagine a standard (right-handed) screw through the origin, perpendicular to the plane spanned by $\vec{r}$ and $\vec{s}$. If you turn this screw from $\vec{r}$ to $\vec{s}$, then the direction of its resulting movement (either in to or out from the plane) is the direction of $\vec{r} \times \vec{s}$ (Figure 2.7).

OR

Cup your right-hand with the thumb outstretched, as if you are about to hitchhike. Place it so that the little finger lies flat on the plane spanned by $\vec{r}$ and $\vec{s}$ and so that the direction from $\vec{r}$ to $\vec{s}$ is the same as the direction of the curl of your fingers as you travel from the wrist around to the finger tips. The thumb is pointing in the direction of $\vec{r} \times \vec{s}$ (Figure 2.8).

OR

Stretch the thumb, index finger, and middle finger of your right-hand so that each is pointing in a different direction. If you line up the thumb with $\vec{r}$ and the index finger with $\vec{s}$, then the middle finger points in the direction of $\vec{r} \times \vec{s}$ (Figure 2.9).

The first observation that we need to make is that the cross product is *not* commutative. If we switch $\vec{r}$ and $\vec{s}$, then we change the direction in which we are turning the screw, and we wind up going in the opposite direction:

$$\vec{s} \times \vec{r} = -\vec{r} \times \vec{s}. \tag{2.24}$$

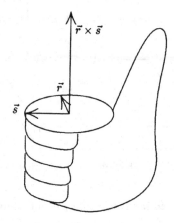

FIGURE 2.8. The right-hand rule: second version.

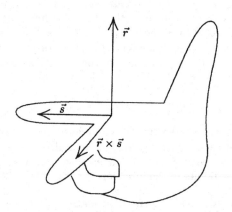

FIGURE 2.9. The right-hand rule: third version.

It is left as an exercise to show that the cross product is also *not* associative.

Applying the definition of the cross product to the basis vectors $\vec{\imath}, \vec{\jmath}$, and $\vec{k}$, we see that

$$\vec{\imath} \times \vec{\imath} = \vec{\jmath} \times \vec{\jmath} = \vec{k} \times \vec{k} = \vec{0}, \tag{2.25}$$

$$\vec{\imath} \times \vec{\jmath} = \vec{k}, \quad \vec{\jmath} \times \vec{\imath} = -\vec{k}, \tag{2.26}$$

$$\vec{\jmath} \times \vec{k} = \vec{\imath}, \quad \vec{k} \times \vec{\jmath} = -\vec{\imath}, \tag{2.27}$$

$$\vec{k} \times \vec{\imath} = \vec{\jmath}, \quad \vec{\imath} \times \vec{k} = -\vec{\jmath}. \tag{2.28}$$

If we multiply the magnitude of one of our vectors by a positive real number, then the magnitude of the product is multiplied by the same positive constant and the direction of the cross product remains the same. If we reverse the direction of one of our vectors, that is, multiply it by -1, that reverses the direction of the resulting cross product. We thus have the following relationship for any real number a:

$$a(\vec{r} \times \vec{s}) = a\vec{r} \times \vec{s} = \vec{r} \times a\vec{s}. \tag{2.29}$$

The Distributive Law

The cross product is also distributive:

$$(\vec{r} + \vec{s}) \times \vec{t} = (\vec{r} \times \vec{t}) + (\vec{s} \times \vec{t}), \tag{2.30}$$

but it will take a little work to see why this is so. We can assume that $\vec{r}$, $\vec{s}$, and $\vec{t}$ are given in the order of the right-hand rule, for if not, then $\vec{r}, \vec{s}$, and $-\vec{t}$ are in the order of the right-hand rule. If

$$(\vec{r} + \vec{s}) \times -\vec{t} = (\vec{r} \times -\vec{t}) + (\vec{s} \times -\vec{t}),$$

then we can divide through by -1 to obtain Equation (2.30).

We begin with the one case on which we can build, namely, when $\vec{r}, \vec{s}$, and $\vec{t}$ are *mutually orthogonal*, each is perpendicular to the other two. The vectors $\vec{r}$ and $\vec{s}$ both lie in the plane perpendicular to $\vec{t}$, and since they are not parallel, they must span this plane. This implies that $\vec{r} + \vec{s}$ also lies in this plane, and so $\vec{r} + \vec{s}$ is perpendicular to $\vec{t}$. The magnitude of our cross product in this case is just the product of the magnitudes:

$$|(\vec{r} + \vec{s}) \times \vec{t}| = |\vec{r} + \vec{s}||\vec{t}|.$$

Since our cross product is perpendicular to $\vec{t}$, it lies in the plane spanned by $\vec{r}$ and $\vec{s}$. It is readily verified (by taking the dot product with $\vec{r} + \vec{s}$ and remembering that $\vec{r} \cdot \vec{s} = 0$) that the vector

$$\vec{v} = \frac{|\vec{s}|}{|\vec{r}|}\vec{r} - \frac{|\vec{r}|}{|\vec{s}|}\vec{s}$$

is perpendicular to $\vec{r} + \vec{s}$ and lies in this plane. The magnitude $|\vec{r} + \vec{s}|$ is

$$|\vec{r} + \vec{s}| = \sqrt{(\vec{r} + \vec{s}) \cdot (\vec{r} + \vec{s})} = \sqrt{|\vec{r}|^2 + 2\vec{r} \cdot \vec{s} + |\vec{s}|^2} = \sqrt{|\vec{r}|^2 + |\vec{s}|^2}.$$

But this is also the magnitude of $\vec{v}$ and so the unit vector in the direction of $\vec{v}$ is $\vec{v}/|\vec{r} + \vec{s}|$. That we want this vector and not its negative can be checked using the right-hand rule or by making certain that the signs are correct if we let $|\vec{r}|$ or $|\vec{s}|$ approach 0.

Since we know the magnitude and direction of our cross product, we have shown that in this special case where $\vec{r}$, $\vec{s}$, and $\vec{t}$ are mutually orthogonal:

$$
\begin{aligned}
(\vec{r} + \vec{s}) \times \vec{t} &= |\vec{r} + \vec{s}||\vec{t}| \frac{\vec{v}}{|\vec{r} + \vec{s}|} \\
&= |\vec{t}|\vec{v} \\
&= \frac{|\vec{s}||\vec{t}|}{|\vec{r}|} \vec{r} - \frac{|\vec{r}||\vec{t}|}{|\vec{s}|} \vec{s} \\
&= -|\vec{r}||\vec{t}| \frac{\vec{s}}{|\vec{s}|} + |\vec{s}||\vec{t}| \frac{\vec{r}}{|\vec{r}|} \\
&= (\vec{r} \times \vec{t}) + (\vec{s} \times \vec{t}),
\end{aligned}
$$

where the last equality holds because $\vec{r} \times \vec{t}$ is in the direction of $-\vec{s}$ and $\vec{s} \times \vec{t}$ is in the direction of $\vec{r}$.

If all we know about $\vec{r}$ and $\vec{s}$ is that they are both perpendicular to $\vec{t}$ (but not necessarily to each other), then we can decompose $\vec{s}$ into

$$\vec{s} = \vec{s}_r + \vec{s}_{r\perp} = \frac{\vec{s} \cdot \vec{r}}{\vec{r} \cdot \vec{r}} \vec{r} + \vec{s}_{r\perp}.$$

Using the special case proved previously, we see that

$$
\begin{aligned}
(\vec{r} + \vec{s}) \times \vec{t} &= \left(\vec{r} + \frac{\vec{s} \cdot \vec{r}}{\vec{r} \cdot \vec{r}} \vec{r} + \vec{s}_{r\perp} \right) \times \vec{t} \\
&= \left(\vec{r} + \frac{\vec{s} \cdot \vec{r}}{\vec{r} \cdot \vec{r}} \vec{r} \right) \times \vec{t} + \vec{s}_{r\perp} \times \vec{t} \\
&= \left(1 + \frac{\vec{s} \cdot \vec{r}}{\vec{r} \cdot \vec{r}} \right) (\vec{r} \times \vec{t}) + \vec{s}_{r\perp} \times \vec{t} \\
&= (\vec{r} \times \vec{t}) + (\vec{s}_r \times \vec{t}) + (\vec{s}_{r\perp} \times \vec{t}) \\
&= (\vec{r} \times \vec{t}) + (\vec{s} \times \vec{t}).
\end{aligned}
$$

What if $\vec{r}$ and $\vec{s}$ are not perpendicular to $\vec{t}$? If θ is the angle between $\vec{r}$ and $\vec{t}$, then the magnitude of $\vec{r}_{t\perp}$ is $|\vec{r}||\sin\theta|$. The plane defined by $\vec{r}$ and $\vec{t}$ is the same as the plane defined by $\vec{r}_{t\perp}$ and $\vec{t}$, and so

$$\vec{r} \times \vec{t} = \vec{r}_{t\perp} \times \vec{t}. \tag{2.31}$$

We can now prove the distributive law for three arbitrary nonzero vectors $\vec{r}$, $\vec{s}$, and $\vec{t}$:

$$
\begin{aligned}
(\vec{r} + \vec{s}) \times \vec{t} &= (\vec{r}_t + \vec{r}_{t\perp} + \vec{s}_t + \vec{s}_{t\perp}) \times \vec{t} \\
&= ((\vec{r}_t + \vec{s}_t) + (\vec{r}_{t\perp} + \vec{s}_{t\perp})) \times \vec{t} \\
&= (\vec{r}_{t\perp} + \vec{s}_{t\perp}) \times \vec{t}.
\end{aligned}
$$

We can use our distributive law for the case where the vectors in the summation are both perpendicular to $\vec{t}$:

$$
\begin{aligned}
(\vec{r} + \vec{s}) \times \vec{t} &= (\vec{r}_{t\perp} \times \vec{t}) + (\vec{s}_{t\perp} \times \vec{t}) \\
&= (\vec{r} \times \vec{t}) + (\vec{s} \times \vec{t}).
\end{aligned}
$$

This concludes the proof that the distributive law [Equation (2.30)] holds for any three nonzero vectors.

Equations (2.25) – (2.28) and (2.30) provide us with a rule for taking the cross product of two vectors given in coordinate form:

$$
\begin{aligned}
(a, b, c) \times (f, g, h) &= (a\vec{i} + b\vec{j} + c\vec{k}) \times (f\vec{i} + g\vec{j} + h\vec{k}) \\
&= af\,(\vec{i} \times \vec{i}) \;+\; ag\,(\vec{i} \times \vec{j}) \;+\; ah\left(\vec{i} \times \vec{k}\right) \\
&\quad +\; bf\,(\vec{j} \times \vec{i}) \;+\; bg\,(\vec{j} \times \vec{j}) \;+\; bh\left(\vec{j} \times \vec{k}\right) \\
&\quad +\; cf\left(\vec{k} \times \vec{i}\right) \;+\; cg\left(\vec{k} \times \vec{j}\right) \;+\; ch\left(\vec{k} \times \vec{k}\right) \\
&= ag\vec{k} - ah\vec{j} - bf\vec{k} + bh\vec{i} + cf\vec{j} - cg\vec{i} \\
&= (bh - cg)\vec{i} + (cf - ah)\vec{j} + (ag - bf)\vec{k} \\
&= (bh - cg, cf - ah, ag - bf). \qquad (2.32)
\end{aligned}
$$

Example

As an example, the plane spanned by $(1, -2, 3)$ and $(5, 0, -1)$ is perpendicular to the vector formed by the cross product of these two vectors:

$$(1, -2, 3) \times (5, 0, -1) = (2 - 0, 15 + 1, 0 + 10) = (2, 16, 10).$$

Thus, the plane spanned by these vectors and containing the origin has the equation

$$2x + 16y + 10z = 0,$$

or equivalently,

$$x + 8y + 5z = 0.$$

One trick for remembering how to take the cross product of two vectors given in coordinate form is to arrange the coordinates as in Figure 2.10.

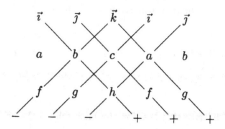

FIGURE 2.10. The cross product $(a, b, c) \times (f, g, h)$.

We take products along the indicated diagonals with a positive sign when moving down from left to right, a negative sign when moving down from right to left. The sum of these products is the cross product. Equivalently, this is the determinant of the 3×3 matrix:

$$(a, b, c) \times (f, g, h) = \begin{vmatrix} \vec{i} & \vec{j} & \vec{k} \\ a & b & c \\ f & g & h \end{vmatrix}.$$

2.4 Using Vector Algebra

For the purposes of this book, it is preferable to concentrate on the geometric meaning of the dot and cross products rather than the rules for formal manipulation of vectors in coordinate form. The power of vector algebra arises from these geometric interpretations. The impetus for adopting vector algebra came precisely from a desire to avoid cumbersome coordinate calculations. Oliver Heaviside, in the introduction to his chapter "The Elements of Vectorial Algebra" in *Electromagnetic Theory*, compares vector algebra with the traditional approach of Cartesian coordinates:

> The mere sight of the arrangement of symbols should call up an immediate picture of the physics symbolised, so that our formulæ may become *alive*, as it were. Now this is possible, and indeed, comparatively easy, in vectorial analysis; but is very difficult in Cartesian analysis, beyond a certain point, owing to the geometrically progressive complexity of the expressions to be interpreted and manipulated. Vectorial algebra is the natural language of vectors, and no one who has ever learnt it (not too late in life, however) will ever care to go back from the vitality of vectors to the bulky inanimateness of the Cartesian system.

Vector algebra is particularly well suited for physical applications because there is no natural system of coordinates for our universe, and so physical

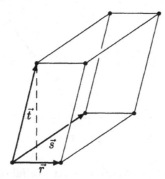

FIGURE 2.11. The scalar triple product: $\vec{r} \times \vec{s} \cdot \vec{t}$.

laws should be invariant of any coordinate frame. This means that we want mathematical models, and ways of manipulating them, that avoid specifying a set of fixed basis vectors. This is precisely what vector algebra does.

The Scalar Triple Product

One attractive consequence of the geometric view of the dot and cross products is the immediate interpretation of the *scalar triple product* that it yields. The scalar triple product of $\vec{r}$, $\vec{s}$, and $\vec{t}$ is defined to be

$$(\vec{r} \times \vec{s}) \cdot \vec{t}$$

or, more simply, just $\vec{r} \times \vec{s} \cdot \vec{t}$, which is well-defined since we must take the cross product first.

If we assume that no two of these vectors are parallel, then they define a parallelepiped (Figure 2.11). The magnitude of the cross product $\vec{r} \times \vec{s}$ is the area of the base, and if $\vec{r}$, $\vec{s}$, and $\vec{t}$ are in the order of the right-hand rule, then $\vec{r} \times \vec{s}$ points in the direction of the altitude of our parallelepiped. The actual altitude is the component of $\vec{t}$ in the direction of $\vec{r} \times \vec{s}$, which is

$$\frac{\vec{t} \cdot (\vec{r} \times \vec{s})}{|\vec{r} \times \vec{s}|}.$$

It follows that the volume of this solid defined by $\vec{r}$, $\vec{s}$, and $\vec{t}$ is

$$\frac{\vec{t} \cdot (\vec{r} \times \vec{s})}{|\vec{r} \times \vec{s}|}|\vec{r} \times \vec{s}| \; = \; \vec{t} \cdot (\vec{r} \times \vec{s}) \; = \; (\vec{r} \times \vec{s}) \cdot \vec{t},$$

the scalar triple product. If $\vec{r}$, $\vec{s}$, and $\vec{t}$ are not in the order of the right-hand rule, then replacing $\vec{t}$ by $-\vec{t}$ rectifies this, and so the scalar triple product will be the negative of the volume of the parallelepiped.

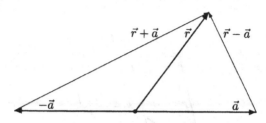

FIGURE 2.12. Theorem 2.1.

We have just proved a most curious and powerful result. There are 12 distinct scalar triple products that can be constructed using the three vectors $\vec{r}$, $\vec{s}$, and $\vec{t}$, namely,

$$\vec{r} \times \vec{s} \cdot \vec{t} \qquad\qquad \vec{t} \times \vec{s} \cdot \vec{r}$$
$$\vec{s} \times \vec{t} \cdot \vec{r} \qquad\qquad \vec{s} \times \vec{r} \cdot \vec{t}$$
$$\vec{t} \times \vec{r} \cdot \vec{s} \qquad\qquad \vec{r} \times \vec{t} \cdot \vec{s}$$
$$\vec{r} \cdot \vec{s} \times \vec{t} \qquad\qquad \vec{t} \cdot \vec{s} \times \vec{r}$$
$$\vec{s} \cdot \vec{t} \times \vec{r} \qquad\qquad \vec{s} \cdot \vec{r} \times \vec{t}$$
$$\vec{t} \cdot \vec{r} \times \vec{s} \qquad\qquad \vec{r} \cdot \vec{t} \times \vec{s}.$$

But, these 12 scalar triple products can take on only two possible values, either the volume of the parallelepiped defined by $\vec{r}$, $\vec{s}$, and $\vec{t}$, or its negative. If $\vec{r}$, $\vec{s}$, and $\vec{t}$ are in the order of the right-hand rule, then each of the scalar triples on the left is equal to the volume of the parallelpiped; each of the six on the right equals the negative of the volume of our solid.

Applications to Euclidean Geometry

Vector algebra also provides a powerful tool for proving the results of Euclidean geometry, and this power comes precisely from the distributive laws that we have struggled to prove. As examples, consider the following theorems.

Theorem 2.1 *Let $\vec{r}$ be any point on a circle with its center at the origin and let the segment from $-\vec{a}$ to $\vec{a}$ be a diameter of this circle (so that $|\vec{a}| = |\vec{r}|$). Then, the vector from $\vec{a}$ to $\vec{r}$ is perpendicular to the vector from $-\vec{a}$ to $\vec{r}$ (Figure 2.12).*

Proof: The vector from $\vec{a}$ to $\vec{r}$ is $\vec{r} - \vec{a}$, the vector from $-\vec{a}$ to $\vec{r}$ is $\vec{r} + \vec{a}$. The dot product of these vectors is

$$\begin{aligned}
(\vec{r} - \vec{a}) \cdot (\vec{r} + \vec{a}) &= (\vec{r} \cdot \vec{r}) + (\vec{r} \cdot \vec{a}) - (\vec{a} \cdot \vec{r}) - (\vec{a} \cdot \vec{a}) \\
&= |\vec{r}|^2 - |\vec{a}|^2 = 0,
\end{aligned}$$

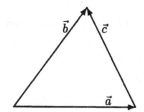

FIGURE 2.13. Theorem 2.2.

and so these vectors are perpendicular.

Q.E.D.

Theorem 2.2 *If a, b, and c are the lengths of the sides of a triangle, and if θ is the angle between the sides of lengths a and b (Figure 2.13), then*

$$c^2 = a^2 + b^2 - 2ab\cos\theta.$$

Proof: Let the vector $\vec{a}$ represent the side of length a, $|\vec{a}| = a$, $\vec{b}$ the side of length b, $|\vec{b}| = b$, and $\vec{c} = \vec{b} - \vec{a}$ the side of length c, $|\vec{c}| = c$. We then see that

$$\begin{aligned}
c^2 &= |\vec{c}|^2 = |\vec{b} - \vec{a}|^2 = (\vec{b} - \vec{a}) \cdot (\vec{b} - \vec{a}) \\
&= |\vec{b}|^2 - \left(2\vec{b} \cdot \vec{a}\right) + |\vec{a}|^2 = b^2 - 2ab\cos\theta + a^2.
\end{aligned}$$

Q.E.D.

Finally, we consider the equation in polar coordinates of an ellipse with one focus at the origin [Equation (1.18)]:

$$r(1 + \varepsilon\cos\theta) = c.$$

Using the notation of vector algebra, we can rewrite this in a form independent of any coordinate frame:

$$|\vec{r}| + \vec{r} \cdot \vec{\varepsilon} = c, \tag{2.33}$$

where $\vec{\varepsilon}$ is a constant vector of length less than 1 and c is a positive constant.

2.5   Exercises

1. For each of the following pairs of vectors, find their sum, dot product, and cross product.

   (a) $(2, -3, 1)$, $(6, 2, -3)$
   (b) $(5, -6, 1)$, $(3, 2, -3)$
   (c) $(3, 0, -2)$, $(-6, 0, 4)$
   (d) $(-2, 5, 1)$, $(3, 0, 6)$

2. Which pairs in Exercise 1 are perpendicular; which are parallel?

3. Illustrate Equations (2.3) and (2.4) by drawing vectors in a plane.
   In Exercises 4 to 8, use $\vec{r} = (-1, 2, -2)$, $\vec{s} = (3, -5, 4)$.

4. Find $|\vec{r}|$, $|\vec{s}|$, and $|\vec{r} + \vec{s}|$. Verify that the sum of any two of these quantities is greater than the third.

5. Find $\vec{r}_s$, $\vec{r}_{s\perp}$, $\vec{s}_r$, and $\vec{s}_{r\perp}$.

6. Find the cosine of the angle between $\vec{r}$ and $\vec{s}$. Find the sine of the angle between $\vec{r}$ and $\vec{s}$ (assuming the angle lies between 0 and π).

7. Find the equation of the plane through the origin spanned by $\vec{r}$ and $\vec{s}$.

8. Find the equation of the plane through $(3, -1, 2)$ spanned by $\vec{r}$ and $\vec{s}$.

9. Consider the plane containing the origin and spanned by the vectors $\vec{r} = (2, 0, -1)$ and $\vec{s} = (-3, 1, 0)$. Which of the following points lie on this plane? For those that do, express them as a linear combination of $\vec{r}$ and $\vec{s}$.

   (a) $(5, 1, -1)$
   (b) $(6, -2, 0)$
   (c) $(1, 1, -2)$
   (d) $(1, -1, -1)$

   In Exercises 10 to 14, the plane in question is the plane whose equation is
   $$x - 4y + 7z = 3.$$

10. Find a vector perpendicular to this plane.

11. Find two vectors that span this plane.

12. Find a parametric representation for this plane.

13. Find the perpendicular distance from $(1, 1, 1)$ to this plane.

14. Find the equation of the plane parallel to this one but passing through $(2, 0, 3)$.

15. Find the distance from $(0, -2, -3)$ to the line passing through $(-1, 2, 0)$ and $(2, -4, 2)$.

16. Find the cosine of the angle between $\vec{i} + \vec{j}$ and $\vec{i} + \vec{j} + \vec{k}$.

17. Let $\vec{a}$ be a nonzero vector. If $\vec{a} \cdot \vec{x} = \vec{a} \cdot \vec{y}$ and $\vec{a} \times \vec{x} = \vec{a} \times \vec{y}$, does it necessarily follow that $\vec{x} = \vec{y}$? Justify your answer.

18. Let $\vec{r} = (2, 1, 1)$. Find the vector $\vec{x}$ that satisfies

$$\vec{r} \cdot \vec{x} = 2 \quad \text{and} \quad \vec{r} \times \vec{x} = (-1, 3, -1).$$

19. Given that $\vec{r} \cdot \vec{s} = 0$, $\vec{r} \cdot \vec{x} = c$, and $\vec{r} \times \vec{x} = \vec{s}$, find the component of $\vec{x}$ in each of the three mutually orthogonal directions: $\vec{r}$, $\vec{s}$, and $\vec{r} \times \vec{s}$.

20. Prove that the cross product is nonassociative. That is, find three vectors $\vec{r}$, $\vec{s}$, and $\vec{t}$ for which

$$(\vec{r} \times \vec{s}) \times \vec{t} \neq \vec{r} \times (\vec{s} \times \vec{t}).$$

21. Prove that
$$|\vec{r} \times \vec{s}|^2 + (\vec{r} \cdot \vec{s})^2 = |\vec{r}|^2 |\vec{s}|^2. \tag{2.34}$$

22. Prove that if $\vec{a} \times \vec{b} = \vec{0}$ and $\vec{a} \cdot \vec{b} = 0$, then at least one of these vectors must be $\vec{0}$.

23. Prove that
$$\vec{a} \cdot \vec{a} \times \vec{b} = 0$$
for any two vectors $\vec{a}$ and $\vec{b}$.

24. Let $\vec{r}$ be a point on a circle with diameter from $-\vec{a}$ to $\vec{a}$ (so that $|\vec{r}| = |\vec{a}|$). Draw the chord through $\vec{r}$ perpendicular to $\vec{a}$ (the length and direction of this chord is $2\vec{r}_{a\perp}$). Let $\vec{t} = \vec{r}_a$ be the point of intersection of this chord with the diameter. Prove that the square of half the length of the chord is the product of the distance from $-\vec{a}$ to $\vec{t}$ times the distance from $\vec{t}$ to $\vec{a}$:

$$|\vec{r}_{a\perp}|^2 = (\vec{t} + \vec{a}) \cdot (\vec{a} - \vec{t}).$$

25. Consider a parallelogram defined by the vectors $\vec{a}$ and $\vec{b}$. Describe the diagonals in terms of these two vectors. Show that the diagonals are perpendicular if and only if $|\vec{a}| = |\vec{b}|$. Use vector algebra to show that the diagonals bisect each other.

26. Prove that
$$|\vec{a} + \vec{b}|^2 + |\vec{a} - \vec{b}|^2 = 2|\vec{a}|^2 + 2|\vec{b}|^2.$$

What geometric property of parallelograms does this imply?

Exercises 27 to 31 use vector algebra to derive *Heron's formula* for computing the area S of an arbitrary triangle with sides of length a, b, and c:

$$S = \sqrt{s(s-a)(s-b)(s-c)}, \qquad (2.35)$$

where $s = (a + b + c)/2.$

27. Let the vectors $\vec{a}$, $\vec{b}$, and $\vec{c}$ represent the three sides of our triangle, $\vec{c} = \vec{b} - \vec{a}$. Show that

$$4S^2 = (\vec{a} \times \vec{b}) \cdot (\vec{a} \times \vec{b}). \qquad (2.36)$$

28. Using Equation (2.34), show that Equation (2.36) can be written as

$$4S^2 = (ab + \vec{a} \cdot \vec{b})(ab - \vec{a} \cdot \vec{b}). \qquad (2.37)$$

29. Show that
$$\vec{a} \cdot \vec{b} = (a^2 + b^2 - c^2)/2. \qquad (2.38)$$

30. Combining Equations (2.37) and (2.38), show that

$$16S^2 = (a + b + c)(a + b - c)(a - b + c)(-a + b + c).$$

31. Complete the proof of Heron's formula.

3

Celestial Mechanics

3.1 The Calculus of Curves

We are not quite ready to prove that Newton's law of gravitational attraction implies Kepler's second law. We need to take a closer look at the Calculus of vector functions in the light of the vector algebra described in the last chapter.

As in Section 1.5, let $\vec{r}(t)$ denote the position of a moving particle at time t. The *derivative* of $\vec{r}(t)$ is defined as it is for scalar functions:

$$\frac{d}{dt}\vec{r}(t) = \lim_{h \to 0} \frac{\vec{r}(t+h) - \vec{r}(t)}{h}. \tag{3.1}$$

Observe that $\vec{r}(t+h) - \vec{r}(t)$ is the vector from $\vec{r}(t)$ to $\vec{r}(t+h)$, which can be viewed as a chord of the curve traced by our moving particle (Figure 3.1). In the limit, this becomes a tangent to the curve. The velocity vector is defined to be this derivative:

$$\vec{v}(t) = \frac{d}{dt}\vec{r}(t). \tag{3.2}$$

The acceleration vector is the derivative of the velocity:

$$\vec{a}(t) = \frac{d}{dt}\vec{v}(t) = \frac{d^2}{dt^2}\vec{r}(t). \tag{3.3}$$

Throughout this chapter, we shall assume that the first two derivatives of $\vec{r}(t)$ exist.

Rules of Differentiation

Derivatives of vector functions satisfy the same basic rules of differentiation that hold for scalar functions.

Theorem 3.1 *If $\vec{r}(t)$ is constant, then*

$$\frac{d}{dt}\vec{r}(t) = \vec{0}.$$

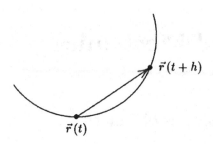

FIGURE 3.1. $\vec{r}(t+h) - \vec{r}(t)$.

Proof: If $\vec{r}(t)$ is constant, then

$$\vec{r}(t+h) - \vec{r}(t) = \vec{0},$$

so that

$$\frac{d}{dt}\vec{r}(t) = \lim_{h \to 0} \frac{\vec{0}}{h} = \vec{0}.$$

Q.E.D.

Theorem 3.2 *Let $\vec{r}(t)$ and $\vec{s}(t)$ be differentiable vector functions, $\lambda(t)$ a differentiable scalar function, and c a real constant. We have*

$$\frac{d}{dt}(c\vec{r}) = c\frac{d\vec{r}}{dt}, \tag{3.4}$$

$$\frac{d}{dt}(\vec{r}+\vec{s}) = \frac{d\vec{r}}{dt} + \frac{d\vec{s}}{dt}, \tag{3.5}$$

$$\frac{d}{dt}(\lambda\vec{r}) = \frac{d\lambda}{dt}\vec{r} + \lambda\frac{d\vec{r}}{dt}, \tag{3.6}$$

$$\frac{d}{dt}(\vec{r}\cdot\vec{s}) = \frac{d\vec{r}}{dt}\cdot\vec{s} + \vec{r}\cdot\frac{d\vec{s}}{dt}, \tag{3.7}$$

$$\frac{d}{dt}(\vec{r}\times\vec{s}) = \frac{d\vec{r}}{dt}\times\vec{s} + \vec{r}\times\frac{d\vec{s}}{dt}. \tag{3.8}$$

Proof: The proofs exactly mimic those for scalar functions. To prove Equation (3.4) we observe that

$$\frac{d}{dt}(c\vec{r}) = \lim_{h \to 0} \frac{c\vec{r}(t+h) - c\vec{r}(t)}{h}$$

$$= \lim_{h \to 0} c\frac{\vec{r}(t+h) - \vec{r}(t)}{h}$$

$$= c \lim_{h \to 0} \frac{\vec{r}(t+h) - \vec{r}(t)}{h}$$

$$= c \frac{d\vec{r}}{dt}.$$

The proof of Equation (3.5) is similarly straightforward and is left as an exercise. For Equation (3.6), we insert

$$-\lambda(t)\,\vec{r}(t+h) + \lambda(t)\,\vec{r}(t+h) = 0$$

into the numerator of the definition of the derivative:

$$\frac{d}{dt}(\lambda \vec{r})$$

$$= \lim_{h \to 0} \frac{\lambda(t+h)\vec{r}(t+h) - \lambda(t)\vec{r}(t)}{h}$$

$$= \lim_{h \to 0} \frac{\lambda(t+h)\vec{r}(t+h) - \lambda(t)\vec{r}(t+h) + \lambda(t)\vec{r}(t+h) - \lambda(t)\vec{r}(t)}{h}$$

$$= \lim_{h \to 0} \left(\frac{\lambda(t+h) - \lambda(t)}{h}\vec{r}(t+h) + \lambda(t)\frac{\vec{r}(t+h) - \vec{r}(t)}{h} \right)$$

$$= \left(\lim_{h \to 0} \frac{\lambda(t+h) - \lambda(t)}{h} \right) \left(\lim_{h \to 0} \vec{r}(t+h) \right)$$

$$+ \lambda(t) \left(\lim_{h \to 0} \frac{\vec{r}(t+h) - \vec{r}(t)}{h} \right)$$

$$= \frac{d\lambda}{dt}\vec{r} + \lambda \frac{d\vec{r}}{dt}.$$

What makes this proof work is the distributive law:

$$\lambda(t+h)\vec{r}(t+h) - \lambda(t)\vec{r}(t+h) = (\lambda(t+h) - \lambda(t))\,\vec{r}(t+h),$$
$$\lambda(t)\vec{r}(t+h) - \lambda(t)\vec{r}(t) = \lambda(t)\,(\vec{r}(t+h) - \vec{r}(t)).$$

Notice that we need both forms of this law, Equations (2.3) and (2.4). We have proven that the distributive law also holds for dot products and cross products, and so exactly the same argument yields Equations (3.7) and (3.8).

Q.E.D.

One corollary to this theorem is a result we used in Section 1.5.

Corollary 3.1 *If* $\vec{r}(t) = (x(t), y(t), z(t))$, *then*

$$\frac{d\vec{r}}{dt} = \left(\frac{dx}{dt}, \frac{dy}{dt}, \frac{dz}{dt} \right).$$

That is, differentiation of a vector function expressed in coordinate form is performed by differentiating each of the coordinates.

Proof: We use the additive rule [Equation (3.5)], the rule for the derivative of a product [Equation (3.6)], and the fact that $\vec{\imath}, \vec{\jmath},$ and $\vec{k}$ are constants so that

$$\frac{d\vec{\imath}}{dt} = \frac{d\vec{\jmath}}{dt} = \frac{d\vec{k}}{dt} = \vec{0}.$$

It follows that

$$\frac{d\vec{r}}{dt} = \frac{d}{dt}(x\vec{\imath} + y\vec{\jmath} + z\vec{k}) = \frac{dx}{dt}\vec{\imath} + \frac{dy}{dt}\vec{\jmath} + \frac{dz}{dt}\vec{k} = \left(\frac{dx}{dt}, \frac{dy}{dt}, \frac{dz}{dt}\right).$$

Q.E.D.

Corollary 3.2 *If $d\vec{r}/dt$ is zero for all t, then $\vec{r}(t)$ is a constant vector.*

Proof: We can write

$$\vec{r}(t) = (x(t), y(t), z(t)),$$

so that $d\vec{r}/dt = 0$ if and only if

$$\frac{dx}{dt} = \frac{dy}{dt} = \frac{dz}{dt} = 0.$$

From single variable calculus, this implies that x, y, and z are constants.

Q.E.D.

The next corollary may be slightly surprising at first glance, but a little consideration of what it means should convince you that you already know it.

Corollary 3.3 *If $|\vec{r}(t)|$ is a constant independent of t, then $\vec{r}(t)$ is perpendicular to $\vec{v}(t)$.*

Proof: By our hypothesis,

$$\vec{r} \cdot \vec{r} = |\vec{r}|^2 = c,$$

where c is a constant. Differentiating both sides with respect to t yields

$$0 = \frac{d}{dt}(\vec{r} \cdot \vec{r}) = \frac{d\vec{r}}{dt} \cdot \vec{r} + \vec{r} \cdot \frac{d\vec{r}}{dt} = 2\vec{r} \cdot \vec{v},$$

and therefore, $\vec{r}$ and $\vec{v}$ are perpendicular.

Q.E.D.

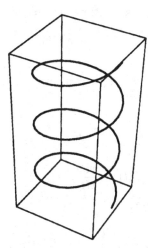

FIGURE 3.2. The curve $\vec{r}(t) = (\cos t, \sin t, t)$.

What we have just proved is that the tangent to a circle is always perpendicular to the radius.

Example

As an example, we take a path that spirals up a vertical cylinder of radius 1 (Figure 3.2):

$$\vec{r}(t) = (\cos t, \sin t, t).$$

The velocity is

$$\vec{v}(t) = (-\sin t, \cos t, 1).$$

The magnitude of this velocity is

$$|\vec{v}(t)| = \sqrt{\sin^2 t + \cos^2 t + 1} = \sqrt{2},$$

a constant, and so Corollary 3.3 implies that the acceleration will always be perpendicular to the velocity. In fact, the acceleration is

$$\vec{a}(t) = (-\cos t, -\sin t, 0),$$

and it is easily seen that

$$\vec{v} \cdot \vec{a} = 0.$$

Arc Length and Tangents

We now define the following functions. The distance from the origin is

$$r(t) = |\vec{r}(t)|, \tag{3.9}$$

the speed is

$$v(t) = |\vec{v}(t)|, \tag{3.10}$$

the arc length is

$$s(t_1) = \int_0^{t_1} v(t)\, dt, \tag{3.11}$$

the unit tangent is

$$\vec{T}(t) = \frac{\vec{v}(t)}{v(t)}, \quad v(t) \neq 0, \tag{3.12}$$

and the principal normal is

$$\vec{N}(t) = \frac{\vec{T}'(t)}{|\vec{T}'(t)|}, \quad \vec{T}'(t) = \frac{d\vec{T}}{dt} \neq \vec{0}. \tag{3.13}$$

We have already met $r(t)$. The speed is the absolute value of the velocity. Integrating the speed or distance traveled per unit time over an interval of time gives the total distance traveled in that time, denoted by $s(t)$. Starting the integral at 0 is an arbitrary convention since we usually treat the arc length not as a function of one time variable but of two: the arc length from $t = t_0$ to $t = t_1$:

$$s(t_1) - s(t_0) = \int_{t_0}^{t_1} v(x)\, dx. \tag{3.14}$$

The unit vector in the direction of $\vec{v}(t)$ is called the *unit tangent* or simply the *tangent*. The principal normal has two important properties given in the next theorem.

Theorem 3.3 *The principal normal, $\vec{N}(t)$, is perpendicular to the tangent, $\vec{T}(t)$, and, if $\vec{v}(t)$ and $\vec{a}(t)$ are not parallel, then $\vec{N}(t)$ lies in the plane spanned by $\vec{v}(t)$ and $\vec{a}(t)$.*

Proof: Since $|\vec{T}(t)| = 1$, Corollary 3.3 implies that $\vec{T}'(t)$, and thus $\vec{N}(t)$, is perpendicular to $\vec{T}(t)$. Since $\vec{v}(t)$ and $\vec{a}(t)$ are not parallel, neither of them is identically $\vec{0}$. We have

$$\vec{a}(t) = \frac{d}{dt}\vec{v} = \frac{d}{dt}(v\vec{T}) = v'\vec{T} + v\vec{T}' = \frac{v'}{v}\vec{v} + v|\vec{T}'|\vec{N}. \tag{3.15}$$

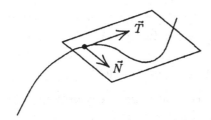

FIGURE 3.3. An osculating plane for $\vec{r}(t) = (\cos t, \sin t, t)$.

Since $\vec{a}$ and $\vec{v}$ are not parallel, $v|\vec{T'}|$ is not zero. Therefore, we have

$$\vec{N} = \frac{1}{v|\vec{T'}|}\vec{a} - \frac{v'}{v^2|\vec{T'}|}\vec{v}.$$

Q.E.D.

What we have demonstrated is that $\vec{T}$ and $\vec{N}$ are perpendicular unit vectors spanning the plane defined by $\vec{v}$ and $\vec{a}$, and thus $\vec{T}$ and $\vec{N}$ provide a convenient basis for describing points in this plane. If we translate this plane so that it passes through $\vec{r}(t)$,

$$\{\vec{r} + \alpha\vec{T} + \beta\vec{N} \mid \alpha, \beta \in \mathbf{R}\},$$

we obtain what is called the *osculating* (or kissing) *plane* (see Figure 3.3). Remember that $\vec{r}$, $\vec{T}$, and $\vec{N}$ are all functions of time, t, so that our plane changes over time. The significance of the osculating plane is that if our acceleration were constant, then our curve would lie in this plane. It thus provides us with a plane we can consider to be tangent to the curve.

Example

Returning to our spiral,

$$\vec{r}(t) = (\cos t, \sin t, t),$$

we have

$$r(t) = \sqrt{\cos^2 t + \sin^2 t + t^2} = \sqrt{1 + t^2},$$

$$v(t) = \sqrt{\sin^2 t + \cos^2 t + 1} = \sqrt{2},$$

$$s(t) = \int_0^t \sqrt{2}\,dx = t\sqrt{2},$$

$$\vec{T}(t) = \frac{\sqrt{2}}{2}(-\sin t, \cos t, 1),$$

$$\vec{N}(t) = (-\cos t, -\sin t, 0).$$

A perpendicular to the osculating plane is given by

$$\vec{v} \times \vec{a} = (\sin t, -\cos t, 1).$$

The dot product of this perpendicular with $\vec{r}$ is

$$\vec{v} \times \vec{a} \cdot \vec{r} = t,$$

and so the equation of the osculating plane passing through $\vec{r}$ is

$$\vec{v} \times \vec{a} \cdot (x, y, z) - \vec{v} \times \vec{a} \cdot \vec{r} = 0,$$

which is

$$(\sin t)x - (\cos t)y + z - t = 0.$$

For example, the osculating planes at $t = 0, \pi/2$, and $5\pi/4$ are, respectively,

$$-y + z = 0,$$
$$x + z - \pi/2 = 0,$$
$$-\frac{\sqrt{2}}{2}x + \frac{\sqrt{2}}{2}y + z - \frac{5\pi}{4} = 0.$$

Curvature

We next investigate the notion of *curvature*, finding the radius of the circle that best approximates our curve. Let us start by assuming that the curve traced by $\vec{r}(t)$ is in fact an arc of a circle of radius ρ lying in some plane. We specify some fixed direction in that plane and let $\alpha(t)$ be the angle between the tangent, $\vec{T}$, and our fixed direction (Figure 3.4). We view our plane so that the path travels counterclockwise around the center of the circle.

We consider the derivative $d\alpha/ds$, the rate at which α changes with respect to the arc length. On a circle, this is constant, and the value of this constant can be determined by considering what happens if we go

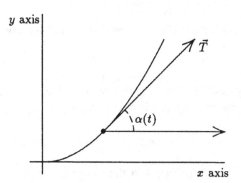

FIGURE 3.4. The angle $\alpha(t)$ between $\vec{T}$ and the chosen direction.

completely around the circle: α will have changed by 2π while the arc length traversed is $2\pi\rho$,

$$\frac{d\alpha}{ds} = \frac{2\pi}{2\pi\rho} = \frac{1}{\rho}. \tag{3.16}$$

For a curve lying in a plane, $d\alpha/ds$ is well defined, and we can define ρ as the reciprocal of this derivative at the point in question. To calculate ρ, we observe that

$$\frac{d\alpha}{ds} = \frac{d\alpha/dt}{ds/dt}.$$

From Equation (3.11), we see that

$$\frac{ds}{dt} = v(t).$$

If we standardize our plane so that $\vec{\imath}$ is the unit vector in our chosen direction and $\vec{\jmath}$ is the perpendicular unit vector in the plane, then

$$\vec{T}(t) = (\cos\alpha)\,\vec{\imath} + (\sin\alpha)\,\vec{\jmath}.$$

Differentiating both sides with respect to t gives us

$$\vec{T}'(t) = \alpha'(t)\left[-(\sin\alpha)\,\vec{\imath} + (\cos\alpha)\,\vec{\jmath}\right].$$

The vector in parentheses is a unit vector and so

$$|\vec{T}'(t)| = |\alpha'(t)|.$$

Since the particle is moving counterclockwise, the angle $\alpha(t)$ is increasing, and so $\alpha'(t)$ is positive:

$$\frac{d\alpha}{dt} = |\vec{T}'(t)|.$$

We have shown that

$$\frac{d\alpha}{ds} = \frac{d\alpha/dt}{ds/dt} = \frac{|\vec{T}'(t)|}{v(t)}.$$

This gives us a definition of curvature that is valid for any curve for which $\vec{T}'$ is well defined.

Definition: The *curvature* of the path $\vec{r}(t)$, denoted by $\kappa(t)$, is defined to be the reciprocal of the radius of the circle that best approximates the curve at $\vec{r}(t)$. Specifically, this is defined to be

$$\kappa(t) = \frac{|\vec{T}'(t)|}{v(t)}. \tag{3.17}$$

The radius $\rho(t) = 1/\kappa(t)$ is called the *radius of curvature*. Note that the curvature is zero if and only if the path is a straight line.

Computing $|\vec{T}'(t)|$ from the definition is often difficult. The following theorem provides us with a more direct approach to computing κ.

Theorem 3.4 *If $\vec{r}(t)$ is twice differentiable and the first derivative is not $\vec{0}$, then the curvature of the path traced by $\vec{r}(t)$ is given by*

$$\kappa = \frac{|\vec{a} \times \vec{v}|}{v^3}. \tag{3.18}$$

Proof: From Equations (3.15) and (3.17) we see that

$$\vec{a}(t) = v'\vec{T} + v|\vec{T}'|\vec{N} = v'\vec{T} + v^2\kappa\vec{N},$$

and so, since $\vec{T} \times \vec{v} = \vec{0}$,

$$\vec{a} \times \vec{v} = v'\vec{T} \times \vec{v} + v^2\kappa\vec{N} \times \vec{v} = v^2\kappa\vec{N} \times \vec{v} = v^3\kappa\vec{N} \times \vec{T}.$$

Since $|\vec{N} \times \vec{T}| = 1$, we finally arrive at

$$|\vec{a} \times \vec{v}| = v^3\kappa.$$

$$\textbf{Q.E.D.}$$

Examples

The curvature of $\vec{r}(t) = (\cos t, \sin t, t)$ is

$$\begin{aligned}
\kappa &= \frac{|(-\sin t, \cos t, 1) \times (-\cos t, -\sin t, 0)|}{2^{3/2}} \\
&= \frac{|(\sin t, -\cos t, 1)|}{2^{3/2}} = \frac{\sqrt{2}}{2^{3/2}} = \frac{1}{2},
\end{aligned}$$

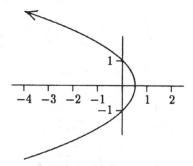

FIGURE 3.5. The curve $r + r \cos \theta = 1$.

and the radius of curvature is 2.

As a second example, consider the problem of finding the velocity, acceleration, and curvature of the path of a particle following the parabola given in polar coordinates by

$$r(1 + \cos \theta) = 1, \qquad (3.19)$$

which is traversed at a constant speed,

$$v(t) = 2,$$

in a counterclockwise direction about the origin (Figure 3.5).

If we differentiate both sides of Equation (3.19) with respect to t and use the relationship

$$1 + \cos \theta = r^{-1}, \qquad (3.20)$$

we see that

$$\frac{dr}{dt}(1 + \cos \theta) - r \sin \theta \frac{d\theta}{dt} = 0,$$

$$\frac{dr}{dt} r^{-1} - r \sin \theta \frac{d\theta}{dt} = 0,$$

$$\frac{d\theta}{dt} = \frac{1}{r^2 \sin \theta} \frac{dr}{dt}. \qquad (3.21)$$

We now recall from Equation (1.15) that

$$\vec{v} = \frac{dr}{dt} \vec{u}_r + r \frac{d\theta}{dt} \vec{u}_\theta,$$

where $\vec{u}_r$ and $\vec{u}_\theta$ are perpendicular unit vectors, and so

$$v = \sqrt{\left(\frac{dr}{dt}\right)^2 + \left(r \frac{d\theta}{dt}\right)^2}. \qquad (3.22)$$

Combining this with the expression for $d\theta/dt$ given in Equation (3.21) and recalling that we have a constant speed of 2, we see that

$$2 = v = \left|\frac{dr}{dt}\right|\sqrt{1 + \frac{1}{r^2\sin^2\theta}} = \left|\frac{dr}{dt}\right|\frac{\sqrt{1 + r^2\sin^2\theta}}{|r\sin\theta|}.$$

Using the fact that our curve is traversed counterclockwise, we can choose the proper sign, and we see that

$$\frac{dr}{dt} = \frac{2r\sin\theta}{\sqrt{1 + r^2\sin^2\theta}},$$

$$\frac{d\theta}{dt} = \frac{2}{r\sqrt{1 + r^2\sin^2\theta}}.$$

We can simplify the square root by using the fact that $\sin^2\theta = 1 - \cos^2\theta$ and, from Equation (3.20), that $\cos\theta = r^{-1} - 1$:

$$1 + r^2\sin^2\theta = 1 + r^2 - r^2\cos^2\theta = 1 + r^2 - r^2(r^{-1} - 1)^2 = 2r.$$

We have shown that

$$\frac{dr}{dt} = \sin\theta\sqrt{2r}, \tag{3.23}$$

$$\frac{d\theta}{dt} = r^{-3/2}\sqrt{2}, \tag{3.24}$$

$$\vec{v} = \sqrt{2}\left(r^{1/2}\sin\theta\,\vec{u}_r + r^{-1/2}\vec{u}_\theta\right). \tag{3.25}$$

If we differentiate Equation (3.25), remembering that r, θ, $\vec{u}_r$, and $\vec{u}_\theta$ are all functions of t, we get the acceleration

$$\vec{a} = \sqrt{2}\left(\frac{1}{2}r^{-1/2}\frac{dr}{dt}\sin\theta\,\vec{u}_r + r^{1/2}\cos\theta\frac{d\theta}{dt}\vec{u}_r + r^{1/2}\sin\theta\frac{d\theta}{dt}\vec{u}_\theta\right.$$
$$\left. - \frac{1}{2}r^{-3/2}\frac{dr}{dt}\vec{u}_\theta - r^{-1/2}\frac{d\theta}{dt}\vec{u}_r\right).$$

Substituting our values of dr/dt and $d\theta/dt$ from Equations (3.23) and (3.24) yields

$$\vec{a} = (\sin^2\theta)\,\vec{u}_r + 2r^{-1}(\cos\theta)\,\vec{u}_r + 2r^{-1}(\sin\theta)\,\vec{u}_\theta$$
$$- r^{-1}(\sin\theta)\,\vec{u}_\theta - 2r^{-2}\vec{u}_r$$
$$= (\sin^2\theta + 2r^{-1}(\cos\theta - r^{-1}))\,\vec{u}_r + r^{-1}(\sin\theta)\,\vec{u}_\theta$$
$$= (\sin^2\theta - 2r^{-1})\,\vec{u}_r + r^{-1}(\sin\theta)\,\vec{u}_\theta, \tag{3.26}$$

where in the last line we have used Equation (3.20) once again. It follows that

$$\vec{v} \times \vec{a} = \left(\sqrt{2}r^{-1/2}\sin^2\theta - \sqrt{2}r^{-1/2}(\sin^2\theta - 2r^{-1})\right)\vec{u}_r \times \vec{u}_\theta$$
$$= 2\sqrt{2}r^{-3/2}\,\vec{u}_r \times \vec{u}_\theta,$$

$$|\vec{v} \times \vec{a}| = 2\sqrt{2}r^{-3/2},$$

$$\kappa = \frac{2\sqrt{2}r^{-3/2}}{8} = (2r)^{-3/2}.$$

3.2 Exercises

1. Prove that
$$\frac{d}{dt}(\vec{r} + \vec{s}) = \frac{d\vec{r}}{dt} + \frac{d\vec{s}}{dt}.$$

2. Prove that
$$\frac{d}{dt}(\vec{r} \cdot \vec{s}) = \frac{d\vec{r}}{dt} \cdot \vec{s} + \vec{r} \cdot \frac{d\vec{s}}{dt}.$$

3. Prove that
$$\frac{d}{dt}(\vec{r} \times \vec{s}) = \frac{d\vec{r}}{dt} \times \vec{s} + \vec{r} \times \frac{d\vec{s}}{dt}.$$

In Exercises 4 through 12, use each of the following trajectories:

(a) $\vec{r}(t) = t^2\vec{i} - 4t\vec{j} - t^2\vec{k},$

(b) $\vec{r}(t) = (\cosh t)\,\vec{i} + (\sinh t)\,\vec{j} + t\vec{k},$
$$(\cosh t = (e^t + e^{-t})/2, \sinh t = (e^t - e^{-t})/2)$$

(c) $\vec{r}(t) = t\cos t\,\vec{i} + t\sin t\,\vec{j} + \vec{k}.$

4. Sketch the curve traced over the interval from $t = 0$ to 2.

5. Find $\vec{v}(t)$ and $\vec{a}(t)$.

6. Find $r(t)$ and $v(t)$.

7. Find the cosine of the angle between $\vec{r}$ and $\vec{v}$. For what values of t is $\vec{r}$ perpendicular to $\vec{v}$? When is it parallel to $\vec{v}$?

8. Find the cosine of the angle between $\vec{v}$ and $\vec{a}$. For what values of t is $\vec{v}$ perpendicular to $\vec{a}$? When is it parallel to $\vec{a}$?

9. Find the definite integral that expresses the arc length from $t = 0$ to 2.

10. Find $\vec{v} \times \vec{a}$.

11. Find the equation of the osculating plane at time t.

12. Find the curvature at time t.

13. Prove that

$$\vec{N} = \frac{\vec{a}_{v\perp}}{|\vec{a}_{v\perp}|}.$$

14. Does $\vec{r}(t) \cdot \vec{v}(t) = 0$ for all t imply that $r(t)$ is constant?

15. Consider a particle whose path is the ellipse

$$r(2 + \cos\theta) = 2,$$

traversed in a counterclockwise direction about the origin, and that sweeps out one unit of area per unit time:

$$\frac{dA}{dt} = \frac{1}{2}r^2\frac{d\theta}{dt} = 1.$$

Find the velocity and acceleration expressed in terms of the local coordinates $\vec{u}_r$ and $\vec{u}_\theta$.

16. Consider a particle whose path is the spiral

$$r = e^{3\theta},$$

sweeping out one unit of area per unit time. Find the velocity and acceleration expressed in terms of the local coordinates $\vec{u}_r$ and $\vec{u}_\theta$.

17. A missile traveling at constant speed is homing in on a target at the origin. Due to an error in its circuitry, it is consistently misdirected by a constant angle α. Find its path. Show that if $|\alpha| < 90°$, then it will eventually hit its target, taking $1/\cos\alpha$ times as long as if it were correctly aimed. (Hint: use local coordinates $\vec{u}_r$ and $\vec{u}_\theta$.)

3.3   Orbital Mechanics

Equipped with calculus and vector algebra, we can now make short work of Newton's result that the law of gravity implies Kepler's second law.

Lemma 3.1 *Let $\vec{r}(t)$ be the position of a particle at time t, $\vec{v}(t)$ its velocity, and $\vec{a}(t)$ its acceleration. If $\vec{a}$ is radial (always parallel to $\vec{r}$), then*

$$\vec{r} \times \vec{v} = \vec{K}, \tag{3.27}$$

where $\vec{K}$ is a constant vector of magnitude

$$K = |\vec{K}| = 2\frac{dA}{dt} \tag{3.28}$$

$$= rv\sin\phi, \tag{3.29}$$

where dA/dt is the rate at which area is swept out and ϕ is the angle between $\vec{r}$ and $\vec{v}$.

Proof: From Equation (3.8) and the fact that $\vec{r}$ and $\vec{a}$ are parallel, we have

$$\frac{d}{dt}(\vec{r}\times\vec{v}) = \vec{v}\times\vec{v}+\vec{r}\times\vec{a} = \vec{0}+\vec{0} = \vec{0},$$

and thus, by Corollary 3.2, $\vec{r}\times\vec{v}$ is a constant vector that we shall call $\vec{K}$.

Equation (3.29) follows from Equation (3.27) and the definition of the cross product. To prove Equation (3.28), we use the representations of $\vec{r}$ and $\vec{v}$ in terms of local coordinates [Equations (1.14) and (1.15)]:

$$\vec{K} = \vec{r}\times\vec{v} = r\vec{u}_r \times \left(\frac{dr}{dt}\vec{u}_r + r\frac{d\theta}{dt}\vec{u}_\theta\right) = r^2\frac{d\theta}{dt}(\vec{u}_r \times \vec{u}_\theta), \tag{3.30}$$

so that

$$K = r^2\frac{d\theta}{dt}. \tag{3.31}$$

Lemma 1.9 now concludes the proof.

Q.E.D.

Kepler's Second Law

The full law of gravity says that the force of gravitational attraction is inversely proportional to the square of the distance and directly proportional to each of the masses:

$$\vec{F} = -G\frac{Mm}{r^2}\vec{u}_r, \tag{3.32}$$

where G is a gravitational constant, M and m are the respective masses, and r is the distance. If m is the mass of our orbiting particle, then its acceleration satisfies

$$\vec{F} = m\vec{a}, \tag{3.33}$$

or

$$\vec{a} = -\frac{GM}{r^2}\vec{u}_r. \tag{3.34}$$

Theorem 3.5 *Let $\vec{r}(t)$ denote the position at time t of a moving particle whose acceleration is given by Equation (3.34) and that sweeps out area at the constant rate $K/2$. There then exists a constant vector $\vec{\varepsilon}$ such that*

$$|\vec{r}| + \vec{r} \cdot \vec{\varepsilon} = \frac{K^2}{GM}. \qquad (3.35)$$

Equivalently, if (r, θ) is the position in polar coordinates, then

$$r(1 + \varepsilon \cos \theta) = \frac{K^2}{GM}. \qquad (3.36)$$

We recognize Equation (3.36) as the equation of a conic section: an ellipse, parabola, or hyperbola (Lemma 1.10 and Exercise 12 of Section 1.5). In particular, if $|\vec{\varepsilon}| < 1$, then it is the equation of an ellipse with one focus at the origin.

Proof: We shall prove this by using the identity for scalar triple products:

$$\vec{r} \times \vec{v} \cdot \vec{K} = \vec{v} \times \vec{K} \cdot \vec{r}. \qquad (3.37)$$

By Equation (3.27), the left side is

$$\vec{r} \times \vec{v} \cdot \vec{K} = \vec{K} \cdot \vec{K} = K^2. \qquad (3.38)$$

To evaluate the right side, we use our definition of $\vec{a}$ [Equation (3.34)], the representation of $\vec{K}$ in local coordinates [Equation (3.30)], and the fact that $(d/dt)\vec{u}_r = (d\theta/dt)\vec{u}_\theta$ [Equation (1.11)]:

$$\frac{d}{dt}(\vec{v} \times \vec{K}) = \vec{a} \times \vec{K} = \left(-\frac{GM}{r^2}\vec{u}_r\right) \times \left(r^2 \frac{d\theta}{dt}\vec{u}_r \times \vec{u}_\theta\right)$$

$$= -GM \frac{d\theta}{dt}[\vec{u}_r \times (\vec{u}_r \times \vec{u}_\theta)] = GM \frac{d\theta}{dt}\vec{u}_\theta = \frac{d}{dt}(GM\vec{u}_r).$$

This means that the derivative of $\vec{v} \times \vec{K} - GM\vec{u}_r$ is $\vec{0}$, and so $\vec{v} \times \vec{K} - GM\vec{u}_r$ is a constant vector which we shall denote by $GM\vec{\varepsilon}$:

$$\vec{v} \times \vec{K} = GM(\vec{u}_r + \vec{\varepsilon}), \qquad (3.39)$$

$$\vec{v} \times \vec{K} \cdot \vec{r} = GM(\vec{u}_r + \vec{\varepsilon}) \cdot \vec{r} = GM(|\vec{r}| + \vec{r} \cdot \vec{\varepsilon}). \qquad (3.40)$$

Combining this result with Equations (3.37) and (3.38), we see that

$$K^2 = \vec{r} \times \vec{v} \cdot \vec{K} = \vec{v} \times \vec{K} \cdot \vec{r} = GM(|\vec{r}| + \vec{r} \cdot \vec{\varepsilon}),$$

$$|\vec{r}| + \vec{r} \cdot \vec{\varepsilon} = \frac{K^2}{GM}.$$

Equation (3.36) follows from the equalities $|\vec{r}| = r$ and $\vec{r} \cdot \vec{\varepsilon} = r\varepsilon \cos\theta$.

Q.E.D.

I challenge the reader to return to Newton's original proof of Kepler's second law (Proposition XVII) and work through it, comparing it to this proof.

Equation of the Orbit

If we define the positive x axis to be parallel to $\vec{\varepsilon} = \varepsilon\vec{i}$ and set

$$\gamma = \frac{K^2}{GM}, \tag{3.41}$$

then we have an elliptic orbit precisely when $|\varepsilon| < 1$, and the equation of this orbit is, by Lemma 1.10,

$$(1 - \varepsilon^2)(x + \alpha)^2 + y^2 = \frac{\gamma^2}{1 - \varepsilon^2}, \tag{3.42}$$

where

$$\alpha = \frac{\varepsilon\gamma}{1 - \varepsilon^2}. \tag{3.43}$$

The *apogee*, or farthest distance, is

$$|\alpha| + \frac{\gamma}{1 - \varepsilon^2} = \frac{|\varepsilon|\gamma}{1 - \varepsilon^2} + \frac{\gamma}{1 - \varepsilon^2} = \frac{(1 + |\varepsilon|)\gamma}{(1 - |\varepsilon|)(1 + |\varepsilon|)} = \frac{\gamma}{1 - |\varepsilon|}. \tag{3.44}$$

The *perigee*, or nearest distance, is

$$-|\alpha| + \frac{\gamma}{1 - \varepsilon^2} = \frac{-|\varepsilon|\gamma}{1 - \varepsilon^2} + \frac{\gamma}{1 - \varepsilon^2} = \frac{(1 - |\varepsilon|)\gamma}{(1 - |\varepsilon|)(1 + |\varepsilon|)} = \frac{\gamma}{1 + |\varepsilon|}. \tag{3.45}$$

The *mean distance* is the semimajor axis:

$$a = \frac{\gamma}{1 - \varepsilon^2}. \tag{3.46}$$

Note also that if ε is positive, then most of the ellipse lies to the *left* of the y axis, and the orbiting particle reaches its perigee when it crosses the positive x axis. If ε is negative, then most of the ellipse lies to the *right* of the y axis, and the orbiting particle reaches its apogee when it crosses the positive x axis (Figure 3.6).

If the absolute value of ε is 1, then our orbit is a parabola. If it is greater than 1, then the orbit is a hyperbola. What is significant about these cases is that they are *nonperiodic*: our particle sweeps in close to the object it is orbiting and then heads off, never to return. A satellite circling the earth

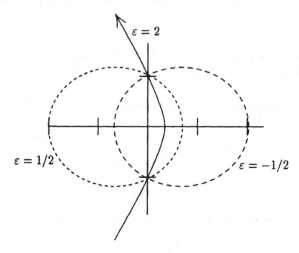

FIGURE 3.6. Orbits for $\varepsilon = -\frac{1}{2}$, $\frac{1}{2}$, and 2.

that achieves a parabolic or hyperbolic orbit is said to reach *escape velocity*. Assuming it is on the outbound arm (and so is in no danger of crashing into the earth), the earth's gravity cannot hold it back.

Eccentricity and Escape Velocity

Let us narrow our focus to satellites orbiting the earth where the value of GM is approximately

$$GM = 4 \times 10^{14} \ \text{m}^3/\text{s}^2. \tag{3.47}$$

To simplify matters, we shall ignore the effects of the moon and other bodies. If we know the position, $\vec{r}$, and velocity, $\vec{v}$, of our satellite at any given time, we can find K and ε and thus compute the orbit. Our first problem will be to find the escape velocity from the earth.

The constant K is easily computed from Equation (3.29): $K = rv \sin \phi$, where ϕ is the angle between $\vec{r}$ and $\vec{v}$. To find ε from r, v, and ϕ is a little trickier. From Equations (3.39) and (3.27), we see that

$$\vec{u}_r + \vec{\varepsilon} = \frac{\vec{v} \times \vec{K}}{GM} = \frac{\vec{v} \times (\vec{r} \times \vec{v})}{GM}. \tag{3.48}$$

We can solve this for $\vec{\varepsilon}$ and then get

$$\varepsilon^2 = \vec{\varepsilon} \cdot \vec{\varepsilon}$$

$$= \left(\frac{\vec{v} \times (\vec{r} \times \vec{v})}{GM} - \vec{u}_r \right) \cdot \left(\frac{\vec{v} \times (\vec{r} \times \vec{v})}{GM} - \vec{u}_r \right)$$

$$= \frac{1}{G^2 M^2} |\vec{v} \times (\vec{r} \times \vec{v})|^2 - \frac{2}{GM} [\vec{v} \times (\vec{r} \times \vec{v}) \cdot \vec{u}_r] + 1. \quad (3.49)$$

Now, $\vec{v}$ is perpendicular to $\vec{r} \times \vec{v}$, and therefore,

$$|\vec{v} \times (\vec{r} \times \vec{v})| = rv^2 |\sin \phi|.$$

We can rearrange our scalar triple product $\vec{v} \times (\vec{r} \times \vec{v}) \cdot \vec{u}_r$ to $[(\vec{u}_r \times \vec{v}) \cdot (\vec{r} \times \vec{v})]$. Since $\vec{u}_r \times \vec{v}$ is parallel to $\vec{r} \times \vec{v}$, the dot product of these vectors is simply the product of their magnitudes:

$$\vec{v} \times (\vec{r} \times \vec{v}) \cdot \vec{u}_r = (v \sin \phi)(rv \sin \phi) = rv^2 \sin^2 \phi.$$

Making these substitutions into the last line of Equation (3.49), we obtain

$$\varepsilon^2 = \frac{r^2 v^4 \sin^2 \phi}{G^2 M^2} - \frac{2rv^2 \sin^2 \phi}{GM} + 1$$

$$= 1 + \frac{rv^2 \sin^2 \phi}{G^2 M^2} (rv^2 - 2GM). \quad (3.50)$$

If, instead of factoring out $rv^2/G^2 M^2$, we replace the 1 in the first line by $\cos^2 \phi + \sin^2 \phi$, we see that we can also write ε^2 as

$$\varepsilon^2 = \sin^2 \phi \left(1 - \frac{rv^2}{GM} \right)^2 + \cos^2 \phi. \quad (3.51)$$

If $\phi = 0$, then our satellite is moving vertically. It either keeps going forever in a straight line or slows down, reverses direction, and crashes back into the earth. Since neither of these cases is particularly interesting, we shall assume that ϕ is not zero.

Because r and v are positive, we have a nonperiodic orbit ($|\varepsilon| \geq 1$) if and only if

$$rv^2 \geq 2GM = 8 \times 10^{14} \text{ m}^3/\text{s}^2,$$

$$v \geq \sqrt{\frac{8 \times 10^{14}}{r}} \text{ m/s}. \quad (3.52)$$

As we get further from the earth, the escape velocity decreases. At the surface of the earth where r is roughly 6.4×10^6 meters, the escape velocity is about 11 200 m/s or 24 800 miles/h. Note that escape velocity does *not* depend on the angle between r and v. In practice, you must check that you are not on a trajectory that will collide with the earth. As long as this will not happen, heading in *any* direction at 25 000 miles/h will launch you toward the ends of the universe.

A circular orbit has an eccentricity of $\varepsilon = 0$, which is achieved if and only if

$$rv^2 = GM \quad \textbf{AND} \quad \phi = \pi/2.$$

In a properly elliptic orbit $(0 < |\varepsilon| < 1)$, the angle between $\vec{r}$ and $\vec{v}$ is $\pi/2$ at precisely two points: the apogee and the perigee. At these points, we have the relationship

$$\varepsilon^2 = \left(1 - \frac{rv^2}{GM}\right)^2,$$

or

$$\varepsilon = \pm\left(1 - \frac{rv^2}{GM}\right). \tag{3.53}$$

The choice of sign is determined by whether the perigee occurs on the positive or negative x axis. The eccentricity is positive when $rv^2 > GM$ and negative when $rv^2 < GM$, and so

$$\varepsilon = \frac{rv^2}{GM} - 1. \tag{3.54}$$

Kepler's Third Law and Geosynchronous Orbit

Kepler's third law now comes for free. The area inside an elliptic orbit is π times the product of the semimajor and semiminor axes, which is

$$\pi\frac{\gamma}{1 - \varepsilon^2}\frac{\gamma}{\sqrt{1 - \varepsilon^2}} = \frac{\pi K^4}{G^2 M^2 (1 - \varepsilon^2)^{3/2}}.$$

Since the area is swept out at the constant rate $K/2$, the *period* of the orbit (the time needed to complete one orbit) is the area divided by the rate:

$$\begin{aligned}
\text{period} &= \frac{2\pi K^3}{G^2 M^2 (1 - \varepsilon^2)^{3/2}} \\
&= \frac{2\pi}{\sqrt{GM}}\frac{\gamma^{3/2}}{(1 - \varepsilon^2)^{3/2}} \\
&= \frac{2\pi}{\sqrt{GM}}a^{3/2}, \tag{3.55} \\
\text{period}^2 &= \frac{4\pi^2}{GM}a^3, \tag{3.56}
\end{aligned}$$

where a is the semimajor axis. Equation (3.56) is Kepler's third law.

An orbit is said to be *geosynchronous* if its period is the same as the time it takes the earth to complete one rotation, that is, 1 day or 86 400 s. Inserting this and the value of GM into Equation (3.55), we get an a value of about 42 000 km, or roughly 35 600 km above the surface of the earth.

Accelerating while in Orbit

A curious phenomenon happens to a vehicle in orbit that fires its rockets to achieve acceleration in a nonradial direction. If two vehicles are traveling in tandem in a circular orbit and one of them produces a brief acceleration in the direction of its velocity, then instead of pulling ahead of its companion, it will swing out into an orbit of greater eccentricity and actually fall behind. The mathematics behind this comes out of our equation for eccentricity [Equation (3.51)].

Initially, since we are in a circular orbit, we have

$$rv^2 = GM,$$

and the semimajor axis is r, so the period is

$$\text{period} = \frac{2\pi}{\sqrt{GM}} r^{3/2}.$$

We let our first vehicle continue in this orbit, but we briefly fire the rockets on the second vehicle. In view of the distances involved, a "burn," or rocket firing, of a few seconds can be viewed as an instantaneous increase in velocity, so that at the moment of the burn the position of our second vehicle, $\vec{r}_2(t)$, is still the same as that of the first vehicle, but the velocity has changed from $\vec{v}$ to

$$\vec{v}_2 = c\vec{v},$$

for some positive constant c. We have assumed that our instantaneous acceleration is parallel to $\vec{v}$, so that initially the angle between $\vec{r}_2$ and $\vec{v}_2$ is still $\pi/2$. It is convenient to define the positive x axis so that the burn occurs as we cross it.

The new value of K is

$$K_2 = |\vec{r}_2 \times \vec{v}_2| = |\vec{r} \times c\vec{v}| = cK. \tag{3.57}$$

Since the angle between $\vec{r}_2$ and $\vec{v}_2$ is $\pi/2$ and $rv^2 = GM$, the new value of ε satisfies

$$\varepsilon_2 = \frac{r_2 v_2^2}{GM} - 1 = c^2 \frac{rv^2}{GM} - 1 = c^2 - 1. \tag{3.58}$$

Since c is assumed to be positive, we have a noncircular orbit unless $c = 1$. Note that we stay in a periodic orbit if and only if $c < \sqrt{2}$. If c is less than 1 (we have decelerated), then we are at our apogee and ε_2 is negative. If c is larger than 1 and less than $\sqrt{2}$, then we are at our perigee and ε_2 is positive.

We have demonstrated that if the second vehicle fires its afterburners, it will swing out into a wider orbit. But as long as c is less than $\sqrt{2}$, it will continue to return to its perigee on each orbit. Which vehicle gets back first? The semimajor axis for the second vehicle is

$$a_2 \;=\; \frac{K_2^2}{GM(1-\varepsilon_2^2)} \;=\; \frac{c^2 K^2}{GM(2c^2 - c^4)} \;=\; \frac{K^2}{GM(2 - c^2)} \;=\; \frac{r}{2 - c^2},$$

and so its period is

$$\frac{2\pi}{\sqrt{GM}} \frac{r^{3/2}}{(2 - c^2)^{3/2}},$$

as opposed to the period of the first vehicle,

$$\frac{2\pi}{\sqrt{GM}} r^{3/2}.$$

If the second vehicle has accelerated, $1 < c < \sqrt{2}$, then it will take it longer to return to the point of the burn. To beat the first rocket back to that point, it must *decelerate*, $0 < c < 1$. Care is required, however, as deceleration puts you into an eccentric orbit passing closer to the earth, and it is desirable to avoid colliding with it.

Caveat

Before using the mathematics of this chapter to send a satellite into orbit, be aware that in practice we cannot ignore the moon's influence. For a few orbits staying relatively close to the earth, the moon will not have much effect, but over time it will modify the orbit considerably. In fact, in time, the sun and each of the planets, even each of the asteroids, will exert a measurable sway over the satellite. The mathematics we have developed is incomplete as an exact model of our universe because our universe consists of more than two objects.

Our model is a good approximation, and the influence of the other bodies can be calculated to almost any degree of accuracy. But, we are placed in a position uncomfortably close to that of pre-Keplerian astronomers: we possess a beautiful and simple theory that is only a first approximation. To make it agree with observational accuracy, we need to complicate it considerably.

This is not to suggest that we are no better off than our medieval predecessors. It is Newton's laws that tell us how to make most of the corrections. There is no need to resort to convoluted inventions to account for them. Yet, there is something basically dissatisfying about the present state of affairs. One wishes for a model that combines elegance, utility, and simplicity, probably in forms we would not yet recognize, in an explanation of the intricate dance of many bodies under gravitational attraction. There is some indication that the collection of results now being grouped under the heading of chaos theory is groping in this direction.

3.4   Exercises

1. Prove that $\vec{u}_r \times (\vec{u}_r \times \vec{u}_\theta) = -\vec{u}_\theta$.

2. Prove that in a properly elliptic orbit, the angle between $\vec{r}$ and $\vec{v}$ is $\pi/2$ only at the apogee and the perigee.

3. Mars has a radius of approximately 3300 km and a mass 0.15 times that of Earth. Find the escape velocity on the surface of Mars.

   For Exercises 4 through 7, we are considering a rocket that is fired to 300 km above the surface of the earth, 6.7×10^6 m from the center of the earth. At this point, the engines are cut off and the rocket enters orbit. The angle between $\vec{r}$ and $\vec{v}$ is denoted by ϕ.

4. What velocity must it have attained if it is to remain in a circular orbit at this height? What is the period of this orbit?

5. If its speed is 9000 m/s and $\phi = \pi/2$, what are the values of the apogee and perigee of the resulting orbit? What is the period of this orbit?

6. If its speed is 9000 m/s and $\phi = \pi/3$, what are the values of the apogee and perigee of the resulting orbit? What is the period of this orbit?

7. If its speed is 9000 m/s, find the angle ϕ that will result in an orbit whose perigee is 6.5×10^6 m. What are the values of the eccentricity, apogee, and period of this orbit?

8. A rocket has attained a circular orbit around the earth at 6.6×10^6 m from the center of the earth. It is traveling at a speed of 7785 m/s. We want to move it out to a circular orbit of $r = 7.0 \times 10^6$ m by executing a burn, increasing its speed to v_1 so that it enters an eccentric orbit whose apogee is 7.0×10^6 m. When it reaches this apogee, we perform a second burn to increase its speed from v_2, the speed of the eccentric orbit at the apogee, to 7560 m/s, the speed needed to maintain a circular orbit at this height. Find v_1 and v_2.

9. Show that the absolute value of the eccentricity is the difference between the apogee and the perigee divided by their sum. Find the absolute value of the eccentricity of an orbit whose apogee is 12×10^6 meters and whose perigee is 8×10^6 m.

   For Exercises 10 through 14, we consider the New York to Tokyo space shuttle now being planned. Our shuttle accelerates until it is 160 km above New York (6.56×10^6 m from the center of the earth). At that point, the engines are cut, and

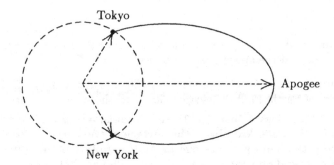

FIGURE 3.7. The New York to Tokyo shuttle, Exercises 10 to 14.

the shuttle enters an orbital glide until it is 160 km above Tokyo, at which time it decelerates for the landing. For the purposes of simplification, we shall ignore the rotation of the earth until the last problem in this set.

10. New York is at 70° W, 41° N; Tokyo is at 140° E, 36° N. Find the angle between the lines connecting the center of the earth, O, to New York and Tokyo, respectively. All of our calculations are on the plane defined by these three points, and we take the bisector of this angle to be the positive x axis (Figure 3.7).

11. Find the speed needed to achieve a circular orbit at $r = 6.56 \times 10^6$. How many minutes will it take for the orbital glide between New York and Tokyo?

12. What speed must it reach if instead of a circular orbit it is to enter an elliptic orbit with apogee at

    (a) 7.0×10^6 m,
    (b) 7.5×10^6 m,
    (c) 8.0×10^6 m,
    (d) 9.0×10^6 m?

    In each of these cases, how many minutes will it take for the orbital glide?

13. Find the value of the apogee that *minimizes* the speed we need when we enter the glide. What is this minimal speed, and how long will the glide last?

14. Redo Exercise 11 taking into consideration the fact that the rotation of the earth is moving Tokyo eastward at the rate of 15°/h.

4

Differential Forms

4.1 Some History

With this chapter we begin the study of functions whose domain and range consist of several variables:

$$\vec{F} : \mathbf{R}^n \longrightarrow \mathbf{R}^m.$$

If the domain has more than one dimension, such functions are often referred to as *fields*. It is a *scalar field* if the range is one dimensional, a *vector field* if the range has more than one dimension. Examples of vector fields in the physical world include fluid flows, where the function is defined on points in space, mapping them to vectors representing the velocity of the flow at that point, and force fields, where the function is again defined on points in space, mapping them to vectors representing the force exerted at that point. It is customary in the second year of a calculus course to use the language of vector algebra in extending the tools of calculus to these phenomena, giving rise to what is known as *vector analysis*. Here I shall break with tradition and move beyond vector analysis to the yet more powerful terminology of *differential forms*, equivalent in its essential aspects to what is known to physicists as *tensor analysis*.

As explained in Chapter 2, vector analysis draws its strength from its invariance under rigid motions: translations and rotations. The notion of absolute position, or frame of reference, should not play a role in the statement of physical laws. In general, physical properties *are* affected by other transformations. For example, a *shear* (Figure 4.1), such as that given by the transformation from the u,v plane to the x,y plane by

$$x = u + v,$$
$$y = v,$$

distorts distances and angles. But, there are some physical laws and properties that are invariant under *any* continuous deformation of space. A good example of this involves the lines of magnetic force. These form closed loops and have the property that for any closed volume that does not include the magnet, the number of lines of magnetic force entering this volume is equal to the number of lines of magnetic force that leave it. This property is known as *incompressibility*, and it remains no matter how we stretch, twist, or deform our space.

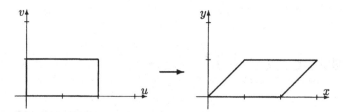

FIGURE 4.1. The shear: $x = u + v, y = v$.

Albert Einstein (1879–1955) based his "Die Grundlage der allgemeinen Relativitätstheorie" or "The Foundation of the General Theory of Relativity" on the assumption that the description of gravity must incorporate just such an invariance. In his words (the italics are his),

> *The general laws of nature are to be expressed by equations which hold good for all systems of coordinates, that are covariant with respect to any substitution whatever (generally covariant).*

The problem is to find such an invariant (or, in the terminology of tensor analysis, covariant) language. Fortunately, by 1916, when Einstein published his treatise, such a language was already well developed, the language of differential forms.

We owe the terminology of differentials to Leibniz, but a real understanding of them did not begin until the mid-1800s. Calculus can be used to compute geometric quantities, such as length, surface area, and volume. The study of differential forms arose out of the nineteenth-century project to use the formulas of calculus as the foundation for geometry. This had its roots in work of Carl Friedrich Gauss (1777–1855) and Gabriel Lamé (1795–1870), among others, but it really began in 1854 with Georg Friedrich Bernhard Riemann (1826–1866) and his thesis "Über die Hypothesen welche der Geometrie zu Grunde liegen" or "On the Hypotheses which Lie at the Foundation of Geometry." A key feature of this new geometry, today called Riemannian geometry, is the invariance of the basic definitions under continuous deformations. In the years after Riemann's death, his ideas were developed by many mathematicians, most notably Eugenio Beltrami (1832–1900), Elwin Bruno Christoffel (1829–1900), and Rudolf Lipschitz (1832–1903). Differential forms as we shall regard them were first described in 1899 in an article by Élie Cartan (1869–1951) and in the third volume of *Les Méthodes Nouvelles de la Mécanique Céleste* or *New Methods of Celestial Mechanics* by Henri Poincaré (1854–1912). In the closing years of the nineteenth century and into the twentieth, Gregorio Ricci-Curbastro (1853–1925) of the University of Padua and his student Tullio Levi-Civita (1873–1941) began the task of restating the laws of physics in terms of differential forms and the invariants of Riemannian geometry.

It was this work that Einstein drew upon in formulating general relativity. The term tensor analysis was his invention. The word tensor goes back to Hamilton's quaternions, out of which vector analysis arose, and the sense in which it is used there is analogous to the manner in which we shall view differential forms. Einstein's designation suggests the strong ties that in fact exist between tensor and vector analysis. The distinction between tensors and differential forms is largely a matter of notation. I have chosen the language of differential forms because I find it simple and direct and because the rules governing the manipulation of differential forms or tensors are mysterious unless one keeps in mind the origins of these creatures in differential and integral calculus.

4.2   Differential 1-Forms

We begin with the simplest of all differential forms, the *differential*

$$dx.$$

More precisely, this is a *constant differential 1-form*, or simply a *1-form*. First-year calculus has problems with this object. The definition in Thomas and Finney's *Calculus and Analytic Geometry* is typical: "We define dx to be an independent variable with domain $(-\infty, \infty)$." Why, if it is only an independent variable, use such a distinctive notation? The reason is that dx satisfies a curious relationship: if $y = f(x)$, then the differentials of x and y are linked by

$$dy = f'(x)\, dx. \tag{4.1}$$

Where does this relationship arise? The answer: in integration. If $y_0 = f(x_0)$ and $y_1 = f(x_1)$, then

$$\int_{y_0}^{y_1} dy = \int_{x_0}^{x_1} f'(x)\, dx.$$

The lesson is that *differential forms exist to be integrated*. This is where we shall look for the definition of a differential.

Given a differential form, such as $3x^2\, dx$, we are only missing the limits of the integral. Once those are supplied, we can evaluate the integral over the line segment between those points. Thus, our differential form $3x^2\, dx$ can be viewed as a *mapping* from the set of all finite intervals to the set of real numbers:

$$[a, b] \longrightarrow \int_a^b 3x^2\, dx = b^3 - a^3.$$

If we introduce a change of variables, for example $x = 2t + 1$, then Equation (4.1) gives the relationship

$$3x^2\, dx = 3(2t + 1)^2\, 2\, dt = (24t^2 + 24t + 6)\, dt,$$

which implies that this differential form is also defined on the t axis, sending $[(a-1)/2, (b-1)/2]$ to precisely the same real number:

$$\left[\frac{a-1}{2}, \frac{b-1}{2}\right] \longrightarrow \int_{(a-1)/2}^{(b-1)/2} (24t^2 + 24t + 6)\, dt = b^3 - a^3.$$

The differential form is thus invariant under invertible differentiable transformations.

Definition: A *differential 1-form in one variable*, call it $g(x)\, dx$, is a mapping from the set of finite intervals to the real numbers defined by

$$g(x)\, dx : [a, b] \longrightarrow \int_a^b g(x)\, dx \in \mathbf{R}.$$

Furthermore, it is invariant under invertible differentiable transformations as follows. If $x = f(t)$, then $g(x)\, dx = g(f(t))f'(t)\, dt$, and

$$g(f(t))f'(t)\, dt : [f^{-1}(a), f^{-1}(b)] \longrightarrow \int_{f^{-1}(a)}^{f^{-1}(b)} g(f(t))f'(t)\, dt$$

$$= \int_a^b g(x)\, dx.$$

This may all be a bit confusing, especially as nothing really new has been said, and yet the usual way of looking at integrals has been turned on its head. You are accustomed to viewing integration as something you do to a function. We are now going to regard it as the action of a differential on a finite interval. The key to remember is that we are focusing on the differential inside the integral and viewing it as a mapping from finite intervals to real numbers.

Constant 1-Forms in Several Variables

In one dimension, dx is a mapping from the interval $[a, b]$ to the number $b - a$. In two dimensions, the x, y plane, we define dx to be a mapping from a directed line segment to the the change in the x coordinate. We thus have, for example,

$$\int_{(1,5)}^{(3,2)} dx = 3 - 1 = 2,$$

$$\int_{(1,5)}^{(3,2)} dy = 2 - 5 = -3.$$

As we move to higher dimensions, the differential form begins to take on physical significance. Consider a constant two-dimensional force field:

at any point in the x, y plane the force exerted on a particle is the same constant vector, for example, (2,3). We want to compute the amount of work done by this force field as it moves a particle from $\vec{a}$ to $\vec{b}$ along a straight line.

The displacement as we move from $\vec{a}$ to $\vec{b}$ is

$$\vec{d} = \vec{b} - \vec{a} = (b_1 - a_1, b_2 - a_2).$$

If we let Δx denote the change in the x coordinate,

$$\Delta x = b_1 - a_1,$$

and Δy the change in the y coordinate,

$$\Delta y = b_2 - a_2,$$

then

$$\vec{d} = (\Delta x, \Delta y).$$

From Section 2.2, we know that the work done by our force field is

$$\begin{aligned}
\text{Work} \;&=\; \vec{F} \cdot \vec{d} \\
&=\; (2, 3) \cdot (\Delta x, \Delta y) \\
&=\; 2\Delta x + 3\Delta y.
\end{aligned}$$

For this reason, it is natural to associate to our constant force field the constant differential 1-form

$$2\, dx + 3\, dy$$

and to define

$$\begin{aligned}
\int_{\vec{a}}^{\vec{b}} 2\, dx + 3\, dy \;&=\; 2 \int_{\vec{a}}^{\vec{b}} dx + 3 \int_{\vec{a}}^{\vec{b}} dy \\
&=\; 2\,\Delta x + 3\,\Delta y.
\end{aligned}$$

The differential 1-form corresponding to a constant force field is therefore a mapping from directed line segments to the amount of work done by the field as it moves a particle along this line segment.

Similarly, in three dimensions, a constant force field such as $(4, -1, 3)$ is described by a 1-form:

$$4\, dx - dy + 3\, dz.$$

The value of the integral of this 1-form over the directed line segement from $\vec{a}$ to $\vec{b}$, where

$$\vec{b} - \vec{a} = (\Delta x, \Delta y, \Delta z),$$

is

$$\int_{\vec{a}}^{\vec{b}} 4\, dx - dy + 3\, dz = 4\,\Delta x - \Delta y + 3\,\Delta z.$$

As an example,

$$\int_{(2,5,-3)}^{(-1,7,4)} 4\,dx - dy + 3\,dz = 4(-3) - (2) + 3(7) = 7.$$

Note that changing the direction of our line segment so that we travel from $\vec{b}$ to $\vec{a}$ changes the sign on the amount of work done, and thus changes the sign of the integral:

$$\int_{(-1,7,4)}^{(2,5,-3)} 4\,dx - dy + 3\,dz = 4(3) - (-2) + 3(-7) = -7.$$

Evaluating 1-Forms by Pullbacks

We can also evaluate the 1-form $4\,dx - dy + 3\,dz$ by using a parametrization of the line segment from $\vec{a}$ to $\vec{b}$:

$$(x, y, z) = \vec{a} + t(\vec{b} - \vec{a}) = (1-t)\vec{a} + t\vec{b}.$$

As t ranges from 0 to 1, (x, y, z) goes from $\vec{a}$ to $\vec{b}$. If $\vec{a} = (2, 5, -3)$ and $\vec{b} = (-1, 7, 4)$, then the parametrization is

$$(x, y, z) = (2, 5, -3) + t(-3, 2, 7),$$

or,

$$\begin{aligned} x &= 2 - 3t \\ y &= 5 + 2t \\ z &= -3 + 7t. \end{aligned} \tag{4.2}$$

The differentials of x, y, and z can be expressed in terms of the differential of t:

$$\begin{aligned} dx &= -3\,dt, \\ dy &= 2\,dt, \\ dz &= 7\,dt, \end{aligned}$$

and so our 1-form can be expressed as a 1-form in the single variable t:

$$\begin{aligned} 4\,dx - dy + 3\,dz &= -12\,dt - 2\,dt + 21\,dt \\ &= 7\,dt. \end{aligned}$$

We call $7\,dt$ the *pullback* of $4\,dx - dy + 3\,dz$. The system of Equations (4.2) took us from t space to x,y,z space. We have used it to *pull* a differential form in x,y,z space *back* to a differential form in t space.

In t space, our integral is from 0 to 1, and so we have

$$\int_{(2,5,-3)}^{(-1,7,4)} 4\,dx - dy + 3\,dz = \int_0^1 7\,dt = 7.$$

Nonconstant 1-Forms

The advantage of using a pullback to evaluate a 1-form is most evident when the coefficients of the differentials are not constant. In most cases, the force vector at any given point in a three-dimensional force field is going to depend on where we are. That is, the x, y, and z components of our force will each be functions of x, y, and z:

$$\vec{F}(x,y,z) = (f_1(x,y,z), f_2(x,y,z), f_3(x,y,z)).$$

The corresponding 1-form in three dimensions is

$$f_1(x,y,z)\,dx + f_2(x,y,z)\,dy + f_3(x,y,z)\,dz.$$

Given the directed line segment from $\vec{a}$ to $\vec{b}$, we define

$$\int_{\vec{a}}^{\vec{b}} f_1(x,y,z)\,dx + f_2(x,y,z)\,dy + f_3(x,y,z)\,dz$$

to be the work done by this force field as it moves a particle along the directed line segment from $\vec{a}$ to $\vec{b}$.

As we shall demonstrate in Chapter 5, we can evaluate this 1-form by parametrizing our line segment, substituting for x, y, and z the appropriate functions of t. For example, let $\vec{a} = (0,3,2)$ and $\vec{b} = (-5,2,0)$. The line segment from $\vec{a}$ to $\vec{b}$ is parametrized by

$$\begin{aligned} x &= -5t, \\ y &= 3 - t, \\ z &= 2 - 2t. \end{aligned}$$

If the force at (x,y,z) is given by $(x^2 + y, -x + 2z, yz)$, then the work done by this field in moving a particle along the directed line segment from $\vec{a}$ to $\vec{b}$ is

$$\int_{(0,3,2)}^{(-5,2,0)} (x^2 + y)\,dx + (-x + 2z)\,dy + yz\,dz$$

$$= \int_0^1 (25t^2 + 3 - t)(-5\,dt) + (5t + 4 - 4t)(-dt)$$

$$+ (2t^2 - 8t + 6)(-2\,dt)$$

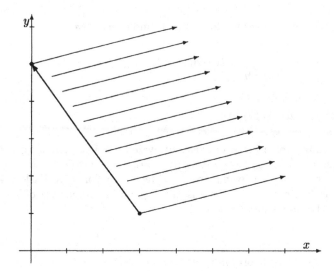

FIGURE 4.2. Constant fluid flow.

$$= \int_0^1 (-129t^2 + 20t - 31) \, dt$$
$$= -43t^3 + 10t^2 - 31t \Big|_0^1$$
$$= -64.$$

Two-Dimensional Fluid Flows

Consider a sheet of liquid flowing over a two-dimensional plane at a uniform velocity; at each point, the velocity is the same, for example, $(4,1)$. Now consider a line segment stretched across the flow, say from $(3,1)$ to $(0,5)$ (Figure 4.2). The problem before us is to compute the rate at which our fluid is crossing this line segment.

The fluid that flows across our line segment in one unit of time is represented by the parallelogram defined by the flow vector, $(4,1)$, and the vector representing our segment, $(0,5) - (3,1) = (-3,4)$. Since we have uniform flow, the quantity of fluid per unit time is the area of this parallelogram. If we imbed our vectors into three-dimensional space, then this area is precisely the magnitude of the cross product:

$$\begin{aligned} \text{area} \ &= \ |(4,1,0) \times (-3,4,0)| \\ &= \ |(0,0,19)| \\ &= \ 19. \end{aligned}$$

The direction of our line segment gives us an orientation. If we decide that the flow that we have described is in the positive direction, then the flow across the line segment from $(0,5)$ to $(3,1)$ would be defined as a negative flow. A convenient convention is to decide that the sign agrees with the sign of the z coordinate in the cross product. If you stand on the directed line segment looking in its direction, then left to right is a positive flow, right to left is negative.

If we define our line segment to be from $\vec{a}$ to $\vec{b}$ where $\vec{b} - \vec{a} = (\Delta x, \Delta y)$, and if the flow velocity is the constant vector (v_1, v_2), then the rate at which the fluid crosses the line segment is the z coordinate of

$$(v_1, v_2, 0) \times (\Delta x, \Delta y, 0) = (0, 0, v_1 \, \Delta y - v_2 \, \Delta x)$$
$$= (0, 0, -v_2 \, \Delta x + v_1 \, \Delta y).$$

This suggests that the 1-form corresponding to a constant fluid flow, (v_1, v_2), should be

$$-v_2 \, dx + v_1 \, dy.$$

With this definition, the rate at which the fluid crosses the directed line segment from $\vec{a}$ to $\vec{b}$ is given by

$$\int_{\vec{a}}^{\vec{b}} -v_2 \, dx + v_1 \, dy = -v_2 \, \Delta x + v_1 \, \Delta y.$$

Examples

The constant flow $(-2, 3)$ crosses the segment from $(-1, 3)$ to $(3,1)$ at the rate of

$$\int_{(-1,3)}^{(3,1)} -3 \, dx - 2 \, dy = -3(4) - 2(-2) = -8.$$

The negative sign implies that the flow is from right to left as seen standing on $(-1, 3)$ and looking toward $(3,1)$. The flow can also be computed by parametrizing our line segment:

$$x = -1 + 4t,$$
$$y = 3 - 2t,$$

$$\int_{(-1,3)}^{(3,1)} -3 \, dx - 2 \, dy = \int_0^1 -8 \, dt$$
$$= -8.$$

The variable flow $(3x + y, x - 2y)$ crosses this same line segment at the rate

$$\int_{(-1,3)}^{(3,1)} (-x + 2y)\, dx + (3x + y)\, dy \;=\; \int_0^1 (7 - 8t)(4\, dt) + (10t)(-2\, dt)$$

$$=\; \int_0^1 (-52t + 28)\, dt$$

$$=\; 2.$$

Note that 2 represents the net flow. From $(-1, 3)$ to $(15/13, 25/13)$, the flow is positive (left to right). From $(15/13, 25/13)$ to $(3,1)$, the flow is negative (right to left).

4.3 Exercises

1. Evaluate the differential form $4\, dx - 2\, dy + 3\, dz$ on each of the following directed line segments:

   (a) from $(-1, 2, 5)$ to $(-3, 3, 2)$,

   (b) from $(-3, 3, 2)$ to $(-1, 2, 5)$,

   (c) from $(2, 0, 1)$ to $(-1, 0, 2)$,

   (d) from $(-1, 2, 5)$ to $(0, 5, 3)$,

   (e) from $(0, 5, 3)$ to $(-3, 3, 2)$.

2. The line segment from $(3, 0, 1)$ to $(1, 4, -2)$ can be parametrized by

$$\begin{aligned} x &= 3 - 2t, \\ y &= 4t, \\ z &= 1 - 3t, \end{aligned}$$

   with t going from 0 to 1, or by

$$\begin{aligned} x &= x, \\ y &= 6 - 2x, \\ z &= -\frac{7}{2} + \frac{3}{2}x, \end{aligned}$$

   with x going from 3 to 1. Show that both parametrizations yield the same result when using the pullback to compute the integral of $6\, dx - 2\, dy + dz$ over the line segment from $(3, 0, 1)$ to $(1, 4, -2)$.

3. Let $k_1\, dx + k_2\, dy + k_3\, dz$ be a constant 1-form and let $\vec{a}, \vec{b}$, and $\vec{c}$ be any three points in x, y, z space. Prove that the integral of this 1-form from $\vec{a}$ to $\vec{c}$ is the sum of the integral from $\vec{a}$ to $\vec{b}$ plus the integral from $\vec{b}$ to $\vec{c}$.

4. Using the result of Exercise 3, prove that the amount of work done by a constant force field in moving a particle from $\vec{a}$ to $\vec{b}$ along a path composed of straight line segments is independent of the path.

5. Show that the evaluation by pullback of a constant 1-form, $k_1 \, dx + k_2 \, dy + k_3 \, dz$, over the directed line segment from $\vec{r}$ to $\vec{s}$ does not depend on which linear parametrization is chosen.

6. Find the amount of work done by the constant force field $dx + 3 \, dy - dz$ as it moves a particle along the intersection of the planes $x + y + z = 1$ and $x - 2y = -2$ from where it intersects the y, z plane ($x = 0$) to where it intersects the z, x plane ($y = 0$).

7. Evaluate the differential form $y \, dx + z \, dy + x \, dz$ on each of the following directed line segments.

   (a) from $(0, 2, -1)$ to $(3, 0, 1)$,
   (b) from $(3, 0, 1)$ to $(0, 2, -1)$,
   (c) from $(3, 1, 2)$ to $(-1, 1, 1)$,
   (d) from $(0, 2, -1)$ to $(1, 3, 2)$,
   (e) from $(1, 3, 2)$ to $(3, 0, 1)$.

8. Prove that in an arbitrary force field the amount of work done in moving from $\vec{a}$ to $\vec{b}$ may depend on the path.

9. Show that the gravitational field generated by a body of mass M at the origin is given by

$$-GM \left(\frac{x}{r^3} \, dx + \frac{y}{r^3} \, dy + \frac{z}{r^3} \, dz \right),$$

where $r = \sqrt{x^2 + y^2 + z^2}$. (Note that this 1-form is not defined at the origin.)

10. Prove that the work done by the gravitational field of Exercise 9 in moving a particle along the directed line segment from $\vec{a}$ to $\vec{b}$ depends only on $|\vec{a}|$ and $|\vec{b}|$.

11. Prove that the work done by the gravitational field of Exercise 9 depends only on the endpoints of the path.

12. Consider a cyclonic force field circling the z axis in a counterclockwise direction with a strength inversely proportional to the distance from the z axis. Show that this force field is described by the 1-form

$$c \left(\frac{-y \, dx}{x^2 + y^2} + \frac{x \, dy}{x^2 + y^2} \right),$$

for some constant c. (This 1-form is not defined on the z axis.)

13. Find the amount of work done by the force field of Exercise 12 over each of the following directed line segments:

    (a) from $(1, -1, 0)$ to $(1, 1, 0)$,

    (b) from $(1, -1, 0)$ to $(-1, 0, 0)$,

    (c) from $(-1, 0, 0)$ to $(1, 1, 0)$.

14. Consider the 2-dimensional flow $(3, -1)$. Find the rate at which it crosses each of the following line segments:

    (a) from $(2, 2)$ to $(3, 5)$,

    (b) from $(3, 5)$ to $(2, 2)$,

    (c) from $(2, 2)$ to $(3, 2)$,

    (d) from $(3, 2)$ to $(3, 5)$,

    (e) from $(-2, 1)$ to $(4, -1)$.

15. Consider the 2-dimensional flow that has the velocity $(x + y, xy)$ at the point (x, y). Find the rate at which it crosses each of the following line segments:

    (a) from $(2, 2)$ to $(3, 5)$,

    (b) from $(3, 5)$ to $(2, 2)$,

    (c) from $(2, 2)$ to $(3, 2)$,

    (d) from $(3, 2)$ to $(3, 5)$,

    (e) from $(-2, 1)$ to $(4, -1)$.

16. Find the rate at which the fluid flow

$$(y - y^2) \, dx + (xy - x) \, dy$$

crosses the line segment from $(2, -2)$ to $(2, 4)$. Where along this segment is the flow positive and where is it negative?

17. Find the rate at which the fluid flow of Exercise 16 crosses the line segment from $(4, -2)$ to $(2, 6)$. Use the pullback to determine where the flow is positive and where it is negative.

18. Consider a fluid flow emerging from the origin at a constant rate and flowing uniformly out in all directions. If we assume that our fluid is incompressible, then the rate at which the fluid flows into any given region must equal the rate at which it flows out of the region. In particular, if we take as our region the annulus with its center at the origin, an inner radius r_1, and an outer radius r_2, then the rate at which the fluid crosses the circle of radius r_1 must equal the rate at which it crosses the circle of radius r_2. Show that this implies that

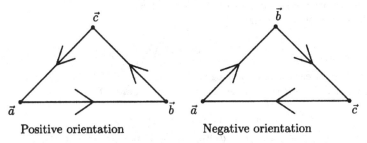

Positive orientation   Negative orientation

FIGURE 4.3. Oriented triangles.

the velocity of the flow is inversely proportional to the distance from
the origin and that the 1-form describing this flow is given by

$$c\left(\frac{-y\,dx}{x^2+y^2}+\frac{x\,dy}{x^2+y^2}\right).$$

4.4 Constant Differential 2-Forms

Differential 1-forms are mappings from directed line segments to the real
numbers. *Differential 2-forms* are mappings from *oriented triangles* to the
real numbers.

Consider a triangle in the x, y plane with vertices at $\vec{a}$, $\vec{b}$, and $\vec{c}$. We
denote this triangle by

$$T = [\vec{a},\,\vec{b},\,\vec{c}]$$

and say that it has a *positive orientation* if traveling from $\vec{a}$ to $\vec{b}$ to $\vec{c}$ and
back to $\vec{a}$ makes the circuit of the boundary in a counterclockwise direction. It has a *negative orientation* if the order of the vertices is clockwise
(Figure 4.3). The differential 2-form

$$dx\,dy$$

is a mapping from oriented triangles in the x, y plane to their *signed area*.
That is, if the orientation is positive, then $dx\,dy$ maps the triangle to its
area. If the orientation is negative, then it maps the triangle to the negative
of its area:

$$dx\,dy : [(0,0),(3,0),(0,4)] \longrightarrow 6,$$
$$dx\,dy : [(0,0),(0,4),(3,0)] \longrightarrow -6.$$

The constant differential 2-form

$$k\,dx\,dy,$$

maps an oriented triangle to k times its signed area.

The area of the triangle

$$T = [\vec{a}, \vec{b}, \vec{c}] = [(a_1, a_2), (b_1, b_2), (c_1, c_2)]$$

is half the area of the parallelogram defined by $\vec{b} - \vec{a}$ and $\vec{c} - \vec{a}$, and so the signed area of our triangle is half the z coordinate of

$$(b_1 - a_1, b_2 - a_2, 0) \times (c_1 - a_1, c_2 - a_2, 0)$$
$$= (0, 0, (b_1 - a_1)(c_2 - a_2) - (b_2 - a_2)(c_1 - a_1)).$$

If we define the integral over T of $dx\,dy$ to be the value of $dx\,dy$ on T, then

$$\int_T dx\,dy = \frac{1}{2}\left((b_1 - a_1)(c_2 - a_2) - (b_2 - a_2)(c_1 - a_1)\right). \qquad (4.3)$$

As an example, let $T = [(1, 2), (4, 0), (0, -2)]$. The value of the integral of $5\,dx\,dy$ over T is

$$\int_T 5\,dx\,dy = \frac{5}{2}[(4 - 1)(-2 - 2) - (0 - 2)(0 - 1)] = -35.$$

The *projection* of a point (x, y, z), onto the x, y plane is the point obtained by setting the z coordinate equal to 0: $(x, y, 0)$. The *projection of the triangle*

$$[\vec{a}, \vec{b}, \vec{c}] = [(a_1, a_2, a_3), (b_1, b_2, b_3), (c_1, c_2, c_3)]$$

onto the x, y plane is the triangle defined by the projections of the vertices:

$$[(a_1, a_2, 0), (b_1, b_2, 0), (c_1, c_2, 0)].$$

We define the differential 2-form, $dx\,dy$, in three-dimensional space to be the mapping from an oriented triangle $T = [\vec{a}, \vec{b}, \vec{c}]$ to the signed area of its projection onto the x, y plane, which is the z coordinate of $\frac{1}{2}(\vec{b} - \vec{a}) \times (\vec{c} - \vec{a})$.

Similarly, $dz\,dx$ maps this triangle to the signed area of its projection onto the z, x plane which is the y coordinate of $\frac{1}{2}(\vec{b} - \vec{a}) \times (\vec{c} - \vec{a})$. The 2-form $dy\,dz$ maps this triangle to the signed area of the projection onto the y, z plane, the x coordinate of $\frac{1}{2}(\vec{b} - \vec{a}) \times (\vec{c} - \vec{a})$. (We put the z before the x in $dz\,dx$ because $\vec{k} \times \vec{i} = \vec{j}$. The positive y axis is in the positive direction relative to the z, x plane, but not to the x, z plane.)

We have shown that

$$\frac{1}{2}(\vec{b} - \vec{a}) \times (\vec{c} - \vec{a}) = \left(\int_T dy\,dz, \int_T dz\,dx, \int_T dx\,dy\right), \qquad (4.4)$$

where $T = [\vec{a}, \vec{b}, \vec{c}]$.

If we take
$$T = [(1,0,2), (-1,1,3), (0,2,2)]$$
to be our oriented triangle, then

$$\frac{1}{2}(\vec{b} - \vec{a}) \times (\vec{c} - \vec{a}) = \frac{1}{2}(-2,1,1) \times (-1,2,0)$$
$$= \left(-1, -\frac{1}{2}, -\frac{3}{2}\right).$$

Using this and Equation (4.4), we see that

$$\int_T dx\, dy = -\frac{3}{2},$$
$$\int_T dz\, dx = -\frac{1}{2},$$
$$\int_T dy\, dz = -1.$$

Three-Dimensional Fluid Flow

We consider a three-dimensional fluid flowing at a constant velocity, $\vec{v} = (v_1, v_2, v_3)$. In three dimensions, it does not make sense to speak of the rate at which this fluid crosses a line segment, but rather the rate at which it crosses a surface. We shall take our surface to be the triangle $T = [\vec{a}, \vec{b}, \vec{c}]$. The positive direction from this triangle is defined to be the direction of the normal vector given by

$$\vec{n} = (\vec{b} - \vec{a}) \times (\vec{c} - \vec{a}). \tag{4.5}$$

Equivalently, if we circle the boundary from $\vec{a}$ to $\vec{b}$ to $\vec{c}$, then the normal vector points in the direction of movement of a right-hand screw turned in this direction (Figure 4.4). A flow across our triangle is positive if and only if its component in the direction of the positive normal is positive; that is, if and only if
$$\vec{v} \cdot \vec{n} > 0.$$

The volume of fluid crossing our triangle per unit time is the volume of the parallel prism whose base is our triangle and whose sides are determined by $\vec{v}$ (Figure 4.5). This is half the volume of the parallelepiped defined by $\vec{v}$, $\vec{b} - \vec{a}$, and $\vec{c} - \vec{a}$, which is half the scalar triple product of these three vectors. Observe that if we take our three vectors in this order, then the sign will be correct: the rate of flow across the triangle is

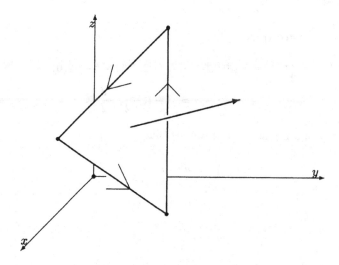

FIGURE 4.4. An oriented triangle in 3-space.

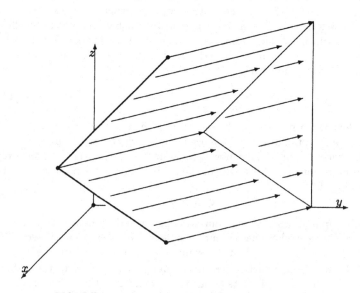

FIGURE 4.5. Three-dimensional flow across a triangle.

$$\frac{1}{2}\vec{v}\cdot(\vec{b}-\vec{a})\times(\vec{c}-\vec{a})$$

$$= v_1\int_T dy\,dz + v_2\int_T dz\,dx + v_3\int_T dx\,dy$$

$$= \int_T v_1\,dy\,dz + v_2\,dz\,dx + v_3\,dx\,dy. \tag{4.6}$$

It therefore makes sense to identify the constant flow $\vec{v} = (v_1, v_2, v_3)$ with the constant differential 2-form

$$v_1\,dy\,dz + v_2\,dz\,dx + v_3\,dx\,dy.$$

Example

To find the rate at which the constant flow $(2, -1, 3)$ crosses the triangle

$$T = [(0, 1, -2), (3, 1, 0), (-2, 2, 1)],$$

we find the signed areas of the projections of this triangle onto the y, z-, z, x-, and x, y planes:

$$\frac{1}{2}(3, 0, 2)\times(-2, 1, 3) = \left(-1, \frac{-13}{2}, \frac{3}{2}\right),$$

and then we evaluate

$$\int_T 2\,dy\,dz - dz\,dx + 3\,dx\,dy = 2(-1) - \left(-\frac{13}{2}\right) + 3\left(\frac{3}{2}\right)$$

$$= 9. \tag{4.7}$$

Pullbacks

We can also evaluate the integral in Equation (4.7) by using a pullback to the triangle

$$U = [(0, 0), (1, 0), (0, 1)]$$

in the u, v plane. A mapping from this triangle to

$$T = [(0, 1, -2), (3, 1, 0), (-2, 2, 1)]$$

is given by

$$\begin{aligned} x &= 3u - 2v, \\ y &= 1 + v, \\ z &= -2 + 2u + 3v. \end{aligned}$$

The problem at hand is to determine how the differentials $dx\,dy$, $dz\,dx$, and $dy\,dz$ are related to $du\,dv$. Our mapping takes the unit square with vertices at $(0,0), (1,0), (1,1)$, and $(0,1)$ in the u,v plane to the parallelogram with vertices at $(0,1), (3,1), (1,2)$, and $(-2,2)$ in the x,y plane. The signed area of this parallelogram is the z coordinate of

$$(3,0,0) \times (-2,1,0) = (0,0,3),$$

which is 3, and so 1 unit of area in the u,v plane corresponds to 3 units of area in the x,y plane which means that

$$dx\,dy = 3\,du\,dv,$$

so that our integrals are related by

$$\int_T dx\,dy = \int_U 3\,du\,dv.$$

In general, the mapping from the triangle

$$U = [(0,0),(1,0),(0,1)]$$

to the triangle with vertices at

$$\begin{aligned} \vec{a} &= (a_1,a_2,a_3), \\ \vec{b} &= (b_1,b_2,b_3), \\ \vec{c} &= (c_1,c_2,c_3) \end{aligned}$$

is given by

$$\begin{aligned} x &= a_1 + (b_1 - a_1)u + (c_1 - a_1)v, \\ y &= a_2 + (b_2 - a_2)u + (c_2 - a_2)v, \\ z &= a_3 + (b_3 - a_3)u + (c_3 - a_3)v. \end{aligned}$$

The unit square in the u,v plane is mapped to the parallelogram in the x,y plane, whose signed area is the z coordinate of

$$(b_1 - a_1, c_1 - a_1, 0) \times (b_2 - a_2, c_2 - a_2, 0)$$
$$= (0, 0, (b_1 - a_1)(c_2 - a_2) - (b_2 - a_2)(c_1 - a_1)),$$

and so

$$dx\,dy = ((b_1 - a_1)(c_2 - a_2) - (b_2 - a_2)(c_1 - a_1))\,du\,dv. \qquad (4.8)$$

Similarly, we have that

$$\begin{aligned} dz\,dx &= ((b_3 - a_3)(c_1 - a_1) - (b_1 - a_1)(c_3 - a_3))\,du\,dv, & (4.9) \\ dy\,dz &= ((b_2 - a_2)(c_3 - a_3) - (b_3 - a_3)(c_2 - a_2))\,du\,dv. & (4.10) \end{aligned}$$

Fortunately, we do not have to memorize these unwieldy formulas. Note the relationships between the 1-forms:

$$\begin{aligned}
dx &= (b_1 - a_1)\, du + (c_1 - a_1)\, dv, \\
dy &= (b_2 - a_2)\, du + (c_2 - a_2)\, dv, \\
dz &= (b_3 - a_3)\, du + (c_3 - a_3)\, dv.
\end{aligned}$$

We need rules for multiplying differentials so that the products $dx\, dy$, $dz\, dx$, and $dy\, dz$ will be correct.

Rules for Multiplying Constant Differential Forms

1. Constants commute with differentials:

$$du(k\, dv) = k\, du\, dv. \qquad (4.11)$$

2. Multiplication of constants and differentials is distributive:

$$\begin{aligned}
(k_1 + k_2)\, du &= k_1\, du + k_2\, du, &\qquad (4.12) \\
k(du + dv) &= k\, du + k\, dv. &\qquad (4.13)
\end{aligned}$$

3. Multiplication of differentials is *anticommutative*:

$$du\, dv = -dv\, du. \qquad (4.14)$$

This implies that the product of a differential with itself is 0:

$$du\, du = 0. \qquad (4.15)$$

Using these rules gives us precisely the results we want. For example,

$$\begin{aligned}
dx\, dy &= [(b_1 - a_1)\, du + (c_1 - a_1)\, dv]\,[(b_2 - a_2)\, du + (c_2 - a_2)\, dv] \\[2mm]
&= (b_1 - a_1)(b_2 - a_2)\, du\, du + (b_1 - a_1)(c_2 - a_2)\, du\, dv \\
&\quad + (c_1 - a_1)(b_2 - a_2)\, dv\, du + (c_1 - a_1)(c_2 - a_2)\, dv\, dv \\[2mm]
&= [(b_1 - a_1)(c_2 - a_2) - (b_2 - a_2)(c_1 - a_1)]\, du\, dv.
\end{aligned}$$

To emphasize the point that multiplication of differentials is *not* commutative, one often sees the special symbol "$\wedge$", called the *wedge* or *exterior product*:

$$du \wedge dv = -dv \wedge du.$$

I have chosen to suppress this symbol because, for our purposes, differentials never multiply in any other way. The anticommutative algebra of differential forms is known as the *exterior algebra*.

Example

Returning to our original example, we have

$$
\begin{aligned}
dx &= 3\,du - 2\,dv,\\
dy &= dv,\\
dz &= 2\,du + 3\,dv,
\end{aligned}
$$

and so

$$
\begin{aligned}
dx\,dy &= (3\,du - 2\,dv)\,dv\\
&= 3\,du\,dv,\\
dz\,dx &= (2\,du + 3\,dv)(3\,du - 2\,dv)\\
&= -13\,du\,dv,\\
dy\,dz &= (dv)(2\,du + 3\,dv)\\
&= -2\,du\,dv.
\end{aligned}
$$

Knowing the pullbacks, we can evaluate the integral over T:

$$
\int_T 2\,dy\,dz - dz\,dx + 3\,dx\,dy
$$

$$
= \int_U 2(-2\,du\,dv) - (-13\,du\,dv) + 3(3\,du\,dv)
$$

$$
= \int_U 18\,du\,dv
$$

$$
= 9.
$$

4.5 Exercises

1. Evaluate the differential form $3\,dx\,dy$ on each of the following oriented triangles:

   (a) $[(0,0),(4,0),(2,1)]$,

   (b) $[(5,2),(1,3),(3,4)]$,

   (c) $[(3,4),(1,3),(5,2)]$,

   (d) $[(1,0,-2),(3,1,5),(-2,1,0)]$.

2. Find the mappings taking the triangle $[(0,0), (1,0), (0,1)]$ in the u, v plane to each of the following oriented triangles in x, y, z space:

   (a) $[(0,1,-3), (2,1,5), (-2,0,6)]$,

   (b) $[(-2,0,6), (2,1,5), (0,1,-3)]$,

   (c) $[(2,7,-1), (-2,3,0), (1,4,1)]$,

   (d) $[(1,-3,2), (2,1,1), (0,-7,5)]$,

   (e) $[(-1,2,3), (0,3,4), (2,5,6)]$.

3. Find the pullback of

$$2 \, dy \, dz + 3 \, dz \, dx - 2 \, dx \, dy$$

   for each of the mappings in Exercise 2.

4. Evaluate the differential form

$$2 \, dy \, dz + 3 \, dz \, dx - 2 \, dx \, dy$$

   for each of the mappings in Exercise 2.

5. Find the rate at which a fluid flowing with constant velocity $\vec{v} = (3,0,-1)$ crosses each of the following triangles:

   (a) $[(-2,1,0), (5,3,-1), (8,-5,2)]$,

   (b) $[(8,-5,2), (5,3,-1), (-2,1,0)]$,

   (c) $[(1,-1,0), (2,0,2), (3,-2,-3)]$,

   (d) $[(0,0,0), (0,1,0), (1,0,0)]$,

   (e) $[(1,0,0), (0,1,0), (0,0,1)]$,

   (f) $[(0,0,0), (0,0,1), (0,1,0)]$,

   (g) $[(0,0,0), (1,0,0), (0,0,1)]$.

6. The four triangles of parts d through g of Exercise 5 are the faces of a tetrahedron (Figure 4.6), oriented so that the positive direction is outward on all four faces. Prove that for any constant flow, the sum of the rates at which the fluid crosses each of these faces is zero.

7. Using Exercise 6 and pullbacks, show that for any constant flow and any tetrahedron whose triangular faces are all oriented outward the net rate at which the flow crosses the surface of the tetrahedron (the sum of the rates at which the flow crosses each of the faces) is zero.

8. Show that the plane containing the triangle $T = [(0,1,-2), (3,1,0), (-2,2,1)]$ is $2x + 13y - 3z = 19$. We can solve for z in terms of x and y. Use this and the rules for multiplying differentials to express $2 \, dy \, dz - dz \, dx + 3 \, dx \, dy$ in terms of just $dx \, dy$.

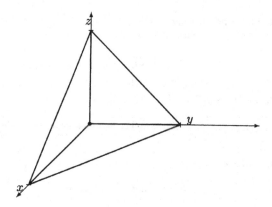

FIGURE 4.6. Tetrahedron.

9. The projection of $T = [(0, 1, -2), (3, 1, 0), (-2, 2, 1)]$ onto the x, y plane is $S = [(0, 1), (3, 1), (-2, 2)]$. Find the area of this triangle. Using Exercise 8, evaluate

$$\int_T 2\, dy\, dz - dz\, dx + 3\, dx\, dy$$

by means of the pullback to the x, y plane. Compare your answer with the example on page 93.

10. Find the rate at which the constant flow $3\, dy\, dz + 2\, dz\, dx - 5\, dx\, dy$ crosses the triangle in the plane

$$x + 2y + 3z = 6$$

with vertices on the x, y, and z axes, oriented so that $(1, 2, 3)$ is the positive direction.

11. If $\vec{a}$, $\vec{b}$, and $\vec{c}$ are on the same line, then

$$\vec{c} = \vec{a} + t(\vec{b} - \vec{a})$$

for some real number t. Find a mapping from $[(0, 0), (1, 0), (0, 1)]$ to $[\vec{a}, \vec{b}, \vec{a} + t(\vec{b} - \vec{a})]$ and show that the pullbacks of $dx\, dy$, $dz\, dx$, and $dy\, dz$ are all 0.

12. Show that the evaluation by pullback of a constant 2-form, $k_1\, dy\, dz + k_2\, dz\, dx + k_3\, dx\, dy$, over an oriented triangle, $T = [\vec{a}, \vec{b}, \vec{c}]$, does not depend on which linear parametrization is chosen.

13. Consider a fluid flow in three dimensions that emerges from the origin at a constant rate and flows uniformly out in all directions. If we

assume that our fluid is incompressible, then the rate at which fluid flows into any given region must equal the rate at which it flows out of that region. In particular, if we take as our region the spherical shell with its center at the origin, an inner radius r_1, and an outer radius r_2, then the rate at which the fluid crosses the spherical surface of radius r_1 must equal the rate at which it crosses the spherical surface of radius r_2. Show that this implies that the velocity of the flow is inversely proportional to the square of the distance from the origin and that the 2-form describing this flow is given by

$$ c\left(\frac{x\,dy\,dz}{r^3} + \frac{y\,dz\,dx}{r^3} + \frac{z\,dx\,dy}{r^3}\right), $$

where $r = \sqrt{x^2 + y^2 + z^2}$.

4.6   Constant Differential k-Forms

An *oriented tetrahedron* or pyramid with a triangular base (Figure 4.6) is defined by four points, $[\vec{a}, \vec{b}, \vec{c}, \vec{d}]$, and has a *positive orientation* if the vectors $\vec{b} - \vec{a}$, $\vec{c} - \vec{a}$, and $\vec{d} - \vec{a}$ are in the order of the right-hand rule. Otherwise, it has a *negative orientation*. We define the constant differential 3-form

$$ dx\,dy\,dz $$

to be a mapping from the oriented tetrahedra to the real numbers defined by taking an oriented tetrahedron to its signed volume. Since the tetrahedron

$$ T = [\vec{a}, \vec{b}, \vec{c}, \vec{d}] $$

has one-sixth the volume of the parallelepiped defined by $\vec{b} - \vec{a}$, $\vec{c} - \vec{a}$, and $\vec{d} - \vec{a}$, we see that

$$ \int_T dx\,dy\,dz = \frac{1}{6}(\vec{b} - \vec{a}) \cdot (\vec{c} - \vec{a}) \times (\vec{d} - \vec{a}). \tag{4.16} $$

A constant 3-form, $k\,dx\,dy\,dz$, maps T to k times the signed volume of T.

If we are in more than three dimensions, for example in 4-dimensional space–time, the oriented tetrahedron is still defined by four points. The differential $dx\,dy\,dz$ maps this tetrahedron to the signed volume of its projection onto x, y, z space.

We can also evaluate 3-forms by pullbacks. Let

$$ U = [(0,0,0), (1,0,0), (0,1,0), (0,0,1)] $$

in u, v, w space and let

$$ T = [(a_1, a_2, a_3), (b_1, b_2, b_3), (c_1, c_2, c_3), (d_1, d_2, d_3)] $$

in x, y, z space. We have a mapping from U to T given by

$$
\begin{aligned}
x &= a_1 + (b_1 - a_1)u + (c_1 - a_1)v + (d_1 - a_1)w, \\
y &= a_2 + (b_2 - a_2)u + (c_2 - a_2)v + (d_2 - a_2)w, \\
z &= a_3 + (b_3 - a_3)u + (c_3 - a_3)v + (d_3 - a_3)w.
\end{aligned} \qquad (4.17)
$$

Under this transformation, the unit cube in u, v, w space is mapped to the parallelepiped spanned by $\vec{b} - \vec{a}$, $\vec{c} - \vec{a}$, and $\vec{d} - \vec{a}$ in x, y, z space, and so the relationship between the respective 3-forms is given by

$$
dx \, dy \, dz = (\vec{b} - \vec{a}) \cdot (\vec{c} - \vec{a}) \times (\vec{d} - \vec{a}) \, du \, dv \, dw. \qquad (4.18)
$$

It is left as Exercise 1 in Section 4.8 to verify that if we take the differential equations

$$
\begin{aligned}
dx &= (b_1 - a_1) \, du + (c_1 - a_1) \, dv + (d_1 - a_1) \, dw, \\
dy &= (b_2 - a_2) \, du + (c_2 - a_2) \, dv + (d_2 - a_2) \, dw, \\
dz &= (b_3 - a_3) \, du + (c_3 - a_3) \, dv + (d_3 - a_3) \, dw
\end{aligned} \qquad (4.19)
$$

and multiply together dx, dy, and dz using the rules of Section 4.4, we get precisely Equation (4.18).

Example

Let $T = [(2, 1, -3), (0, 2, 1), (5, -2, 4), (-3, 1, 0)]$ and U as above. The mapping from U to T is given by

$$
\begin{aligned}
x &= 2 - 2u + 3v - 5w, \\
y &= 1 + u - 3v, \\
z &= -3 + 4u + 7v + 3w.
\end{aligned}
$$

The 3-forms are related by

$$
\begin{aligned}
dx \, dy \, dz &= (-2 \, du + 3 \, dv - 5 \, dw)(du - 3 \, dv)(4 \, du + 7 \, dv + 3 \, dw) \\
&= 18 \, du \, dv \, dw + 9 \, dv \, du \, dw - 35 \, dw \, du \, dv + 60 \, dw \, dv \, du \\
&= (18 - 9 - 35 - 60) \, du \, dv \, dw \\
&= -86 \, du \, dv \, dw.
\end{aligned}
$$

Evaluating $dx \, dy \, dz$ over T yields

$$
\int_T dx \, dy \, dz = \int_U -86 \, du \, dv \, dw = \frac{1}{6}(-86) = -14\frac{1}{3}.
$$

Simplices

A line segment is also known as a *1-simplex*, a triangle as a *2-simplex*, and a tetrahedron as a *3-simplex*. In general, an *oriented k-simplex* lives in at least k dimensions and is defined by $k+1$ ordered points:

$$T = [\vec{a}_0, \vec{a}_1, \vec{a}_2, \ldots, \vec{a}_k],$$

where each $\vec{a}_i$ is a vector of at least k dimensions:

$$\vec{a}_i = (a_{1i}, a_{2i}, \ldots, a_{li}), \quad l \geq k.$$

The *standard k-simplex* in k-dimensional space is given by

$$U = [(0,0,0,\ldots,0), (1,0,0,\ldots,0), (0,1,0,\ldots,0), \ldots, (0,0,0,\ldots,1)],$$

and the mapping from U to T is given by

$$
\begin{aligned}
x_1 &= a_{10} + (a_{11} - a_{10})u_1 + (a_{12} - a_{10})u_2 + \cdots + (a_{1k} - a_{10})u_k, \\
x_2 &= a_{20} + (a_{21} - a_{20})u_1 + (a_{22} - a_{20})u_2 + \cdots + (a_{2k} - a_{20})u_k, \\
&\vdots \\
x_l &= a_{l0} + (a_{l1} - a_{l0})u_1 + (a_{l2} - a_{l0})u_2 + \cdots + (a_{lk} - a_{l0})u_k.
\end{aligned}
$$
$$(4.20)$$

k-Forms

The differential k-form

$$dx_1 \, dx_2 \, \cdots \, dx_k$$

is a mapping from oriented k-simplices to their signed hypervolumes which can be defined by the pullback

$$
\begin{aligned}
dx_1 \, dx_2 \, \cdots \, dx_k &= [(a_{11} - a_{10}) \, du_1 + \cdots + (a_{1k} - a_{10}) \, du_k] \\
&\quad \times \cdots \times [(a_{k1} - a_{k0}) \, du_1 + \cdots + (a_{kk} - a_{k0}) \, du_k] \\
&= D(\vec{a}_1 - \vec{a}_0, \ldots, \vec{a}_k - \vec{a}_0) \, du_1 \, du_2 \, \cdots \, du_k, \quad (4.21)
\end{aligned}
$$

where $D(\vec{a}_1 - \vec{a}_0, \ldots, \vec{a}_k - \vec{a}_0)$ is the coefficient of $du_1 \ldots du_k$ obtained when we use our rules for multiplying differentials.

As is shown in Exercise 13 of Section 4.8, the hypervolume of the standard k-simplex is

$$\frac{1}{k!} = \frac{1}{1 \cdot 2 \cdot \cdots \cdot k},$$

and so

$$\int_T dx_1 \cdots dx_k = \int_U D(\vec{a}_1 - \vec{a}_0, \ldots, \vec{a}_k - \vec{a}_0) \, du_1 \cdots du_k,$$

$$= \frac{1}{k!} D(\vec{a}_1 - \vec{a}_0, \ldots, \vec{a}_k - \vec{a}_0). \tag{4.22}$$

Evaluating $D(\vec{b}_1, \vec{b}_2, \ldots, \vec{b}_k)$

A *permutation* of the set of integers 1 through k is a 1-to-1 and onto mapping, denoted by σ, from this set to itself:

For i in $\{1, 2, \ldots, k\}$, $\sigma(i)$ is also in $\{1, 2, \ldots, k\}$.

If $i \neq j$, then $\sigma(i) \neq \sigma(j)$.

For example, there are 24 permutations of the integers 1 through 4. One of these permutations is

$$\sigma(1) = 3, \quad \sigma(2) = 2, \quad \sigma(3) = 4, \quad \sigma(4) = 1.$$

Let S_k be the set of all permutations of the integers 1 through k. For $i = 1, 2, \ldots, k$, let $\vec{b}_i$ be a k-dimensional vector:

$$\vec{b}_i = (b_{1i}, b_{2i}, \ldots, b_{ki}).$$

You are asked in Exercise 7 of Section 4.8 to verify that the function D of Equation (4.21) satisfies

$$D(\vec{b}_1, \ldots, \vec{b}_k) \, du_1 \cdots du_k = \sum_{\sigma \in S_k} \prod_{i=1}^{k} b_{i\sigma(i)} \, du_{\sigma(i)}. \tag{4.23}$$

The symbol Π denotes the product:

$$\prod_{i=1}^{k} b_{i\sigma(i)} \, du_{\sigma(i)} = \left(b_{1\sigma(1)} \, du_{\sigma(1)} \right) \left(b_{2\sigma(2)} \, du_{\sigma(2)} \right) \cdots \left(b_{k\sigma(k)} \, du_{\sigma(k)} \right).$$

If we define $\mathrm{sgn}(\sigma)$ to be $+1$ or -1 so that

$$\mathrm{sgn}(\sigma) \prod_{i=1}^{k} du_{\sigma(i)} = du_1 \cdots du_k,$$

then

$$D(\vec{b}_1, \ldots, \vec{b}_k) = \sum_{\sigma \in S_k} \mathrm{sgn}(\sigma) \prod_{i=1}^{k} b_{i\sigma(i)}. \tag{4.24}$$

Example

In the example given on page 100, we can write our transformation from u, v, w space to x, y, z space as

$$
\begin{array}{ccccccccc}
x & = & 2 & + & -2u & + & 3v & + & -5w, \\
y & = & 1 & + & u & + & -3v & + & 0 \cdot w, \\
z & = & -3 & + & 4u & + & 7v & + & 3w.
\end{array}
$$

The vectors $\vec{b}_1$, $\vec{b}_2$, and $\vec{b}_3$ are the columns of coefficients of u, v, and w, respectively:

$$
\begin{aligned}
\vec{b}_1 & = (-2, 1, 4), \\
\vec{b}_2 & = (3, -3, 7), \\
\vec{b}_3 & = (-5, 0, 3).
\end{aligned}
$$

Our function D must satisfy

$$
dx \, dy \, dz = D\left((-2, 1, 4), (3, -3, 7), (-5, 0, 3)\right) du \, dv \, dw,
$$

and so

$$
D\left((-2, 1, 4), (3, -3, 7), (-5, 0, 3)\right) = -86.
$$

We also get -86 if we use the formula given in Equation (4.24). There are six permutations in S_3:

$$
\begin{aligned}
D\left((-2, 1, 4), (3, -3, 7), (-5, 0, 3)\right) & = b_{11}b_{22}b_{33} - b_{11}b_{23}b_{32} \\
& \quad - b_{12}b_{21}b_{33} + b_{12}b_{23}b_{31} \\
& \quad - b_{13}b_{22}b_{31} + b_{13}b_{21}b_{32} \\
& = (-2)(-3)(3) - (-2)(0)(7) \\
& \quad - (3)(1)(3) + (3)(0)(4) \\
& \quad - (-5)(-3)(4) + (-5)(1)(7) \\
& = -86.
\end{aligned}
$$

0-Forms

We have missed one type of differential form, the *0-form*. A *0-simplex* is a point. A 0-form is a mapping from a point in the given space to a real number. In other words, 0-forms are simply real-valued functions or scalar fields. It may seem strange to even bother defining 0-forms, but as we shall see later, it is convenient to include ordinary functions into the grand scheme of differential forms.

Summary

Most of the differential forms that we shall need are 0-forms, 1-forms, 2-forms, or 3-forms.

0-forms: These are real-valued functions of one or more variables, for example,

$$f(x) = x^3, \quad g(x,y) = x\cos y, \quad h(x,y,z) = \frac{x - y + z}{x^2 + y^2 + z^2}.$$

A physical example of a 0-form is the function that describes the temperature (a real number) at any point in a room (specified by a 3-dimensional vector).

1-forms: These map directed line segments (which are 1-dimensional) to real numbers. In 3-dimensional space, we use 1-forms to describe force fields:

$$\frac{x\,dx + y\,dy + z\,dz}{(x^2 + y^2 + z^2)^{3/2}}$$

is a field whose force vector always points away from the origin and whose strength is the inverse square of the distance from the origin. Its effect on a directed line segment is to compute the amount of work done by the force field in moving a particle along that segment.

In 2-dimensional space, a 1-form can represent either a force field or a fluid flow. In the latter case, it maps a directed line segment to the rate at which fluid is flowing across that segment. The 1-form

$$\frac{x\,dx + y\,dy}{x^2 + y^2}$$

represents a clockwise flow around the origin whose velocity at (x,y) is

$$\left(\frac{y}{x^2 + y^2}, \frac{-x}{x^2 + y^2} \right).$$

In 1-dimensional space, a 1-form such as $3x^2\,dx$ is a mapping from segments of the real number line to real numbers. The 1-form $3x^2\,dx$ takes the interval $[a, b]$ to the real number

$$\int_a^b 3x^2\,dx = b^3 - a^3.$$

2-forms: These map oriented triangles (which are 2-dimensional) to real numbers. In 3-dimensional space, we use 2-forms to describe flows:

$$\frac{x\,dy\,dz + y\,dz\,dx + z\,dx\,dy}{(x^2 + y^2 + z^2)^{3/2}}$$

is a flow whose velocity at (x, y, z) is

$$\left(\frac{x}{(x^2 + y^2 + z^2)^{3/2}}, \frac{y}{(x^2 + y^2 + z^2)^{3/2}}, \frac{z}{(x^2 + y^2 + z^2)^{3/2}} \right).$$

It assigns to the oriented triangle (or more generally to any surface that can be built out of triangles) the rate at which fluid is flowing through the surface.

In 2-dimensional space, a 2-form is a mapping from regions which can be built out of triangles to the real number line. As an example, if we have a thin, flat plate whose density at each point (x, y) is given by the function $\rho(x, y)$, then the 2-form $\rho(x, y)\, dx\, dy$ assigns to any 2-dimensional region R the total mass of that region.

3-forms: These map oriented tetrahedra (which are 3-dimensional) to real numbers. As an example, if $\rho(x, y, z)$ denotes the density at the point (x, y, z), then the 3-form $\rho(x, y, z)\, dx\, dy\, dz$ maps a 3-dimensional region R that can be built out of tetrahedra to the mass of the region, a real number.

4.7   Prospects

First, we take a quick glance back to vector algebra. There is a natural correspondence between vectors and constant 1-forms:

$$(a, b, c) \longleftrightarrow a\, dx + b\, dy + c\, dz.$$

In three dimensions, vectors also correspond to constant 2-forms:

$$(a, b, c) \longleftrightarrow a\, dy\, dz + b\, dz\, dx + c\, dx\, dy.$$

The cross product is simply the product of the vectors considered as 1-forms:

$$(a\, dx + b\, dy + c\, dz)(r\, dx + s\, dy + t\, dz)$$
$$= (bt - cs)\, dy\, dz + (cr - at)\, dz\, dx + (as - br)\, dx\, dy.$$

The dot product is the product of a 1-form and a 2-form:

$$(a\, dx + b\, dy + c\, dz)(r\, dy\, dz + s\, dz\, dx + t\, dx\, dy)$$
$$= (ar + bs + ct)\, dx\, dy\, dz.$$

Note that in three dimensions a scalar can be either a 0-form or a 3-form.

Looking ahead, it is clear that we have barely begun the task of evaluating differential forms. For $k \geq 2$, we have only considered forms with constant coefficients. We shall need to expand this to forms whose coefficients are real-valued functions of position. But more than this, we are going to have to be able to evaluate forms on geometric objects that are more complex than a simplex: on curves, contoured surfaces, and molded volumes. An indication of how this is accomplished can be gained from a description of evaluating 1-forms on differentiable curves.

We shall evaluate the 1-form

$$\omega = f(x, y, z)\, dx + g(x, y, z)\, dy + h(x, y, z)\, dz$$

over the curve

$$C = \{(x(t), y(t), z(t)) \mid t \in [a, b]\},$$

where x, y, and z are differentiable functions of t. The pullback of ω is

$$\omega = \left(f(x, y, z)\frac{dx}{dt} + g(x, y, z)\frac{dy}{dt} + h(x, y, z)\frac{dz}{dt} \right) dt.$$

For example, let

$$
\begin{aligned}
\omega &= y\, dx - x\, dy + z\, dz, \\
C &= \{(\cos t,\ \sin t,\ t) \mid 0 \le t \le 2\pi\}.
\end{aligned}
$$

Since $x = \cos t$, $y = \sin t$, and $z = t$, we see that

$$
\begin{aligned}
dx &= -\sin t\, dt, \\
dy &= \cos t\, dt, \\
dz &= dt.
\end{aligned}
$$

The pullback of ω is

$$
\begin{aligned}
\omega &= (\sin t\,(-\sin t) - \cos t\,(\cos t) + t)\, dt \\
&= (-1 + t)\, dt,
\end{aligned}
$$

and so we have

$$
\begin{aligned}
\int_C \omega &= \int_0^{2\pi} (-1 + t)\, dt \\
&= \left. -t + \frac{t^2}{2} \right|_0^{2\pi} \\
&= 2\pi(\pi - 1).
\end{aligned}
$$

This procedure will be justified in the next chapter, in which we shall also show how to evaluate a k-form in k-dimensional space. In general, we shall evaluate k-forms in spaces with more than k dimensions by parametrizing the object over which we wish to integrate and by finding a pullback to

k-dimensional space. To do this, we shall need to investigate differentiable mappings of higher dimension and to find the effect of a pullback on an arbitrary k-form. Chapters 6 through 9 will address this problem as we study the differential calculus of vector fields.

The next step, taken in Chapter 10, is to investigate the fundamental theorem of calculus in our new setting. For real-valued functions of a single variable, the fundamental theorem carries the implication that the value of an integral depends only on the endpoints of the interval and not on how we get from one endpoint to the other. As revealed in the exercises of Section 4.3, this is not always true in higher dimensions. Part of our program will be to investigate under what conditions it is still true. More generally, we shall see how the fundamental theorem of calculus extends to the evaluation of differential forms in higher dimensions.

Finally, in Chapter 11, we shall take this amazing machinery and apply it to the physical phenomena of electricity and magnetism, showing how it reveals unexpected properties of these phenomena and how it leads to special relativity and the equivalence of matter and energy.

All of this still begs the question of precisely what we mean by the integral of a k-form and more importantly whether the value of our integral depends upon our choice of parametrization. It does not, but this is a result that is beyond the scope of this book. The interested reader is directed to Edwards' *Advanced Calculus*.

4.8 Exercises

1. Verify that if we multiply dx, dy, and dz as given in Equations (4.19), we get Equation (4.18).

2. Find the mappings taking

$$[(0,0,0),(1,0,0),(0,1,0),(0,0,1)]$$

   in u,v,w space to each of the following oriented tetrahedra in x,y,z space:

   (a) $[(-2,0,3),(1,1,-2),(3,5,0),(4,-2,1)]$,

   (b) $[(1,0,0),(0,0,0),(0,1,0),(0,0,1)]$,

   (c) $[(5,-1,2),(3,1,0),(1,0,2),(7,0,0)]$,

   (d) $[(-3,0,0),(2,-1,3),(0,0,2),(5,1,-2)]$.

3. Find the pullback of $dx\,dy\,dz$ for each of the mappings in Exercise 2.

4. Evaluate the 3-form $dx\,dy\,dz$ on each of the oriented tetrahedra of Exercise 2.

5. If $\vec{a}$, $\vec{b}$, $\vec{c}$, and $\vec{d}$ lie on the same plane, then

$$\vec{d} = \vec{a} + r(\vec{b} - \vec{a}) + s(\vec{c} - \vec{a})$$

for some real numbers r and s. Find a mapping from

$$[(0,0,0), (1,0,0), (0,1,0), (0,0,1)]$$

to

$$[\vec{a},\, \vec{b},\, \vec{c},\, \vec{a} + r(\vec{b} - \vec{a}) + s(\vec{c} - \vec{a})]$$

and show that the pullback of $dx\, dy\, dz$ for this mapping is always 0.

6. Consider the system of equations:

$$\begin{aligned}
y_1 &= 3 - 2x_1 + 5x_2 - 3x_3, \\
y_2 &= x_1 - 3x_3 + 2x_4, \\
y_3 &= -1 + 5x_1 - x_2 + 4x_4, \\
y_4 &= 6 + 2x_2 - x_4.
\end{aligned}$$

Find the pullbacks of each of the following differential forms:

(a) $dy_1\, dy_3$,

(b) $dy_1\, dy_4$,

(c) $dy_2\, dy_3\, dy_4$,

(d) $dy_1\, dy_2\, dy_3\, dy_4$.

7. Using the definition of D given in Equation (4.21), prove that D can be written as a sum over permutations as given in Equation (4.23).

8. A convenient way of describing a given permutation is to list the images in order: $\sigma(1)\sigma(2)\cdots\sigma(k)$. Thus, 35142 represents the permutation

$$\begin{aligned}
\sigma(1) &= 3, \\
\sigma(2) &= 5, \\
\sigma(3) &= 1, \\
\sigma(4) &= 4, \\
\sigma(5) &= 2.
\end{aligned}$$

Using this notation, list all permutations for $k = 3$. What is the value of $\text{sgn}(\sigma)$ for each of these permutations?

9. List all permutations for $k = 4$. What is the value of $\text{sgn}(\sigma)$ for each of these permutations?

10. The *inversion number* of a permutation is the number of pairs, (i, j), for which $i < j$ but $\sigma(i) > \sigma(j)$. It is denoted by $\text{inv}(\sigma)$. As an example, the permutation 35142 has six pairs of inversions: $3 > 1$, $3 > 2$, $5 > 1$, $5 > 4$, $5 > 2$, $4 > 2$, and so $\text{inv}(35412) = 6$. Prove that

$$\text{sgn}(\sigma) = (-1)^{\text{inv}(\sigma)}. \qquad (4.25)$$

11. Verify that for $k = 3$

$$D(\vec{b}_1, \vec{b}_2, \vec{b}_3) = \vec{b}_1 \cdot \vec{b}_2 \times \vec{b}_3.$$

12. Let

$$\begin{aligned}
\vec{b}_1 &= (0, -3, 0, 1), \\
\vec{b}_2 &= (1, 0, 0, -1), \\
\vec{b}_3 &= (-4, 0, -2, 0), \\
\vec{b}_4 &= (0, 5, -1, 0).
\end{aligned}$$

    Evaluate $D(\vec{b}_1, \vec{b}_2, \vec{b}_3, \vec{b}_4)$.

13. Using induction and the fact that

$$\int_0^1 \frac{t^{k-1}}{(k-1)!} \, dt = \frac{1}{k!},$$

    justify the statement that the hypervolume of a fundamental k-simplex is $1/k!$.

14. In 4-dimensional space–time $(x, y, z, t$ space), a 4-dimensional vector can be represented by either a constant 1-form or a constant 3-form. Find the appropriate representations so that the product of a 1-form and a 3-form corresponds to the dot product:

$$(a, b, c, d) \cdot (e, f, g, h) = ae + bf + cg + dh.$$

15. What is the analog of the cross product in four dimensions?

16. What is a constant 2-form in 4-dimensional space–time? (Note: it is not a 4-dimensional vector.)

17. Find the work done by the force field

$$(x + y) \, dx - (2z + 1) \, dy + (y - z) \, dz$$

    as it moves a particle along the curve

$$C = \{(t^2, t + 1, t - 1) \mid 0 \le t \le 1\}.$$

18. Let c be a fixed constant. Find the rate at which fluid crosses the curve
$$y = c - cx^2,$$
directed from $(-1, 0)$ to $(1, 0)$, if the fluid velocity at (x, y) is described by the vector $(-2x, 3y^2)$. Find the value of c that maximizes the rate of flow.

19. Why do you think we define 2-forms and 3-forms in terms of triangles and tetrahedra instead of parallelograms and parallelepipeds?

5

Line Integrals, Multiple Integrals

5.1   The Riemann Integral

There is a curious contradiction in the standard presentation of integral calculus. It is an ancient subject rooted in the geometric investigations of Archimedes of Syracuse (287–212 B.C.). His "method of exhaustion" for computing arc lengths, areas, and volumes is recognizably equivalent to our modern understanding of the integral as a limit of the summation

$$\sum f(x_i)\,\Delta x_i \longrightarrow \int f(x)\,dx,$$

as the change in x becomes progressively smaller. Yet, in contraposition to this, the most modern mathematics presented in most first-year calculus courses is the definition of the integral given by Bernhard Riemann in his 1854 paper "Über die Darstellbarkeit einer Funktion durch eine trigonometrische Reihe" or "On the Representation of a Function as a Trigonometric Series."

What has happened is that the concept of the integral as an area under a curve or as a limit of the sums of areas of approximating rectangles is intuitively clear. In fact, the concept is so transparent that until the late 18th or early 19th centuries few mathematicians were led to think deeply about what was involved. But, at this time, science turned to the physical problems of finding models for mechanical vibrations and heat conduction. These problems led to the introduction of infinite sums of trigonometric functions, *Fourier series*, named for Joseph Fourier (1768–1830). His papers of 1807, 1811, and 1822 on heat conduction created great controversy over whether he was justified in integrating an infinite sum of functions by integrating each term in the summation.

In time, it was realized that this is a difficult question. In many cases, there are no complications, but contradictions, paradoxes, and outright errors can arise if such integration is not approached very carefully. The need for a more precise definition of the integral became apparent. It was in the context of integrating Fourier series that Riemann set what has since become the standard definition of the definite integral as presented in first-year calculus.

A second impetus for a careful look at integration came from the investigation of geometric invariants recounted in the previous chapter. Integrals had been conceived in terms of areas and volumes. If they were now to define area and volume, a new characterization was needed in order to avoid circular definitions. Since we shall not deal with Fourier series in this book, most of the techniques presented in this chapter will be justified by an appeal to the reader's intuitive understanding of integrals. For a more rigorous treatment, the reader is directed to texts on real analysis or Edwards' *Advanced Calculus*.

To the student who hopes to pursue mathematics well into the 20th century, let me simply mention that by the end of the 19th, Riemann's understanding of the integral had proven inadequate. Henri Lebesgue (1875–1941) was to redefine the integral so that it could be applied to a far broader class of functions.

Definition of the Riemann Integral

We restrict our attention to the definite integral of a bounded function over a finite interval:

$$\int_a^b f(x)\,dx.$$

We choose a *partition* of this interval,

$$\pi = \{a = x_0 < x_1 < \cdots < x_n = b\},$$

set

$$\Delta x_i = x_i - x_{i-1},$$

and define the *mesh* of π, denoted $\|\pi\|$, by

$$\|\pi\| = \max_{1 \le i \le n} \Delta x_i.$$

This is also sometimes called the *norm* of π. Note that n is not fixed but depends on π and that our subintervals do not have to be of equal length. However, as we make the mesh smaller, the number of subintervals, n, is forced to become larger.

Given a partition of $[a, b]$, we select points

$$x_1^* \in [x_0,\, x_1],\ x_2^* \in [x_1,\, x_2], \ldots, x_n^* \in [x_{n-1},\, x_n].$$

A *Riemann sum* is an approximation to the integral given by

$$\sum_{i=1}^n f(x_i^*)\,\Delta x_i.$$

We say that f is *integrable* or *Riemann integrable* over $[a, b]$ if, regardless of how we choose our points $x_1^*, x_2^*, \ldots, x_n^*$, as the mesh of the partition approaches 0, the value of the Riemann sums approaches a fixed limit. Equivalently, f is integrable over $[a, b]$ if we can assign a value to $\int_a^b f(x)\,dx$ so that for any given $\epsilon > 0$ (how close we want the Riemann sum to approximate the integral), we can find a δ (how small the mesh must be) such that $\|\pi\| < \delta$ implies

$$\left| \sum_{i=1}^n f(x_i^*)\,\Delta x_i - \int_a^b f(x)\,dx \right| < \epsilon,$$

regardless of how we choose $x_1^*, x_2^*, \ldots, x_n^*$.

If $f(x)$ is continuous on $[a, b]$ or has only finitely many points of discontinuity, then it will be integrable over $[a, b]$. The proof of this can be found in Edwards' *Advanced Calculus*.

5.2   Line Integrals

As we saw in Chapter 4, the integral of a 1-form over a curve is defined in terms of a parametrization of that curve and the pullback to 1-dimensional space. As we shall now see, that is a reasonable definition consistent with the physical explanation of the integral as the amount of work done by the force field described by our 1-form as it moves a particle along the curve.

Initially, we shall restrict our attention to curves in $\mathbf{R}^2$. Higher dimensions present no additional difficulties. Let

$$f(x, y)\,dx + g(x, y)\,dy$$

be the 1-form to be evaluated on the curve

$$C = \{(x(t), y(t)) \mid a \leq t \leq b\},$$

where $x(t)$ and $y(t)$ are functions of t with continuous derivatives in (a, b). Let

$$F(t) = f(x(t), y(t)), \quad G(t) = g(x(t), y(t)).$$

We have defined the integral of our 1-form in terms of the pullback:

$$\int_C f(x, y)\,dx + g(x, y)\,dy = \int_a^b \left(F(t)\frac{dx}{dt} + G(t)\frac{dy}{dt} \right) dt. \tag{5.1}$$

To justify this, let us divide our curve into small arcs by partitioning the interval $[a, b]$ into n subintervals:

$$[t_0, t_1], [t_1, t_2], \ldots, [t_{n-1}, t_n],$$

where $a = t_0 < t_1 < \cdots < t_n = b$. Let C_i be the piece of C from $(x(t_{i-1}), y(t_{i-1}))$ to $(x(t_i), y(t_i))$ (Figure 5.1) and let

FIGURE 5.1. Approximating the curve by short lines.

$$\Delta t_i = t_i - t_{i-1}.$$

Let us focus on the integral of the 1-form $f(x, y)\, dx$ as we move along C_i. If $f(x, y)$ is a continuous function on C_i, then $F(t)$ is continuous on $[t_{i-1}, t_i]$, and so for any $t_i^* \in [t_{i-1}, t_i]$, $F(t_i^*)$ will be a good approximation for $f(x, y)$ over C_i. More precisely, we can guarantee that $f(x, y)$ will be within any prespecified amount of $F(t_i^*)$ if we have taken Δt_i sufficiently small. The effect of this is to imply that if f and g are continuous and our pieces of arc are short enough, then on each particular C_i we can treat our force field as if it were constant:

$$\int_{C_i} f(x,\, y)\, dx + g(x,\, y)\, dy \simeq \int_{C_i} F(t_i^*)\, dx + G(t_i^*)\, dy,$$

with a manageable error that can be made arbitrarily small.

From Exercise 4 of Section 4.3, the integral of a constant 1-form over any path from $\vec{a}$ to $\vec{b}$ is the same as the integral over the line segment from $\vec{a}$ to $\vec{b}$. Since $F(t_i^*)$ and $G(t_i^*)$ are independent of x and y, we see that

$$
\begin{aligned}
\int_{C_i} F(t_i^*)\, dx + G(t_i^*)\, dy &= \int_{(x_{i-1}, y_{i-1})}^{(x_i, y_i)} F(t_i^*)\, dx + G(t_i^*)\, dy \\
&= F(t_i^*)(x_i - x_{i-1}) + G(t_i^*)(y_i - y_{i-1}).
\end{aligned}
$$

We combine this result with the fundamental theorem of calculus for real-valued functions of a single variable:

$$\int_{t_{i-1}}^{t_i} x'(t)\, dt = x(t_i) - x(t_{i-1}),$$

and we see that

$$\int_{C_i} F(t_i^*)\, dx + G(t_i^*)\, dy$$

$$= F(t_i^*)[x(t_i) - x(t_{i-1})] + G(t_i^*)[y(t_i) - y(t_{i-1})]$$

$$= F(t_i^*) \int_{t_{i-1}}^{t_i} x'(t)\, dt \;+\; G(t_i^*) \int_{t_{i-1}}^{t_i} y'(t)\, dt$$

$$= \int_{t_{i-1}}^{t_i} [F(t_i^*)x'(t) + G(t_i^*)y'(t)]\, dt.$$

Using the continuity of F, G, x', and y', this last integrand can be made arbitrarily close to $F(t)x'(t) + G(t)y'(t)$ over $[t_{i-1}, t_i]$ by controlling the size of Δt_i. We thus have

$$\int_C f(x, y)\, dx + g(x, y)\, dy \;=\; \sum_{i=1}^{n} \int_{C_i} f(x, y)\, dx + g(x, y)\, dy$$

$$\simeq \sum_{i=1}^{n} \int_{C_i} F(t_i^*)\, dx + G(t_i^*)\, dy$$

$$\simeq \sum_{i=1}^{n} \int_{t_{i-1}}^{t_i} (F(t_i^*)x'(t) + G(t_i^*)y'(t))\, dt$$

$$= \int_a^b (F(t)x'(t) + G(t)y'(t))\, dt.$$

This has not been a proof of Equation (5.1), but rather a sketch of a justification. Hopefully, it has been a convincing argument that when f and g are continuous it makes sense to evaluate the integral in terms of the pullbacks.

Integration with Respect to Arc Length

Recall from Chapter 3 that if we have a curve parametrized by

$$\vec{r}(t) = (x(t), y(t), z(t)),$$

then the velocity vector at time t is

$$\vec{v}(t) = \left(\frac{dx}{dt}, \frac{dy}{dt}, \frac{dz}{dt} \right),$$

the speed at time t is

$$v(t) = \sqrt{\left(\frac{dx}{dt}\right)^2 + \left(\frac{dy}{dt}\right)^2 + \left(\frac{dz}{dt}\right)^2},$$

and the length of the arc from $\vec{r}(t_0)$ to $\vec{r}(t_1)$ is given by

$$s(t_1) - s(t_0) = \int_{t_0}^{t_1} v(t)\, dt.$$

Since

$$\frac{ds}{dt} = v(t),$$

the differentials of s and t are related by

$$ds = v(t)\, dt. \tag{5.2}$$

Example

Let C be the curve $y = \sqrt{x}$ from $x = 2$ to $x = 6$. The integral

$$\int_C y\, ds$$

can be evaluated by using x as our parameter:

$$(x, y) = (x, \sqrt{x}).$$

With this choice of parameter, we have

$$ds = \sqrt{\left(\frac{dx}{dx}\right)^2 + \left(\frac{dy}{dx}\right)^2}\, dx = \sqrt{1 + \frac{1}{4x}}\, dx.$$

We therefore can evaluate our integral:

$$\begin{aligned}
\int_C y\, ds &= \int_2^6 \sqrt{x}\sqrt{1 + \frac{1}{4x}}\, dx \\
&= \frac{1}{2}\int_2^6 \sqrt{4x + 1}\, dx \\
&= \frac{1}{8} \cdot \frac{2}{3}(4x + 1)^{3/2}\Big|_2^6 \\
&= \frac{49}{6}.
\end{aligned}$$

Mass and Center of Mass

It is often appropriate to integrate with respect to arc length. For example, if we have a wire described by

$$\vec{r}(t) = \{(t, t^2, 1) \mid 1 \le t \le 3\},$$

whose density, mass per unit length, at (x, y, z) is

$$\rho(x, y, z) = \frac{yz}{x},$$

then the total mass is calculated by dividing the wire into small pieces where we treat the density as constant, multiplying each constant by the length of the small piece, adding these masses, and then taking the limiting case as the individual lengths approach 0:

$$
\begin{aligned}
\text{Mass} &= \int_{\vec{r}} \rho(x, y, z) \, ds \\
&= \int_1^3 \rho(t, t^2, 1) v(t) \, dt \\
&= \int_1^3 t(1 + 4t^2)^{1/2} \, dt \\
&= \frac{1}{8} \cdot \frac{2}{3} (1 + 4t^2)^{3/2} \Big|_1^3 \\
&= \frac{1}{12} \left(37^{3/2} - 5^{3/2} \right) \\
&= 17.823 \ldots .
\end{aligned}
$$

Given several point masses, m_1 at $\vec{x}_1$, m_2 at $\vec{x}_2, \ldots, m_n$ at $\vec{x}_n$, the *center of mass* is defined as the point $\vec{c}$ satisfying

$$\sum_{i=1}^n m_i(\vec{x}_i - \vec{c}) = \vec{0},$$

or equivalently,

$$\vec{c} = \frac{\sum_{i=1}^n m_i \vec{x}_i}{\sum_{i=1}^n m_i}. \tag{5.3}$$

If instead of point masses we have a solid wire C of known density $\rho(\vec{x})$ at each point $\vec{x}$, then the center of mass is given by

$$\vec{c} = \frac{\int_C \vec{x} \rho(\vec{x}) \, ds}{\int_C \rho(\vec{x}) \, ds} = \left(\frac{\int_C x \rho(\vec{x}) \, ds}{\int_C \rho(\vec{x}) \, ds}, \frac{\int_C y \rho(\vec{x}) \, ds}{\int_C \rho(\vec{x}) \, ds}, \frac{\int_C z \rho(\vec{x}) \, ds}{\int_C \rho(\vec{x}) \, ds} \right). \tag{5.4}$$

For our example, if $\vec{c} = (\bar{x}, \bar{y}, \bar{z})$, then

$$
\begin{aligned}
\bar{x} &= \frac{\int_1^3 t^2 (1 + 4t^2)^{1/2} \, dt}{(37^{3/2} - 5^{3/2})/12} \\
&= \frac{3 \left[x(1 + 2x^2)\sqrt{1 + x^2} - \sinh^{-1} x \right]_2^6}{16 (37^{3/2} - 5^{3/2})} \\
&= \frac{3 \left(438\sqrt{37} - 18\sqrt{5} - \sinh^{-1} 6 + \sinh^{-1} 2 \right)}{16 (37^{3/2} - 5^{3/2})} = 2.299 \ldots ,
\end{aligned}
$$

$$\bar{y} = \frac{\int_1^3 t^3(1+4t^2)^{1/2}\,dt}{(37^{3/2}-5^{3/2})/12}$$

$$= \frac{t^2(1+4t^2)^{3/2}\big|_1^3 - 2\int_1^3 t(1+4t^2)^{3/2}dt}{(37^{3/2}-5^{3/2})}$$

$$= \frac{333\sqrt{37} - 5\sqrt{5} - \left[\frac{1}{10}(1+4t^2)^{5/2}\right]_1^3}{(37^{3/2}-5^{3/2})}$$

$$= \frac{1961\sqrt{37} - 25\sqrt{5}}{10\,(37^{3/2}-5^{3/2})} = 5.550\ldots\,,$$

$$\bar{z} = \frac{\int_1^3 t(1+4t^2)^{1/2}\,dt}{(37^{3/2}-5^{3/2})/12} = 1.$$

An Alternate Notation

The line integral

$$\int_C f\,dx + g\,dy + h\,dz$$

can also be expressed as an integral with respect to arc length. Let $\vec{F}$ be the force field described by our 1-form:

$$\vec{F} = (f, g, h).$$

If $\vec{r}(t)$ parametrizes C, then the unit tangent to C is

$$\vec{T} = \frac{\vec{v}}{v}.$$

The component of $\vec{F}$ in the direction of the unit tangent is

$$\vec{F} \cdot \frac{\vec{v}}{v}.$$

The amount of work done by the force field in moving a particle along C can be expressed as

$$\int_C \vec{F} \cdot \frac{\vec{v}}{v}\,ds.$$

Using the fact that $ds = v\,dt$ and defining $d\vec{r} = \vec{v}\,dt$, we can write

$$\int_C f\,dx + g\,dy + h\,dz = \int_C \frac{\vec{F}\cdot\vec{v}}{v}\,ds$$

$$= \int_C \vec{F}\cdot\vec{v}\,dt$$

$$= \int_C \vec{F}\cdot d\vec{r}. \tag{5.5}$$

This last is really no more than an alternate notation for the integral of the 1-form. You should familiarize yourself with it as the line integral is often written in this manner.

5.3 Exercises

1. Integrate the 1-form $xy^2\,dx + y\,dy$ along each of the following paths from $(0,0)$ to $(1,1)$:

   (a) the straight line from $(0,0)$ to $(1,1)$,

   (b) the line from $(0,0)$ to $(1,0)$ followed by the line from $(1,0)$ to $(1,1)$,

   (c) the lines from $(0,0)$ to $(0,1)$ to $(1,1)$,

   (d) the curve $y = x^2$,

   (e) the curve $x = y^2$,

   (f) the lines from $(0,0)$ to $(2,0)$ to $(2,1)$ to $(1,1)$.

2. Draw each path of Exercise 1 and compare the values of $\int_C xy^2\,dx + y\,dy$. Find a path C from $(0,0)$ to $(1,1)$ for which

$$\int_C xy^2\,dx + y\,dy = 0.$$

3. Repeat Exercise 1 for the 1-form $xy^2\,dx + x^2y\,dy$.

4. Integrate the 1-form $yz\,dx + zx\,dy + xy\,dz$ over each of the following paths from $(0,1,0)$ to $(2,1,1)$:

   (a) the straight line from $(0,1,0)$ to $(2,1,1)$,

   (b) the lines from $(0,1,0)$ to $(0,1,1)$ to $(2,1,1)$,

   (c) the lines from $(0,1,0)$ to $(2,1,0)$ to $(2,1,1)$,

   (d) the arc $(2t,(2t-1)^2,t)$, $0 \le t \le 1$.

5. Repeat Exercise 4 for the 1-form $xy\,dx + yz\,dy + zx\,dz$.

6. Integrate the 1-form

$$\frac{-y\,dx + x\,dy}{x^2 + y^2}$$

   along each of the following paths from $(-1,0)$ to $(1,0)$:

   (a) the lines from $(-1,0)$ to $(-1,1)$ to $(1,1)$ to $(1,0)$,

   (b) the lines from $(-1,0)$ to $(-1,-1)$ to $(1,-1)$ to $(1,0)$,

   (c) the lines from $(-1,0)$ to $(0,1)$ to $(1,0)$,

(d) the lines from $(-1,0)$ to $(0,-1)$ to $(1,0)$,

(e) the curve $(-\cos t, \sin t)$, $0 \le t \le \pi$,

(f) the curve $(-\cos t, -\sin t)$, $0 \le t \le \pi$.

7. Draw each path in Exercise 6. Can you find a path from $(-1,0)$ to $(1,0)$ for which

$$\int_C \frac{-y\,dx + x\,dy}{x^2 + y^2} = 0?$$

Justify your answer.

8. Compare the results obtained in Exercises 1, 3, 4, 5, and 6. What can you say about whether or not the value of the line integral depends only on the endpoints of the curve over which you integrate? Ponder what is different about these 1-forms.

9. Evaluate $\int_C (x^2 - 2xy + y^2)\,ds$, where $C = \{(2\cos t, 2\sin t)|0 \le t \le \pi\}$.

10. Evaluate $\int_C (x^3 - yz)\,ds$, where C is the intersection of the planes $x + y - z = 1$ and $z = 3x$ from $x = 0$ to $x = 1$.

11. Evaluate $\int_C x^2\,ds$, where C is the curve $y = x^{5/2}$ from $x = 1$ to $x = 4$.

12. Find the mass of a wire that is parametrized by $C = \{(\frac{3}{2}t^2, (1 + 2t)^{3/2})|0 \le t \le 2\}$ and that has a density given by $\rho(x, y) = 2x + 1$.

13. Find the center of mass of a semicircular wire of uniform density and radius r.

14. Find the center of mass in terms of a of the spring coil of uniform density described by

$$\{(\cos t, \sin t, t)\,|\,0 \le t \le a\}.$$

5.4   Multiple Integrals

The integrals of a 2-form over a region in $\mathbf{R}^2$, of a 3-form over a solid in $\mathbf{R}^3$, or more generally of a k-form over a portion of $\mathbf{R}^k$ are *multiple integrals* and they form the backbone of integral calculus. The integral of a 2-form over a piece of surface in $\mathbf{R}^3$ will be defined in terms of the pullback to a double integral in $\mathbf{R}^2$. The integral of a k-form over a k-dimensional "manifold" in $\mathbf{R}^n$ will be defined in terms of the pullback to a k-fold integral in $\mathbf{R}^k$.

The origins of multiple integration are hazy. It is at least implicit in Newton's *Principia* in which the total force exerted by a large body is calculated by summing the contributions from each small volume of mass and then passing to the limit where these small volumes approach points. Multiple integrals were freely used and occasionally misused in the eighteenth century. Augustin-Louis Cauchy (1789–1857) was among the first to recognize their potential problems. Karl J. Thomae (1840–1921) and Paul Du Bois-Reymond (1831–1889) did much of the work of extending Riemann's notion of the integral to higher dimensions.

Double Integrals

We restrict our attention to bounded 2-forms over bounded rectangular regions:

$$\int_R f(x,y) \, dx \, dy, \quad R = \{(x,y) \mid a_1 \leq x \leq b_1, \ a_2 \leq y \leq b_2\}.$$

Just as it is helpful to think of an integral of a 1-form,

$$\int_a^b f(x) \, dx,$$

as the area of a region between $y = f(x)$ and the x axis, so we can think of the integral of a 2-form,

$$\int_R f(x,y) \, dx \, dy,$$

as the *volume* of a region above the x, y plane whose height is given by $f(x,y)$. Thus, if R is the rectangle

$$R = \{(x,y) \mid 0 \leq x \leq 1, \ 2 \leq y \leq 3\},$$

then

$$\int_R xy^2 \, dx \, dy$$

is the volume of the solid lying above R, bounded above by the surface $z = xy^2$ (see Figure 5.2).

If the region over which we are integrating, call it R^*, is not a rectangle, then we embed it inside a rectangle (Figure 5.3),

$$R^* \subseteq R,$$

and define $f(x,y)$ to be 0 if (x,y) is in R but not in R^*. If the orientation of R is not specified, then it is assumed to be positive. That is, positive flow is in the direction of the positive z axis.

We approximate this volume with a Riemann sum. We partition R by partitioning the segments on the x and y axes:

$$a_1 = x_0 < x_1 < \cdots < x_m = b_1, \quad a_2 = y_0 < y_1 < \cdots < y_n = b_2.$$

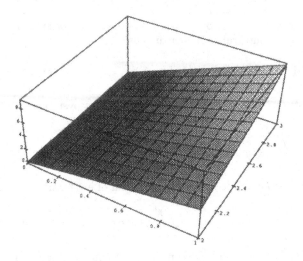

FIGURE 5.2. The solid bounded above by $z = xy^2$.

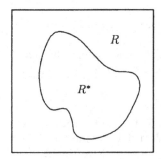

FIGURE 5.3. Embedding in a rectangle R.

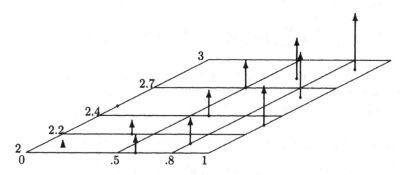

FIGURE 5.4. Chosen values of $f(x, y) = xy^2$.

The *mesh* of this partition is the maximum over all i and j of $\Delta x_i = x_i - x_{i-1}$ and $\Delta y_j = y_j - y_{j-1}$. Our partition gives us an $n \times m$ grid. Inside each subrectangle, we choose x_{ij}^* and y_{ij}^*:

$$x_{i-1} \leq x_{ij}^* \leq x_i, \quad y_{j-1} \leq y_{ij}^* \leq y_j.$$

A Riemann sum for this partition is given by the double sum

$$\sum_{i=1}^{m} \sum_{j=1}^{n} f(x_{ij}^*, y_{ij}^*) \, \Delta x_i \, \Delta y_j,$$

which is the area of the base, $\Delta x_i \, \Delta y_j$, times the height at (x_{ij}^*, y_{ij}^*).

As an example, let

$$f(x, y) = xy^2, \quad R = \{(x, y) \mid 0 \leq x \leq 1, \ 2 \leq y \leq 3\}.$$

Let us take for our partition:

$$x_0 = 0, \quad x_1 = .5, \quad x_2 = .8, \quad x_3 = 1,$$
$$y_0 = 2, \quad y_1 = 2.2, \quad y_2 = 2.4, \quad y_3 = 2.7, \quad y_4 = 3.$$

The mesh is 0.5. The values of (x_{ij}^*, y_{ij}^*) are plotted in Figure 5.4, with vertical arrows representing the values of $f(x_{ij}^*, y_{ij}^*)$. The corresponding Riemann sum is

$$
\begin{aligned}
&(.1)(2.1)^2(.5)(.2) + (.4)(2.2)^2(.5)(.2) + (0)(2.5)^2(.5)(.3) \\
&+ (.5)(2.7)^2(.5)(.3) + (.6)(2)^2(.3)(.2) + (.6)(2.4)^2(.3)(.2) \\
&+ (.5)(2.7)^2(.3)(.3) + (.7)(2.8)^2(.3)(.3) + (.8)(2.1)^2(.2)(.2) \\
&+ (1)(2.3)^2(.2)(.2) + (.9)(2.6)^2(.2)(.3) + (.9)(2.9)^2(.2)(.3) \\
&= 3.12968.
\end{aligned}
$$

The double integral exists if and only if there is a real number, denoted by $\int_R f(x, y) \, dx \, dy$, such that

$$\lim_{||\pi|| \to 0} \sum_{i=1}^{m} \sum_{j=1}^{n} f(x_{ij}^*, y_{ij}^*) \, \Delta x_i \, \Delta y_j = \int_R f(x, y) \, dx \, dy.$$

Most of the time it does not matter how we proceed in reducing the size of the mesh. Let π_x be the partition of the x's, π_y the partition of the y's. Let x_{ij}^* depend only on i:

$$x_{ij}^* = x_i^*,$$

and let y_{ij}^* depend only on j:

$$y_{ij}^* = y_j^*.$$

If we let $||\pi_y||$ approach 0 first, we see that

$$
\begin{aligned}
\int_R f(x, y) \, dx \, dy
&= \lim_{||\pi_x|| \to 0} \left(\lim_{||\pi_y|| \to 0} \sum_{i=1}^{m} \sum_{j=1}^{n} f(x_i^*, y_j^*) \, \Delta x_i \, \Delta y_j \right) \\
&= \lim_{||\pi_x|| \to 0} \sum_{i=1}^{m} \left(\lim_{||\pi_y|| \to 0} \sum_{j=1}^{n} f(x_i^*, y_j^*) \, \Delta y_j \right) \Delta x_i \\
&= \lim_{||\pi_x|| \to 0} \sum_{i=1}^{m} \left(\int_{a_2}^{b_2} f(x_i^*, y) \, dy \right) \Delta x_i \\
&= \int_{a_1}^{b_1} \left(\int_{a_2}^{b_2} f(x, y) \, dy \right) dx. \qquad (5.6)
\end{aligned}
$$

Reversing the order of our limits yields

$$
\begin{aligned}
\int_R f(x, y) \, dx \, dy
&= \lim_{||\pi_y|| \to 0} \sum_{j=1}^{n} \left(\lim_{||\pi_x|| \to 0} \sum_{i=1}^{m} f(x_i^*, y_j^*) \, \Delta x_i \right) \Delta y_j \\
&= \int_{a_2}^{b_2} \left(\int_{a_1}^{b_1} f(x, y) \, dx \right) dy. \qquad (5.7)
\end{aligned}
$$

For our example,

$$f(x, y) = xy^2, \quad R = \{(x, y) \mid 0 \leq x \leq 1, \ 2 \leq y \leq 3\},$$

we have

$$\int_R xy^2 \, dx \, dy = \int_0^1 \left(\int_2^3 xy^2 \, dy \right) dx$$

$$= \int_0^1 \left(\frac{1}{3}xy^3 \Big|_{y=2}^{y=3} \right) dx$$

$$= \int_0^1 \frac{19}{3} x \, dx$$

$$= \frac{19}{6}.$$

In the other order, we also have that

$$\int_R xy^2 \, dx \, dy = \int_2^3 \left(\int_0^1 xy^2 \, dx \right) dy$$

$$= \int_2^3 \left(\frac{1}{2}x^2y^2 \Big|_{x=0}^{x=1} \right) dy$$

$$= \int_2^3 \frac{1}{2}y^2 \, dy$$

$$= \frac{19}{6}.$$

The name "multiple integral" reflects the fact that higher dimensional integrals can usually be evaluated by doing an iterated integration as described previously. But be careful: there are two potential traps lying in wait for the unwary.

First Warning

The first trap is notational. It is common practice to drop the parentheses in the iterated integral:

$$\int_R xy^2 \, dx \, dy = \int_2^3 \int_0^1 xy^2 \, dx \, dy \qquad (5.8)$$

$$= \int_0^1 \int_2^3 xy^2 \, dy \, dx. \qquad (5.9)$$

It is tempting when looking at Equations (5.8) and (5.9) to say that the integral of $xy^2 \, dx \, dy$ is the same as the integral of $xy^2 \, dy \, dx$. We know that this is not true because

$$xy^2 \, dx \, dy = -xy^2 \, dy \, dx$$

by the anticommutativity of differentials. The problem is that on the right-hand sides of Equations (5.8) and (5.9) we are *not* multiplying our differentials. Rather, we are iterating our integrals using two 1-forms.

This is a strong argument for using the special notation, the wedge $\wedge$ mentioned in Section 4.4, to signify the product of differentials:

$$\int_R xy^2 \, dx \wedge dy = -\int_R xy^2 \, dy \wedge dx.$$

But since the confusion can only occur inside an integral, we shall adopt the convention that a single integral signifies that we are integrating a differential form; more than one integral means that we iterate the integration:

$$\int_R xy^2 \, dx \, dy \;=\; -\int_R xy^2 \, dy \, dx, \tag{5.10}$$

$$\int_2^3 \int_0^1 xy^2 \, dx \, dy \;=\; \int_0^1 \int_2^3 xy^2 \, dy \, dx. \tag{5.11}$$

If this is still confusing, put the parentheses back into the iterated integral:

$$\int_2^3 \left(\int_0^1 xy^2 \, dx \right) dy = \int_0^1 \left(\int_2^3 xy^2 \, dy \right) dx.$$

Second Warning

Not all double integrals can be evaluated by iterating single integrals, and order of integration will sometimes matter. Fortunately, for a continuous function in a bounded region whose boundary can be expressed as a finite set of differentiable curves, the integral always exists and can be found by iterating in either order. The kinds of situations that lead to problems can be found in Exercises 19 and 20 of Section 5.6. These are the only examples in this book for which the order of integration is a problem.

Higher Dimensions

What is true in two dimensions carries over with very little modification to higher dimensions. Let $f(\vec{x})$ be a bounded scalar field defined on a bounded rectangular region $R \subset \mathbf{R}^k$:

$$R = \{(x_1, x_2, \ldots, x_k) \mid a_i \leq x_i \leq b_i\}.$$

We partition each interval $[a_i, b_i]$:

$$\pi_{x_i} = (a_i = x_{i0} < x_{i1} < \cdots < x_{in_i} = b_i),$$

and for each k-tuple, $\vec{j} = (j_1, j_2, \ldots, j_k), 1 \leq j_i \leq n_i$, choose $x_{i\vec{j}}^*$ such that

$$x_{i(j_i-1)} \leq x_{i\vec{j}}^* \leq x_{ij_i}.$$

We have divided our region R into small hypercubes and chosen one point, $(x_{1\bar{j}}^*, x_{2\bar{j}}^*, \ldots, x_{k\bar{j}}^*)$, inside each. From this partition and these points, we obtain the Riemann sum:

$$\sum_{j_1=1}^{n_1} \sum_{j_2=1}^{n_2} \cdots \sum_{j_k=1}^{n_k} f(x_{1\bar{j}}^*, x_{2\bar{j}}^*, \ldots, x_{k\bar{j}}^*) \, \Delta x_{1j_1} \, \Delta x_{2j_2} \cdots \Delta x_{kj_k},$$

where

$$\Delta x_{ij_i} = x_{ij_i} - x_{i(j_i-1)}.$$

Again, if $f(\vec{x})$ is continuous over a bounded region whose boundary is made up of finitely many differentiable surfaces, then the limit of the Riemann sums as $||\pi||$ approaches zero exists, and we can find this limit by successively evaluating the one-dimensional integrals in any order:

$$\int_R f(\vec{x}) \, dx_1 \, dx_2 \cdots dx_k = \int_{a_1}^{b_1} \int_{a_2}^{b_2} \cdots \int_{a_k}^{b_k} f(\vec{x}) \, dx_k \, dx_{k-1} \cdots dx_1. \quad (5.12)$$

Example

Let

$$
\begin{aligned}
f(x,y,z) &= 3x^2y + 2y \sin z, \\
R &= \{(x,y,z) \mid -1 \le x \le 1,\ 0 \le y \le 2,\ 0 \le z \le \pi/2\},
\end{aligned}
$$

then

$$
\begin{aligned}
\int_R f(x,y,z) \, dx\, dy\, dz &= \int_0^{\pi/2} \int_0^2 \int_{-1}^1 (3x^2y + 2y \sin z) \, dx\, dy\, dz \\
&= \int_{-1}^1 \int_0^2 \int_0^{\pi/2} (3x^2y + 2y \sin z) \, dz\, dy\, dx \\
&= \int_{-1}^1 \int_0^2 \left. (3x^2yz - 2y \cos z) \right|_{z=0}^{z=\pi/2} dy\, dx \\
&= \int_{-1}^1 \int_0^2 \left(\frac{3\pi}{2} x^2y + 2y \right) dy\, dx \\
&= \int_{-1}^1 \left. \left(\frac{3\pi}{4} x^2y^2 + y^2 \right) \right|_{y=0}^{y=2} dx \\
&= \int_{-1}^1 (3\pi x^2 + 4) \, dx \\
&= 2\pi + 8.
\end{aligned}
$$

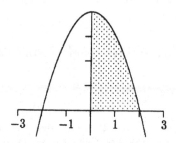

FIGURE 5.5. $R = \{(x,y) \mid x \geq 0,\ 0 \leq y \leq 4 - x^2\}$.

Integration over Nonrectangular Regions

If we are to integrate $f(x,y) = xy^2$ over the region

$$R = \{(x,y) \mid x \geq 0,\ 0 \leq y \leq 4 - x^2\}$$

(Figure 5.5), we can define a new function that agrees with $f(x,y)$ inside R and is 0 everywhere else:

$$g(x,y) = \begin{cases} xy^2, & 0 \leq x \leq 2, \quad 0 \leq y \leq 4 - x^2, \\ 0, & \text{otherwise.} \end{cases}$$

Integrating g over the rectangle $0 \leq x \leq 2$, $0 \leq y \leq 4$ is the same as integrating f over R:

$$\int_R f(x,y)\,dx\,dy = \int_0^2 \int_0^4 g(x,y)\,dy\,dx.$$

As it stands, this appears of little practical use, but if we isolate the integral with respect to y, we see that x is being held constant and that

$$
\begin{aligned}
\int_0^4 g(x,y)\,dy &= \int_0^{4-x^2} xy^2\,dy + \int_{4-x^2}^4 0\,dy \\
&= \int_0^{4-x^2} xy^2\,dy \\
&= \frac{1}{3}x(4 - x^2)^3 \\
&= -\frac{x^7}{3} + 4x^5 - 16x^3 + \frac{64x}{3},
\end{aligned}
$$

and so

$$\int_0^2 \left(\int_0^4 g(x,y)\,dy \right) dx = \int_0^2 \left(-\frac{x^7}{3} + 4x^5 - 16x^3 + \frac{64x}{3} \right) dx$$

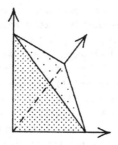

FIGURE 5.6. $T = \{(x, y, z) \mid x, y, z \geq 0,\ x + y + z \leq 1\}.$

$$= \frac{-256}{24} + \frac{256}{6} - \frac{256}{4} + \frac{256}{6}$$

$$= \frac{32}{3}.$$

We could equally well have integrated over x first:

$$\int_0^4 \int_0^2 g(x, y)\, dx\, dy = \int_0^4 \left(\int_0^{\sqrt{4-y}} xy^2\, dx \right) dy$$

$$= \int_0^4 \frac{4y^2 - y^3}{2}\, dy$$

$$= \frac{128}{3} - 32$$

$$= \frac{32}{3}.$$

A Nonrectangular Region in $\mathbb{R}^3$

Let us integrate $f(x, y, z) = xyz$ over the tetrahedron (see Figure 5.6)

$$T = \{(x,\, y,\, z) \mid x,\, y,\, z \geq 0,\ x + y + z \leq 1\}.$$

We change the problem to integrating $g(x, y, z)$ over the cube $0 \leq x \leq 1$, $0 \leq y \leq 1$, $0 \leq z \leq 1$, where

$$g(x, y, z) = \begin{cases} xyz & \text{if } x + y + z \leq 1, \\ 0 & \text{if } x + y + z > 1, \end{cases}$$

and so

$$\int_T xyz\, dx\, dy\, dz = \int_0^1 \int_0^1 \int_0^1 g(x, y, z)\, dx\, dy\, dz.$$

Recall that an iterated integral corresponds to a Riemann sum. The innermost integral is the innermost sum; this is the one that we compute first. If the integral on x is innermost, then we are fixing values of z and y and integrating with respect to x. After this integration is accomplished, we move out one layer to the integral over y, where we treat z as constant. Finally, we integrate over z. To rewrite our integral in terms of the original function, we work our way inward, finding the maximum possible range of the outermost variables first and then expressing succeeding variables in terms of those whose range has already been delimited.

The outermost variable, z, can take on any value from 0 through 1. Once it has been fixed, y can only range from 0 to $1 - z$. If y were larger than $1 - z$, then $x \geq 0$ and $x \leq 1 - y - z$ could not both be satisfied. Once we have fixed both z and y, x can only range from 0 to $1 - y - z$. This is precisely the region where $g(x, y, z) = xyz$:

$$
\int_0^1 \int_0^1 \int_0^1 g(x, y, z)\, dx\, dy\, dz
$$

$$
= \int_0^1 \int_0^{1-z} \int_0^{1-y-z} xyz\, dx\, dy\, dz
$$

$$
= \int_0^1 \int_0^{1-z} \left(\frac{1}{2}(1 - y - z)^2 yz \right) dy\, dz
$$

$$
= \int_0^1 \int_0^{1-z} \left(\frac{1}{2}yz + \frac{1}{2}y^3 z + \frac{1}{2}yz^3 - y^2 z - yz^2 + y^2 z^2 \right) dy\, dz
$$

$$
= \int_0^1 \left(\frac{1}{24}z - \frac{1}{6}z^2 + \frac{1}{4}z^3 - \frac{1}{6}z^4 + \frac{1}{24}z^5 \right) dz
$$

$$
= \frac{1}{720}.
$$

A Glance Ahead

One very efficient way of handling nonrectangular regions is to change variables. Thus, if R is the unit disc,

$$
R = \{(x, y) \mid x^2 + y^2 \leq 1\},
$$

it is usually much easier to switch to polar coordinates:

$$
x = \rho \cos\theta, \quad y = \rho \sin\theta.
$$

The pullback of this transformation takes our disc in the x, y plane back to a rectangle, $\{(\rho, \theta) \mid 0 \leq \rho \leq 1, \ 0 \leq \theta \leq 2\pi\}$ in the ρ, θ plane. This change of variables will be dealt with in Chapter 8 after results from differential calculus have shown us how to find the pullback of arbitrary k-forms.

5.5 Using Multiple Integrals

We can use our double integral to compute the value of a nonconstant 2-form over an oriented triangle in $\mathbf{R}^3$. If a fluid has velocity $(y+z, x+z, x-y)$ at (x, y, z), then the rate at which it crosses the triangle

$$T = [(1, 2, -1), (3, 1, 0), (0, 1, 1)]$$

is

$$\int_T (y + z)\, dy\, dz + (x + z)\, dz\, dx + (x - y)\, dx\, dy.$$

We find a pullback to 2-dimensional space, for example by using the parametric equations:

$$\begin{aligned}
x &= 1 + 2u - v, \\
y &= 2 - u - v, \\
z &= -1 + u + 2v,
\end{aligned}$$

which pull T back to $U = [(0, 0), (1, 0), (0, 1)]$:

$$\begin{aligned}
\int_T (y &+ z) dy\, dz + (x + z) dz\, dx + (x - y) dx\, dy \\
&= \int_U (1 + v)(-du - dv)(du + 2\,dv) \\
&\quad + (3u + v)(du + 2\,dv)(2\,du - dv) \\
&\quad + (-1 + 3u)(2\,du - dv)(-du - dv) \\
&= \int_U (1 + v)(-du\, dv) + (3u + v)(-5\,du\, dv) + (-1 + 3u)(-3\,du\, dv) \\
&= \int_U (2 - 24u - 6v)\, du\, dv = \int_0^1 \int_0^{1-v} (2 - 24u - 6v)\, du\, dv \\
&= \int_0^1 \left[(2 - 6v)(1 - v) - 12(1 - v)^2 \right] dv \\
&= \int_0^1 (-10 + 16v - 6v^2)\, dv = -10 + 8 - 2 = -4.
\end{aligned}$$

Area

The 2-form $dx\, dy$ maps a rectangle in $\mathbf{R}^2$ to its area. Now that we have defined the double integral for arbitrary bounded regions, we can define the area of a bounded region R by

$$\text{Area}(R) = \int_R dx\, dy. \tag{5.13}$$

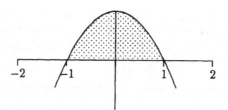

FIGURE 5.7. $R = \{(x, y) \mid 0 \leq y \leq 1 - x^2\}$.

The area of R is well-defined if and only if the integral exists.

Given a thin plate of shape R and density expressed by the function $\rho(x, y)$, we find its total mass by integrating density over area:

$$\text{Mass} = \int_R \rho(x, y) \, dx \, dy. \tag{5.14}$$

As in the case of the 1-dimensional wire, the center of mass, $(\bar{x}, \bar{y})$, satisfies

$$\int_R (x - \bar{x})\rho(x, y) \, dx \, dy = 0,$$

$$\int_R (y - \bar{y})\rho(x, y) \, dx \, dy = 0,$$

so that

$$\bar{x} = \frac{\displaystyle\int_R x\rho(x, y) \, dx \, dy}{\displaystyle\int_R \rho(x, y) \, dx \, dy}, \tag{5.15}$$

$$\bar{y} = \frac{\displaystyle\int_R y\rho(x, y) \, dx \, dy}{\displaystyle\int_R \rho(x, y) \, dx \, dy}. \tag{5.16}$$

As an example, let (Figure 5.7)

$$R = \{(x, y) \mid 0 \leq y \leq 1 - x^2\},$$
$$\rho(x, y) = \frac{y}{1 + x^2},$$

so that

$$\text{Area} = \int_{-1}^{1} \int_{0}^{1-x^2} dy\, dx$$

$$= \int_{-1}^{1} (1 - x^2)\, dx = \frac{4}{3},$$

$$\text{Mass} = \int_{-1}^{1} \int_{0}^{1-x^2} \frac{y}{1+x^2}\, dy\, dx$$

$$= \int_{-1}^{1} \frac{(1-x^2)^2}{2(1+x^2)}\, dx$$

$$= \int_{-1}^{1} \left(\frac{x^2 - 3}{2} + \frac{2}{1+x^2} \right) dx = \pi - \frac{8}{3},$$

$$\left(\pi - \frac{8}{3} \right) \bar{x} = \int_{-1}^{1} \int_{0}^{1-x^2} \frac{xy}{1+x^2}\, dy\, dx$$

$$= \int_{-1}^{1} \left(\frac{x^3 - 3x}{2} + \frac{2x}{1+x^2} \right) dx = 0,$$

$$\left(\pi - \frac{8}{3} \right) \bar{y} = \int_{-1}^{1} \int_{0}^{1-x^2} \frac{y^2}{1+x^2}\, dy\, dx$$

$$= \int_{-1}^{1} \frac{(1-x^2)^3}{3(1+x^2)}\, dx$$

$$= \int_{-1}^{1} \left(\frac{-x^4 + 4x^2 - 7}{3} + \frac{8}{3(1+x^2)} \right) dx$$

$$= \frac{4\pi}{3} - \frac{176}{45},$$

$$(\bar{x}, \bar{y}) = \left(0, \frac{60\pi - 176}{45\pi - 120} \right) \simeq (0, 0.58).$$

Volume

We define the volume of a bounded region $R \subset \mathbf{R}^3$ to be the image of R under the 3-form $dx\, dy\, dz$:

$$\text{Volume}(R) = \int_{R} dx\, dy\, dz, \tag{5.17}$$

where the volume is well defined if and only if this integral exists. If the density of R at (x, y, z) is given by $\rho(x, y, z)$, then

$$\text{Mass} = \int_{R} \rho(x, y, z)\, dx\, dy\, dz. \tag{5.18}$$

The coordinates of the center of mass are found by

$$\bar{x} = \frac{\displaystyle\int_R x\rho(x,y,z)\,dx\,dy\,dz}{\displaystyle\int_R \rho(x,y,z)\,dx\,dy\,dz}, \tag{5.19}$$

$$\bar{y} = \frac{\displaystyle\int_R y\rho(x,y,z)\,dx\,dy\,dz}{\displaystyle\int_R \rho(x,y,z)\,dx\,dy\,dz}, \tag{5.20}$$

$$\bar{z} = \frac{\displaystyle\int_R z\rho(x,y,z)\,dx\,dy\,dz}{\displaystyle\int_R \rho(x,y,z)\,dx\,dy\,dz}. \tag{5.21}$$

The hypervolume of a bounded region $R \subset \mathbf{R}^n$ in a higher dimensional space is defined by

$$\text{Hypervolume}(R) = \int_R dx_1\,dx_2\cdots dx_n. \tag{5.22}$$

5.6   Exercises

1. Sketch the region R over which you are integrating and then evaluate the following integrals:

   (a) $\displaystyle\int_R (x^2 + y^2)\,dx\,dy, \quad R = \{(x,y) \mid 1 \le x \le 2,\ -1 \le y \le 1\};$

   (b) $\displaystyle\int_R x\sin y\,dx\,dy, \quad R = \{(x,y) \mid 0 \le x \le 1,\ x^2 \le y \le 2x^2\};$

   (c) $\displaystyle\int_R (xy + 2)\,dx\,dy, \quad R = \{(x,y) \mid 1 \le y \le 2,\ y^2 \le x \le y^3\};$

   (d) $\displaystyle\int_R x\cos y\,dx\,dy, \quad R = \{(x,y) \mid 0 \le y \le \pi/2,\ 0 \le x \le \sin y\};$

   (e) $\displaystyle\int_R xy\,dx\,dy, \quad R = $ triangle with vertices at $(1,0)$, $(2,2)$, and $(1,2)$;

   (f) $\displaystyle\int_R e^{xy}\,dx\,dy, \quad R = \{(x,y) \mid 0 \le x \le 1 + (\log y)/y\},\ 2 \le y \le 3\};$

   (g) $\displaystyle\int_R (x^3 + 2xy)\,dx\,dy, \quad R = $ the parallelogram with vertices at $(1,3)$, $(3,4)$, $(4,6)$, and $(2,5)$.

2. What is the region R over which you integrate when evaluating the double integral

$$\int_0^1 \int_{x^2}^1 x\sqrt{1-y^2}\, dy\, dx?$$

Rewrite this as an iterated integral first with respect to x, then with respect to y. Evaluate this integral. Which order of integration is easier?

3. What is the region R over which you integrate when evaluating the double integral

$$\int_1^2 \int_1^x \frac{x}{\sqrt{x^2+y^2}}\, dy\, dx?$$

Rewrite this as an iterated integral first with respect to x, then with respect to y. Evaluate this integral. Which order of integration is easier?

4. What is the region R over which you integrate when evaluating the double integral

$$\int_1^4 \int_1^{\sqrt{x}} \frac{e^{xy^{-2}}}{y^5}\, dy\, dx?$$

Rewrite this as an iterated integral first with respect to x, then with respect to y. Evaluate this integral. Which order of integration is easier?

5. Find the area of the region bounded by the curves $y = 2x^2$ and $x = 4y^2$.

6. Find the area, mass, and center of mass of a thin plate bounded by $x^2 - y^2 = 1$ and $x = 4$ with density

$$\rho(x,y) = x.$$

7. Find the mass of the elliptic region $x^2 + 4y^2 \le 4$ if its density is given by $\rho(x,y) = x^2 + y^2$.

8. Find the center of mass of a thin plate in the shape of a triangle with vertices at $(1,0,0)$, $(1,2,0)$, $(1,0,3)$ if the density is given by $\rho(x,y,z) = x + y + z$.

9. Find a transformation from u,v space to x,y,z space that takes the triangle $U = [(0,0),(1,0),(0,1)]$ to the triangle

$$T = [(1,0,-2),(-1,2,0),(1,1,2)].$$

10. Using the tranformation of Exercise 9, find the pullback of

$$(y-1)\,dy\,dz + (y+z)\,dz\,dx - x\,dx\,dy$$

to u, v space.

11. Using the pullback of Exercise 10, evaluate

$$\int_T (y-1)\,dy\,dz + (y+z)\,dz\,dx - x\,dx\,dy.$$

12. Using the pullback of Exercise 9, find where the flow across the triangle U is positive and where it is negative. Use this to determine where the flow across T is positive and where it is negative.

13. Let R be the region above the unit disc, $x^2 + y^2 \le 1$, lying between the x, y plane and the surface $z = |xy|$. Find the volume of R.

14. Sketch the region R over which you are integrating and then evaluate the following integrals:

(a) $\displaystyle\int_R (x + xz - y^2)\,dx\,dy\,dz$,

   $R = \{(x, y, z) \mid 0 \le x \le 1,\ -2 \le y \le 0,\ 3 \le z \le 5\}$;

(b) $\displaystyle\int_R (x + z)\,dx\,dy\,dz$,

   $R = \{(x, y, z) \mid x \ge 0,\ y \ge 0,\ z \ge 0,\ x + y + 2z \le 3\}$;

(c) $\displaystyle\int_R xyz\,dx\,dy\,dz$,

   $R = \{(x, y, z) \mid x \ge 0,\ y \ge 0,\ z \ge 0,\ x^2 + y^2 + z^2 \le 1\}$.

15. Describe the region over which each integral is evaluated and find the limits if the order of integration is changed to $dy\,dz\,dx$ or to $dy\,dx\,dz$:

(a)

$$\int_0^2 \int_0^1 \int_0^y f(x,\,y,\,z)\,dx\,dy\,dz.$$

(b)

$$\int_1^2 \int_0^{\sqrt{4-z^2}} \int_0^{\sqrt{4-y^2-z^2}} f(x,\,y,\,z)\,dx\,dy\,dz.$$

(c)

$$\int_{-2}^{2}\int_{0}^{4-z^2}\int_{-(4-y-z^2)}^{4-y-z^2} f(x,\,y,\,z)\,dx\,dy\,dz.$$

16. Find the mass of the solid satisfying

$$x \geq 0, \quad y \geq 0, \quad z \geq 0, \quad z^2 \leq 4x, \quad x^2 + y^2 \leq 16,$$

whose density is given by $\rho(x,y,z) = xyz^3$.

17. Find the mass and center of mass of the solid of constant density 1 satisfying

$$x \geq 0, \quad y \geq 0, \quad x + y \leq 2,$$
$$0 \leq z \leq 1 + x + 2y.$$

18. Using multiple integration, show that the simplex satisfying

$$x_1 \geq 0, \, x_2 \geq 0, \, \ldots, \, x_n \geq 0, \, x_1 + x_2 + \cdots + x_n \leq 1,$$

has the hypervolume

$$\frac{1}{n!} = \frac{1}{1 \cdot 2 \cdot \cdots \cdot n}.$$

19. Let

$$f(x,\,y) = \begin{cases} 1, & \text{if } x \text{ is irrational,} \\ 4y^3, & \text{if not.} \end{cases}$$

Show that

$$\int_{0}^{1}\left(\int_{0}^{1} f(x,y)\,dy\right) dx = 1,$$

but that

$$\int_{0}^{1}\left(\int_{0}^{1} f(x,y)\,dx\right) dy$$

does not exist (and therefore the order of integration *can* be very important).

20. Define the function $f(x,y)$ to be 1 if x and y are both rational numbers with the same denominator (when written in reduced form), and 0 for all other pairs (x,y). For example,

$$f(1/2,5/2) = 1, \quad f(1/3,1/2) = 0, \quad f(1/6,2/6) = 0, \quad f(1,\pi) = 0.$$

In the third example, the denominators are different because 2/6 reduces to 1/3. Let R be the unit square $\{(x,y) \mid 0 \leq x \leq 1, \, 0 \leq y \leq 1\}$. Show that it is possible for the iterated integrals to exist and be the same in either order while the corresponding multiple integral does not exist by showing that

$$\int_0^1 \left(\int_0^1 f(x,\, y)\, dx \right) dy = \int_0^1 \left(\int_0^1 f(x,\, y)\, dy \right) dx = 0,$$

but that

$$\int_R f(x,\, y)\, dx\, dy$$

does not exist.

6

Linear Transformations

6.1 Basic Notions

A *linear transformation* is a function, $\vec{L}$, from one or more real variables to one or more real variables, that satisfies the following two conditions for any values of $\vec{a}$ and $\vec{b}$ in the domain and any real constant c:

$$\vec{L}(\vec{a}+\vec{b}) = \vec{L}(\vec{a})+\vec{L}(\vec{b}), \qquad (6.1)$$
$$\vec{L}(c\vec{a}) = c\vec{L}(\vec{a}). \qquad (6.2)$$

These conditions are extremely restrictive. Equation (6.1) implies that

$$\begin{aligned} \vec{L}(\vec{x}) &= \vec{L}(\vec{x}+\vec{0}) \\ &= \vec{L}(\vec{x})+\vec{L}(\vec{0}), \end{aligned}$$

and so

$$\vec{L}(\vec{0}) = \vec{0}. \qquad (6.3)$$

If $\vec{L}$ is defined on just one variable, then that variable x is real, and so

$$\vec{L}(x) = x\vec{L}(1). \qquad (6.4)$$

In this case, the function is uniquely determined by the value of $\vec{L}(1)$. If $\vec{L}$ is defined in 3-dimensional space, then any point in the domain is uniquely representable as $a\vec{i}+b\vec{j}+c\vec{k}$, where a, b, and c are real, and so

$$\vec{L}(a\vec{i}+b\vec{j}+c\vec{k}) = a\vec{L}(\vec{i})+b\vec{L}(\vec{j})+c\vec{L}(\vec{k}). \qquad (6.5)$$

Here the function is uniquely determined by the values of $\vec{L}(\vec{i})$, $\vec{L}(\vec{j})$, and $\vec{L}(\vec{k})$.

In general, if the domain is an n-dimensional space, then any point in the domain is uniquely representable in terms of the unit basis vectors:

$$\begin{aligned} \vec{i}_1 &= (1,0,0,\ldots,0), \\ \vec{i}_2 &= (0,1,0,\ldots,0), \\ &\vdots \\ \vec{i}_n &= (0,0,0,\ldots,1). \end{aligned}$$

We have

$$\vec{L}(a_1\vec{i}_1 + a_2\vec{i}_2 + \cdots + a_n\vec{i}_n) = a_1\vec{L}(\vec{i}_1) + a_2\vec{L}(\vec{i}_2) + \cdots + a_n\vec{L}(\vec{i}_n). \quad (6.6)$$

Our function is uniquely determined by the values of $\vec{L}(\vec{i}_1), \vec{L}(\vec{i}_2), \ldots, \vec{L}(\vec{i}_n)$.

Note that most of the transformations given in Chapter 4 were *not* linear transformations. The example

$$
\begin{aligned}
x &= 2 - 2u + 3v - 5w, \\
y &= 1 + u - 3v, \\
z &= -3 + 4u + 7v + 3w
\end{aligned}
$$

is not linear because it takes $(0,0,0)$ to $(2,1,-3)$ instead of $(0,0,0)$. On the other hand, the system of equations relating the differentials is a linear transformation

$$
\begin{aligned}
dx &= -2\,du + 3\,dv - 5\,dw, \\
dy &= du - 3\,dv, \\
dz &= 4\,du + 7\,dv + 3\,dw.
\end{aligned}
$$

In fact, if $\vec{y} = \vec{F}(\vec{x})$ and $\vec{F}$ is differentiable, then the transformation from the differential of $\vec{x}$ to the differential of $\vec{y}$ is always linear, just as it is in one dimension: at $(c, f(c))$ on the curve $y = f(x)$, the differentials of x and y are related by the linear transformation:

$$dy = f'(c)\,dx. \quad (6.7)$$

Differentiability

What does it mean to say that $\vec{F}$ is *differentiable*? For real-valued functions of a real variable, $y = f(x)$ is differentiable at c if

$$\lim_{x \to c} \frac{f(x) - f(c)}{x - c}$$

exists. This is equivalent to saying that there exists a number, which we call $f'(c)$, such that

$$\frac{f(x) - f(c)}{x - c} - f'(c)$$

approaches 0 as x approaches c. If we set $\Delta y = f(x) - f(c)$ and $\Delta x = x - c$ and define

$$E(c, \Delta x) = \frac{\Delta y}{\Delta x} - f'(c),$$

then the existence of a derivative at c is equivalent to saying that there exists a number, $f'(c)$, such that

$$\Delta y = f'(c)\Delta x + \Delta x\, E(c, \Delta x), \quad (6.8)$$

where

$$\lim_{\Delta x \to 0} E(c, \Delta x) = 0.$$

If we ignore the error term, $\Delta x\, E(c, \Delta x)$, in Equation (6.8), we are left with a linear function from Δx to Δy. It is this equation, giving us a linear approximation to the relationship between the change in x and the change in y, that provides the power of differential calculus. It is thus this equation that we wish to mimic as we move to higher dimensions.

Definition: Given a vector field

$$\vec{F} : \mathbf{R}^n \longrightarrow \mathbf{R}^m, \quad \vec{y} = \vec{F}(\vec{x}),$$

we say that $\vec{F}$ is *differentiable at* $\vec{c}$ if there exists a linear transformation

$$\vec{L}_c : \mathbf{R}^n \longrightarrow \mathbf{R}^m$$

such that

$$\Delta \vec{y} = \vec{L}_c(\Delta \vec{x}) + |\Delta \vec{x}|\, \vec{E}(\vec{c}, \Delta \vec{x}), \tag{6.9}$$

where

$$\begin{aligned}
\Delta \vec{y} &= \vec{F}(\vec{x}) - \vec{F}(\vec{c}), \\
\Delta \vec{x} &= \vec{x} - \vec{c},
\end{aligned}$$

and

$$\lim_{\Delta \vec{x} \to \vec{0}} \vec{E}(\vec{c}, \Delta \vec{x}) = \vec{0}.$$

The linear transformation $\vec{L}_c$ is *the derivative of* $\vec{F}$ *at* $\vec{c}$.

If we define

$$\begin{aligned}
d\vec{x} &= (dx_1, dx_2, \ldots, dx_n), \\
d\vec{y} &= (dy_1, dy_2, \ldots, dy_m),
\end{aligned}$$

and if $\vec{L}_c$ is the derivative of $\vec{y} = \vec{F}(\vec{x})$ at $\vec{c}$, then we define the relationship between $d\vec{y}$ and $d\vec{x}$ to be

$$d\vec{y} = \vec{L}_c(d\vec{x}). \tag{6.10}$$

The definition of differentiability is not complete because we have not yet defined

$$\lim_{\Delta \vec{x} \to \vec{0}}.$$

This will be dealt with at the beginning of the next chapter. The point I wish to make here is that while you are accustomed to thinking of the derivative of $f(x)$ at $x = c$ as a real number we shall now need to think of it as a linear transformation. How we determine if such a linear transformation exists and how we find it if it does will be the principal tasks of Chapter 7. For now, we shall simply investigate linear transformations.

Matrix Notation

Consider a linear transformation, $\vec{L}$, from u, v, w space to x, y, z, t space and let $\vec{\imath}, \vec{\jmath}$, and $\vec{k}$ be the unit basis vectors in u, v, w space. As we have seen, $\vec{L}$ is uniquely determined by what it does to $\vec{\imath}, \vec{\jmath}$, and $\vec{k}$. As an example, let us assume that

$$
\begin{aligned}
\vec{L}(\vec{\imath}) &= (2, -1, 6, 13), \\
\vec{L}(\vec{\jmath}) &= (-3, 5, 7, -8), \\
\vec{L}(\vec{k}) &= (24, 9, -4, 10).
\end{aligned}
$$

It follows that

$$
\begin{aligned}
&\vec{L}(u\vec{\imath} + v\vec{\jmath} + w\vec{k}) \\
&= u\vec{L}(\vec{\imath}) + v\vec{L}(\vec{\jmath}) + w\vec{L}(\vec{k}) \\
&= (2u, -u, 6u, 13u) + (-3v, 5v, 7v, -8v) + (24w, 9w, -4w, 10w) \\
&= (2u - 3v + 24w, -u + 5v + 9w, 6u + 7v - 4w, 13u - 8v + 10w),
\end{aligned}
$$

and $\vec{L}$ is represented by the system of equations:

$$
\begin{aligned}
x &= 2u - 3v + 24w, \\
y &= -u + 5v + 9w, \\
z &= 6u + 7v - 4w, \\
t &= 13u - 8v + 10w.
\end{aligned} \tag{6.11}
$$

Note that each column of coefficients is the image of the respective basis vector.

In general, if $\vec{L}$ is a mapping from $\mathbf{R}^n$ to $\mathbf{R}^m$ and

$$
\vec{L}(\vec{\imath}_j) = (a_{1j}, a_{2j}, \ldots, a_{mj}), \quad j = 1, 2, \ldots, n,
$$

then $\vec{L}$ is represented by the system of equations:

$$
\begin{aligned}
y_1 &= a_{11}x_1 + a_{12}x_2 + \cdots + a_{1n}x_n, \\
y_2 &= a_{21}x_1 + a_{22}x_2 + \cdots + a_{2n}x_n, \\
&\vdots \\
y_m &= a_{m1}x_1 + a_{m2}x_2 + \cdots + a_{mn}x_n.
\end{aligned}
$$

Since we always have such a representation, it is convenient to record our transformation by just recording the coefficients in a *matrix* or rectangular array and to refer to this as our transformation. To help avoid confusion, we shall change from vector notation to boldface when we are referring to the matrix:

$$\mathbf{L} = \begin{pmatrix} a_{11} & a_{12} & \cdots & a_{1n} \\ a_{21} & a_{22} & \cdots & a_{2n} \\ & \vdots & & \\ a_{m1} & a_{m2} & \cdots & a_{mn} \end{pmatrix}. \tag{6.12}$$

The dimensions of this matrix are $m \times n$ (m rows by n columns). The first subscript is the number of the row, the second subscript is the number of the column: a_{ij} is the entry in the ith row and jth column.

An Example

Consider the mapping that rotates the x, y plane by $60°$ counterclockwise. This is a linear transformation, and so it can be represented by the matrix whose columns are the images of $(1,0)$ and $(0,1)$. Rotating $(1,0)$ by $60°$ yields $(1/2, \sqrt{3}/2)$; rotating $(0,1)$ gives us $(-\sqrt{3}/2, 1/2)$. It follows that the matrix for this transformation is

$$\begin{pmatrix} \frac{1}{2} & \frac{-\sqrt{3}}{2} \\ \frac{\sqrt{3}}{2} & \frac{1}{2} \end{pmatrix}.$$

Composition of Linear Transformations

Let $\vec{L}$ be the linear transformation given in Equation (6.11) and let $\vec{M}$ be the linear transformation from 4-dimensional to 3-dimensional space whose matrix representation is given by:

$$\mathbf{M} = \begin{pmatrix} 2 & 6 & -3 & 1 \\ 10 & -5 & 0 & 2 \\ 3 & -2 & 4 & -6 \end{pmatrix}.$$

The composition of $\vec{L}$ followed by $\vec{M}$, written $\vec{M} \circ \vec{L}$, is a mapping from $\mathbf{R}^3$ to $\mathbf{R}^3$ sending (u, v, w) to (p, q, r) where

$$\begin{aligned} p &= 2x + 6y - 3z + t \\ &= 2(2u - 3v + 24w) + 6(-u + 5v + 9w) \\ &\quad - 3(6u + 7v - 4w) + (13u - 8v + 10w) \\ &= -7u - 5v + 124w, \end{aligned}$$

$$
\begin{aligned}
q \;=\;& 10x - 5y + 2t \\
=\;& 10(2u - 3v + 24w) - 5(-u + 5v + 9w) \\
& + 2(13u - 8v + 10w) \\
=\;& 51u - 71v + 215w, \\
r \;=\;& 3x - 2y + 4z - 6t \\
=\;& 3(2u - 3v + 24w) - 2(-u + 5v + 9w) \\
& + 4(6u + 7v - 4w) - 6(13u - 8v + 10w) \\
=\;& -46u + 57v - 22w.
\end{aligned}
$$

If we define the product of **M** and **L** to be the matrix of their composition, we see that we have:

$$
\begin{pmatrix}
2 & 6 & -3 & 1 \\
10 & -5 & 0 & 2 \\
3 & -2 & 4 & -6
\end{pmatrix}
\begin{pmatrix}
2 & -3 & 24 \\
-1 & 5 & 9 \\
6 & 7 & -4 \\
13 & -8 & 10
\end{pmatrix}
=
\begin{pmatrix}
-7 & -5 & 124 \\
51 & -71 & 215 \\
-46 & 57 & -22
\end{pmatrix}.
$$

Definition: If the number of columns of **M** is equal to the number of rows of **L**, then we define the *matrix product*, **ML**, to be the matrix representation of the composition $\vec{M} \circ \vec{L}$. To compute the matrix product, the entry in the ith row and jth column of **ML** is the dot product of the ith row of **M** with the jth column of **L**.

For our example, the entry in the second row, third column of **ML** is

$$
(10, -5, 0, 2) \cdot (24, 9, -4, 10) = 215.
$$

For the given transformations, the range of $\vec{M}$ is also the domain of $\vec{L}$. We can define the composition in the opposite order, $\vec{L} \circ \vec{M}$. Here, we get a mapping from 4-dimensional space to 4-dimensional space whose matrix representation is given by

$$
\begin{pmatrix}
2 & -3 & 24 \\
-1 & 5 & 9 \\
6 & 7 & -4 \\
13 & -8 & 10
\end{pmatrix}
\begin{pmatrix}
2 & 6 & -3 & 1 \\
10 & -5 & 0 & 2 \\
3 & -2 & 4 & -6
\end{pmatrix}
$$
$$
=
\begin{pmatrix}
46 & -21 & 90 & -148 \\
75 & -49 & 39 & -45 \\
70 & 9 & -34 & 44 \\
-24 & 98 & 1 & -63
\end{pmatrix}.
$$

This product notation is extremely useful. We can employ it to write our linear transformations in a form more directly analogous to the 1-dimensional equation: $y = cx$. If **L** is the matrix representation of our transformation and if we write $\vec{x}$ as an $n \times 1$ matrix denoted by **x**, $\vec{y}$ as an

$m \times 1$ matrix denoted by $\mathbf{y}$, then we can represent our system of equations as

$$
\begin{pmatrix} y_1 \\ y_2 \\ \vdots \\ y_m \end{pmatrix} = \begin{pmatrix} a_{11}x_1 + a_{12}x_2 + \cdots + a_{1n}x_n \\ a_{21}x_1 + a_{22}x_2 + \cdots + a_{2n}x_n \\ \vdots \\ a_{m1}x_1 + a_{m2}x_2 + \cdots + a_{mn}x_n \end{pmatrix}
$$

$$
= \begin{pmatrix} a_{11} & a_{12} & \cdots & a_{1n} \\ a_{21} & a_{22} & \cdots & a_{2n} \\ & \vdots & & \\ a_{m1} & a_{m2} & \cdots & a_{mn} \end{pmatrix} \begin{pmatrix} x_1 \\ x_2 \\ \vdots \\ x_n \end{pmatrix},
$$

or equivalently,

$$
\mathbf{y} = \mathbf{Lx}.
$$

Identity and Inverses

If $\vec{L}$ has an inverse, say $\vec{M}$, then $\vec{M} \circ \vec{L}$ must be the identity transformation taking $\vec{\imath}_1$ to $\vec{\imath}_1$, $\vec{\imath}_2$ to $\vec{\imath}_2$, ..., $\vec{\imath}_m$ to $\vec{\imath}_m$. The matrix representation of the identity transformation is

$$
\mathbf{I} = \begin{pmatrix} 1 & 0 & 0 & & 0 \\ 0 & 1 & 0 & \cdots & 0 \\ 0 & 0 & 1 & & 0 \\ & \vdots & & \ddots & \vdots \\ 0 & 0 & 0 & \cdots & 1 \end{pmatrix}. \tag{6.13}
$$

To say that $\vec{M}$ is the inverse of $\vec{L}$ is equivalent to the statement

$$
\mathbf{ML} = \mathbf{I}.
$$

For example, if

$$
\mathbf{L} = \begin{pmatrix} 2 & -3 \\ -3 & 4 \end{pmatrix},
$$

then the inverse is given by

$$
\mathbf{M} = \begin{pmatrix} -4 & -3 \\ -3 & -2 \end{pmatrix},
$$

because

$$
\begin{pmatrix} -4 & -3 \\ -3 & -2 \end{pmatrix} \begin{pmatrix} 2 & -3 \\ -3 & 4 \end{pmatrix} = \begin{pmatrix} 1 & 0 \\ 0 & 1 \end{pmatrix}.
$$

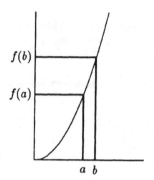

FIGURE 6.1. The derivative as magnification.

6.2   Determinants

In this section, we shall restrict our attention to linear transformations for which the domain and the range have the same dimension. Equivalently, these are transformations whose matrix representation is a square matrix ($m \times m$). There are two related questions that we shall address in this section and again in the last section.

1. When is such a linear transformation invertible?

2. What is the magnification of volume under this transformation? That is, if we start with the unit cube spanned by the unit basis vectors, what is the volume of its image?

Part of the reason for asking this second question is that for real-valued functions of a single variable, supplying this magnification is one of the jobs of the derivative. The function f sends the interval $[a, b]$ to the interval $[f(a), f(b)]$ (see Figure 6.1). The ratio of the lengths of these intervals is the magnification. The derivative

$$f'(c) = \lim_{b \to a} \frac{f(b) - f(a)}{b - a}$$

is an approximation to this ratio and is equal to the limiting value as we take progressively smaller initial intervals. As we shall see, in higher dimensions, we often do not need the full derivative, the linear transformation $\vec{L}_c$, but only the number that tells us how it magnifies volume.

In a linear mapping from $\mathbf{R}^2$ to $\mathbf{R}^2$, as an example

$$\mathbf{L} = \begin{pmatrix} 2 & -3 \\ 5 & 4 \end{pmatrix},$$

the unit square spanned by $(1, 0)$ and $(0, 1)$ is mapped to the parallelogram spanned by $(2, 5)$ and $(-3, 4)$. The signed area of this image parallelogram is the z coordinate of $(2, 5, 0) \times (-3, 4, 0)$, which is

$$2 \times 4 - (-3) \times 5 = 23.$$

In general, if the linear mapping is given by

$$\mathbf{L} = \left(\begin{array}{cc} a_{11} & a_{12} \\ a_{21} & a_{22} \end{array} \right),$$

then the image parallelogram is spanned by (a_{11}, a_{21}) and (a_{12}, a_{22}) and it has signed area

$$a_{11}a_{22} - a_{12}a_{21}.$$

We call this number the *determinant* of $\mathbf{L}$. The notation for the determinant is $\det(\mathbf{L})$ or, if $\mathbf{L}$ is given explicitly as a matrix,

$$\det(\mathbf{L}) = \left| \begin{array}{cc} a_{11} & a_{12} \\ a_{21} & a_{22} \end{array} \right|.$$

Theorem 6.1 *A linear transformation,*

$$\vec{L} : \mathbf{R}^2 \longrightarrow \mathbf{R}^2,$$

is invertible if and only if the determinant of its matrix representation is not zero.

Proof: We begin by assuming that

$$\det(\mathbf{L}) = a_{11}a_{22} - a_{12}a_{21} = 0.$$

If $a_{11} = a_{12} = a_{21} = a_{22} = 0$, then $\vec{L}$ sends everything to $\vec{0}$, and $\vec{L}$ is clearly not invertible. If at least one of our entries is not zero, let us assume that it is a_{11}. The following argument can be modified so that it works for any of the four entries.

We use the fact that $a_{11}a_{22} = a_{12}a_{21}$. If $a_{12} = 0$, then a_{22} must be 0, and so $\vec{L}(0, c) = c\vec{L}(0, 1) = \vec{0}$ for any c. It follows that $\vec{L}$ is not 1 to 1 and so is not invertible. If $a_{12} \neq 0$, then

$$
\begin{aligned}
\vec{L}(0, 1) &= (a_{12}, a_{22}) \\
&= \frac{a_{12}}{a_{11}} \left(a_{11}, \frac{a_{11}a_{22}}{a_{12}} \right) \\
&= \frac{a_{12}}{a_{11}} \left(a_{11}, \frac{a_{12}a_{21}}{a_{12}} \right) \\
&= \frac{a_{12}}{a_{11}} (a_{11}, a_{21}) \\
&= \frac{a_{12}}{a_{11}} \vec{L}(1, 0) \\
&= \vec{L} \left(\frac{a_{12}}{a_{11}}, 0 \right),
\end{aligned}
$$

and therefore $\vec{L}$ is not 1 to 1 and so not invertible. We have thus proved that if $\det(\mathbf{L}) = 0$, then $\vec{L}$ is not invertible.

If the determinant

$$\delta = a_{11}a_{22} - a_{12}a_{21}$$

is not zero, then

$$\mathbf{L} = \begin{pmatrix} a_{11} & a_{12} \\ a_{21} & a_{22} \end{pmatrix}$$

has as inverse:

$$\mathbf{L}^{-1} = \begin{pmatrix} a_{22}/\delta & -a_{12}/\delta \\ -a_{21}/\delta & a_{11}/\delta \end{pmatrix}.$$

This is readily verified by multiplication:

$$
\begin{aligned}
\mathbf{L}^{-1}\mathbf{L} &= \begin{pmatrix} a_{22}/\delta & -a_{12}/\delta \\ -a_{21}/\delta & a_{11}/\delta \end{pmatrix} \begin{pmatrix} a_{11} & a_{12} \\ a_{21} & a_{22} \end{pmatrix} \\
&= \begin{pmatrix} (a_{22}a_{11} - a_{12}a_{21})/\delta & (a_{22}a_{12} - a_{12}a_{22})/\delta \\ (-a_{21}a_{11} + a_{11}a_{21})/\delta & (-a_{21}a_{12} + a_{11}a_{22})/\delta \end{pmatrix} \\
&= \begin{pmatrix} 1 & 0 \\ 0 & 1 \end{pmatrix}.
\end{aligned}
$$

Q.E.D.

Using the example from the beginning of this section, if

$$\mathbf{L} = \begin{pmatrix} 2 & -3 \\ 5 & 4 \end{pmatrix},$$

then

$$\mathbf{L}^{-1} = \begin{pmatrix} 4/23 & 3/23 \\ -5/23 & 2/23 \end{pmatrix}.$$

Three-Dimensional Determinants

In three dimensions, we define the determinant of a linear transformation to be the signed volume of the parallelepiped that is the image of the cube spanned by $\vec{i}$, $\vec{j}$, and $\vec{k}$. We already know that this is the scalar triple product of the images of the three unit basis vectors:

$$\det(\mathbf{L}) = \vec{L}(\vec{i}) \cdot \vec{L}(\vec{j}) \times \vec{L}(\vec{k}), \tag{6.14}$$

$$\begin{vmatrix} a_{11} & a_{12} & a_{13} \\ a_{21} & a_{22} & a_{23} \\ a_{31} & a_{32} & a_{33} \end{vmatrix} = \begin{cases} a_{11}a_{22}a_{33} - a_{11}a_{32}a_{23} \\ \quad + a_{21}a_{32}a_{13} - a_{21}a_{12}a_{33} \\ \quad + a_{31}a_{12}a_{23} - a_{31}a_{22}a_{13}. \end{cases} \tag{6.15}$$

Our three image vectors,

$$(a_{11}, a_{21}, a_{31}), (a_{12}, a_{22}, a_{32}), (a_{13}, a_{23}, a_{33}),$$

span all of $\mathbf{R}^3$ if and only if they do not lie in a single plane containing the origin. This is equivalent to saying that the parallelepiped that they span has a nonzero volume. Thus, $\vec{L}$ is invertible if and only if det($\mathbf{L}$) is not zero.

If our linear transformation is represented by the matrix

$$\mathbf{L} = \begin{pmatrix} 2 & -1 & 0 \\ 3 & 0 & -4 \\ 2 & 1 & 3 \end{pmatrix},$$

which has the determinant

$$0 + 8 + 8 + 9 + 0 - 0 = 25 \neq 0,$$

then it is invertible. The inverse happens to be

$$\mathbf{L}^{-1} = \begin{pmatrix} 4/25 & 3/25 & 4/25 \\ -17/25 & 6/25 & 8/25 \\ 3/25 & -4/25 & 3/25 \end{pmatrix}.$$

In Section 6.5, we shall see how to explicitly construct the inverse.

Properties of Three-Dimensional Volume

In more than three dimensions, we want to define the determinant to be the appropriately signed hypervolume of the parallelepiped spanned by $\vec{L}(\vec{\imath}_1)$, $\vec{L}(\vec{\imath}_2),\ldots,\vec{L}(\vec{\imath}_m)$. The problem before us it how to define this signed hypervolume.

Ordinary volume in three dimensions has three properties that we would like it see carried over to our definition of signed hypervolume:

1. The volume of the unit cube is 1.

2. Volume is linear in each dimension (see Figure 6.2): The volume of the parallelepiped spanned by $\vec{a}$, $\vec{b}$, and $\vec{c}_1 + \vec{c}_2$ is the sum of the volumes of the parallelepipeds spanned by $\vec{a}$, $\vec{b}$, and $\vec{c}_1$ and by $\vec{a}$, $\vec{b}$, and $\vec{c}_2$:

$$\vec{a} \cdot \vec{b} \times (\vec{c}_1 + \vec{c}_2) = (\vec{a} \cdot \vec{b} \times \vec{c}_1) + (\vec{a} \cdot \vec{b} \times \vec{c}_2).$$

The volume of the parallelepiped spanned by $\vec{a}$, $\vec{b}$, and $k\vec{c}$ is k times the volume of the parallelepiped spanned by $\vec{a}$, $\vec{b}$, and $\vec{c}$:

$$\vec{a} \cdot \vec{b} \times k\vec{c} = k(\vec{a} \cdot \vec{b} \times \vec{c}).$$

3. If the three spanning vectors are linearly dependent, that is, if they all lie in a common plane through the origin, then the resulting parallelepiped has zero volume.

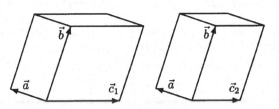

FIGURE 6.2. Volume is linear in each dimension.

Properties of the Determinant

Given a linear transformation

$$\vec{L} : \mathbf{R}^m \longrightarrow \mathbf{R}^m,$$

let $\vec{a}_1$, $\vec{a}_2$, ..., $\vec{a}_m$ be the images of the unit basis vectors:

$$\begin{aligned}
\vec{a}_1 &= (a_{11}, a_{21}, \ldots, a_{m1}) &= \vec{L}(\vec{i}_1), \\
\vec{a}_2 &= (a_{12}, a_{22}, \ldots, a_{m2}) &= \vec{L}(\vec{i}_2), \\
&\quad\vdots \\
\vec{a}_m &= (a_{1m}, a_{2m}, \ldots, a_{mm}) &= \vec{L}(\vec{i}_m).
\end{aligned}$$

The matrix representation of $\vec{L}$ is given by

$$\mathbf{L} = \begin{pmatrix}
a_{11} & a_{12} & \cdots & a_{1m} \\
a_{21} & a_{22} & \cdots & a_{2m} \\
& \vdots & & \\
a_{m1} & a_{m2} & \cdots & a_{mm}
\end{pmatrix}.$$

The *determinant of* $\mathbf{L}$,

$$\det(\mathbf{L}) = \det(\vec{a}_1, \vec{a}_2, \ldots, \vec{a}_m),$$

must satisfy the following conditions:

1. The signed hypervolume of the standard unit hypercube is 1:

$$\det(\vec{i}_1, \vec{i}_2, \ldots, \vec{i}_m) = 1. \tag{6.16}$$

2. The value of the determinant is linear as a function of each column vector:

$$\begin{aligned}
\det(\vec{a}_1, \vec{a}_2, \ldots, \vec{a}_k + \vec{b}_k, \ldots, \vec{a}_m) &= \det(\vec{a}_1, \vec{a}_2, \ldots, \vec{a}_k, \ldots, \vec{a}_m) \\
&\quad + \det(\vec{a}_1, \vec{a}_2, \ldots, \vec{b}_k, \ldots, \vec{a}_m), \\
&\tag{6.17}
\end{aligned}$$

$$\det(\vec{a}_1, \vec{a}_2, \ldots, c\vec{a}_k, \ldots, \vec{a}_m) = c \det(\vec{a}_1, \vec{a}_2, \ldots, \vec{a}_k, \ldots, \vec{a}_m). \tag{6.18}$$

3. If any two distinct basis vectors have the same image, then the value of the determinant is zero:

$$\vec{a}_i = \vec{a}_k \ (i \neq k) \implies \det(\vec{a}_1, \vec{a}_2, \ldots, \vec{a}_m) = 0. \qquad (6.19)$$

It may appear that in the third property we have settled for a weaker condition than what can be demanded. For 3-dimensional volume, the volume is zero whenever the spanning vectors are linearly dependent. Here, we have only required that the hypervolume be zero whenever two of the spanning vectors are identical. However, as the next theorem demonstrates, these restrictions are actually equivalent.

Theorem 6.2 *Let $f(\vec{a}_1, \vec{a}_2, \ldots, \vec{a}_m)$ be any real-valued function that satisfies Equations (6.17) through (6.19), then the following properties hold:*

1. *If any of the vectors in the argument of f is $\vec{0}$, then the value of f is 0:*

$$f(\vec{a}_1, \vec{a}_2, \ldots, \vec{0}, \ldots, \vec{a}_m) = 0. \qquad (6.20)$$

2. *Interchanging any two of the vectors in the argument of f changes the sign:*

$$f(\vec{a}_1, \ldots, \vec{a}_i, \ldots, \vec{a}_k, \ldots, \vec{a}_m) = -f(\vec{a}_1, \ldots, \vec{a}_k, \ldots, \vec{a}_i, \ldots, \vec{a}_m). \qquad (6.21)$$

3. *If the vectors in the argument of f are linearly dependent, then the value of f is 0. That is, if there exist constants $c_1, c_2, \ldots, c_m$, not all zero, such that*

$$\sum_{j=1}^{m} c_j \vec{a}_j = \vec{0},$$

then

$$f(\vec{a}_1, \vec{a}_2, \ldots, \vec{a}_m) = 0.$$

Note: Linear dependence means that at least one of these vectors is a linear combination of the others. For example, if $c_1 \neq 0$, then

$$\vec{a}_1 = -\frac{c_2}{c_1} \vec{a}_2 - \frac{c_3}{c_1} \vec{a}_3 - \cdots - \frac{c_m}{c_1} \vec{a}_m.$$

Proof: Equation (6.20) is simply Equation (6.18) with $c = 0$. Using Equations (6.17) and (6.19), we see that

$$\begin{aligned}
0 &= f(\vec{a}_1, \ldots, \vec{a}_i + \vec{a}_k, \ldots, \vec{a}_i + \vec{a}_k, \ldots, \vec{a}_m) \\
&= f(\vec{a}_1, \ldots, \vec{a}_i, \ldots, \vec{a}_i, \ldots, \vec{a}_m) \\
&\quad + f(\vec{a}_1, \ldots, \vec{a}_i, \ldots, \vec{a}_k, \ldots, \vec{a}_m) \\
&\quad + f(\vec{a}_1, \ldots, \vec{a}_k, \ldots, \vec{a}_i, \ldots, \vec{a}_m) \\
&\quad + f(\vec{a}_1, \ldots, \vec{a}_k, \ldots, \vec{a}_k, \ldots, \vec{a}_m) \\
&= f(\vec{a}_1, \ldots, \vec{a}_i, \ldots, \vec{a}_k, \ldots, \vec{a}_m) \\
&\quad + f(\vec{a}_1, \ldots, \vec{a}_k, \ldots, \vec{a}_i, \ldots, \vec{a}_m),
\end{aligned}$$

which is Equation (6.21). Finally, let us assume that c_1 is not zero in the equation of linear dependence. We have

$$
\begin{aligned}
0 &= f(\vec{0}, \vec{a}_2, \ldots, \vec{a}_m) \\
&= f\left(\sum_{j=1}^{m} c_j \vec{a}_j, \vec{a}_2, \ldots, \vec{a}_m\right) \\
&= \sum_{j=1}^{m} c_j f(\vec{a}_j, \vec{a}_2, \ldots, \vec{a}_m).
\end{aligned}
$$

Equation (6.19) implies that for each $j > 1$,

$$
f(\vec{a}_j, \vec{a}_2, \ldots, \vec{a}_m) = 0.
$$

Since $c_1 \neq 0$, it follows that

$$
0 = f(\vec{a}_1, \vec{a}_2, \ldots, \vec{a}_m).
$$

Q.E.D.

Uniqueness of Hypervolume

We are now prepared to prove that there is at most one real-valued function of $\vec{a}_1, \vec{a}_2, \ldots, \vec{a}_m$ that satisfies Equations (6.16) through (6.19), and therefore these equations actually define the determinant.

Theorem 6.3 *For any positive dimension m, let $f(\vec{a}_1, \vec{a}_2, \ldots, \vec{a}_m)$ be a real-valued function satisfying Equations (6.17) through (6.19), then*

$$
f(\vec{a}_1, \vec{a}_2, \ldots, \vec{a}_m) = \det(\vec{a}_1, \vec{a}_2, \ldots, \vec{a}_m) f(\vec{\imath}_1, \vec{\imath}_2, \ldots, \vec{\imath}_m). \tag{6.22}
$$

Proof: Let F be the difference of these functions:

$$
F(\vec{a}_1, \ldots, \vec{a}_m) = f(\vec{a}_1, \ldots, \vec{a}_m) - \det(\vec{a}_1, \ldots, \vec{a}_m) f(\vec{\imath}_1, \ldots, \vec{\imath}_m).
$$

The function F satisfies Equations (6.17) through (6.19), and so it also satisfies all of the properties given in Theorem 6.2. When the vectors in the argument of F are the unit basis vectors in order, we have

$$
F(\vec{\imath}_1, \vec{\imath}_2, \ldots, \vec{\imath}_m) = f(\vec{\imath}_1, \ldots, \vec{\imath}_m) - f(\vec{\imath}_1, \ldots, \vec{\imath}_m) = 0.
$$

Any rearrangement of these unit basis vectors changes the sign of both the determinant and f in exactly the same way, and so F continues to have the value 0. Furthermore, if any of the unit basis vectors repeat, then by

Equation (6.19) the value of F is again 0. What this implies is that for *any* choice of unit basis vectors, we have

$$F(\vec{i}_{j_1}, \vec{i}_{j_2}, \ldots, \vec{i}_{j_m}) = 0.$$

Now, we use the fact that F is linear in each coordinate. Let

$$\vec{a}_1, \vec{a}_2, \ldots, \vec{a}_m$$

be any set of m m-dimensional vectors. For any set of values of $j_2, j_3, \ldots, j_m$ taken from $[1, m]$ we have

$$
\begin{aligned}
F(\vec{a}_1, \vec{i}_{j_2}, \ldots, \vec{i}_{j_m}) &= \sum_{j_1=1}^{m} a_{j_1 1} F(\vec{i}_{j_1}, \vec{i}_{j_2}, \ldots, \vec{i}_{j_m}) \\
&= \sum_{j_1=1}^{m} 0 = 0.
\end{aligned}
$$

We proceed by induction. Assume that we have demonstrated that for any set of values of $j_t, j_{t+1}, \ldots, j_m$ taken from $[1, m]$ we have

$$F(\vec{a}_1, \ldots, \vec{a}_{t-1}, \vec{i}_{j_t}, \ldots, \vec{i}_{j_m}) = 0.$$

It then follows that

$$
\begin{aligned}
&F(\vec{a}_1, \ldots, \vec{a}_{t-1}, \vec{a}_t, \vec{i}_{j_{t+1}}, \ldots, \vec{i}_{j_m}) \\
&= \sum_{j_t=1}^{m} a_{j_t t} F(\vec{a}_1, \ldots, \vec{a}_{t-1}, \vec{i}_{j_t}, \vec{i}_{j_{t+1}}, \ldots, \vec{i}_{j_m}) \\
&= \sum_{j_t=1}^{m} 0 = 0.
\end{aligned}
$$

Therefore, $F(\vec{a}_1, \vec{a}_2, \ldots, \vec{a}_m) = 0$.

Q.E.D.

Corollary 6.1 *There is at most one real-valued function satisfying all of the conditions on the determinant: Equations (6.16) through (6.19). Equivalently, Equations (6.16) through (6.19) uniquely define the determinant.*

Proof: If $f(\vec{i}_1, \ldots, \vec{i}_m) = 1$, then Equation (6.22) states

$$f(\vec{a}_1, \vec{a}_2, \ldots, \vec{a}_m) = \det(\vec{a}_1, \vec{a}_2, \ldots, \vec{a}_m).$$

Q.E.D.

Definition of the Determinant

Is there a function that satisfies these four equations? Yes, and we have already met it. We recall from Section 4.6 the function

$$D(\vec{a}_1, \vec{a}_2, \ldots, \vec{a}_m) = \sum_{\sigma \in S_m} \text{sgn}(\sigma) \prod_{i=1}^{m} a_{i\sigma(i)}, \qquad (6.23)$$

where

$$\vec{a}_j = (a_{1j}, a_{2j}, \ldots, a_{mj}).$$

When $m = 2$, $D(\vec{a}_1, \vec{a}_2)$ is equal to the determinant of the 2×2 matrix:

$$\begin{vmatrix} a_{11} & a_{12} \\ a_{21} & a_{22} \end{vmatrix} = a_{11}a_{22} - a_{12}a_{21}.$$

When $m = 3$, $D(\vec{a}_1, \vec{a}_2, \vec{a}_3)$ is equal to the determinant of the 3×3 matrix [see Equation (6.15)].

Theorem 6.4 *Equations (6.16) through (6.19) are satisfied by our function* D*, and therefore,*

$$\det(\vec{a}_1, \vec{a}_2, \ldots, \vec{a}_m) = D(\vec{a}_1, \vec{a}_2, \ldots, \vec{a}_m). \qquad (6.24)$$

Proof: For Equation (6.16), we have that

$$D(\vec{\imath}_1, \vec{\imath}_2, \ldots, \vec{\imath}_m) = \sum_{\sigma \in S_m} \text{sgn}(\sigma) \prod_{j=1}^{m} i_{j\sigma(j)},$$

where

$$i_{jk} = \begin{cases} 1 & \text{if} \quad j = k, \\ 0 & \text{if} \quad j \neq k. \end{cases}$$

The only nonzero term in the summation corresponds to the identity permutation:

$$\sigma(j) = j, \quad \text{for all } j.$$

The term corresponding to the identity permutation is 1.
For Equation (6.17), we have that

$$D(\vec{a}_1, \ldots, \vec{a}_k + \vec{b}_k, \ldots, \vec{a}_m)$$

$$= \sum_{\sigma \in S_m} \text{sgn}(\sigma) \left(\prod_{\substack{j=1 \\ \sigma(j) \neq k}}^{m} a_{j\sigma(j)} \right) \left(a_{\sigma^{-1}(k)k} + b_{\sigma^{-1}(k)k} \right)$$

$$= \sum_{\sigma \in S_m} \text{sgn}(\sigma) \left(\prod_{\substack{j=1 \\ \sigma(j) \neq k}}^{m} a_{j\sigma(j)} \right) a_{\sigma^{-1}(k)k}$$

$$+ \sum_{\sigma \in S_m} \text{sgn}(\sigma) \left(\prod_{\substack{j=1 \\ \sigma(j) \neq k}}^{m} a_{j\sigma(j)} \right) b_{\sigma^{-1}(k)k}$$

$$= D(\vec{a}_1, \ldots, \vec{a}_k, \ldots, \vec{a}_m) + D(\vec{a}_1, \ldots, \vec{b}_k, \ldots, \vec{a}_m).$$

The notation $\sigma^{-1}(k)$ signifies the inverse image of k. For our purposes, it is the row number that σ associates to column k.

For Equation (6.18), we have that

$$D(\vec{a}_1, \ldots, c\vec{a}_k, \ldots, \vec{a}_m)$$

$$= \sum_{\sigma \in S_m} \text{sgn}(\sigma) \left(\prod_{\substack{j=1 \\ \sigma(j) \neq k}}^{m} a_{j\sigma(j)} \right) c a_{\sigma^{-1}(k)k}$$

$$= c \sum_{\sigma \in S_m} \text{sgn}(\sigma) \prod_{j=1}^{m} a_{j\sigma(j)}$$

$$= cD(\vec{a}_1, \ldots, \vec{a}_k, \ldots, \vec{a}_m).$$

Finally, for Equation (6.19), we return to the original definition of D. If we are given the linear relationships:

$$\begin{aligned}
dy_1 &= a_{11}\, dx_1 + a_{12}\, dx_2 + \cdots + a_{1m}\, dx_m, \\
dy_2 &= a_{21}\, dx_1 + a_{22}\, dx_2 + \cdots + a_{2m}\, dx_m, \\
&\vdots \\
dy_m &= a_{m1}\, dx_1 + a_{m2}\, dx_2 + \cdots + a_{mm}\, dx_m,
\end{aligned} \qquad (6.25)$$

then using our rules for multiplying differentials leads us to the relationship

$$dy_1\, dy_2 \cdots dy_m = D(\vec{a}_1, \vec{a}_2, \ldots, \vec{a}_m)\, dx_1\, dx_2 \ldots dx_m. \qquad (6.26)$$

If $\vec{a}_i = \vec{a}_k$, then our system of differential equations (6.25) is symmetric in dx_i and dx_k. In other words, interchanging dx_i and dx_k does not change these equations. Since Equation (6.26) is defined by these equations, it also must be symmetric in dx_i and dx_k. But in order to interchange dx_i and dx_k, $i < k$, we need to move dx_k $k-i$ places to the left, resulting in $k-i$ changes of sign, and then move dx_i $k-i-1$ places to the right, resulting in an additional $k-i-1$ changes of sign:

$$dx_1 \cdots dx_{i-1} \, dx_i \quad \cdots \quad dx_{k-1} \, dx_k \, dx_{k+1} \cdots dx_m,$$

$$k - i \text{ spaces}$$

$$dx_1 \cdots dx_{i-1} \, dx_k \, dx_i \quad \cdots \quad dx_{k-1} \, dx_{k+1} \cdots dx_m.$$

$$k - i - 1 \text{ spaces}$$

The total number of changes of sign is $2k - 2i - 1$, which is odd, and so

$$dx_1 \ldots dx_i \ldots dx_k \ldots dx_m = -dx_1 \ldots dx_k \ldots dx_i \ldots dx_m.$$

What we have shown is that the right-hand side of Equation (6.26) remains unchanged when we interchange dx_i and dx_k and that it changes sign when we interchange dx_i and dx_k. The only way this can happen is if the right-hand side is 0.

Q.E.D.

An Example

The determinant of the 4×4 matrix

$$\begin{pmatrix} 1 & 0 & -2 & 3 \\ 5 & -7 & 4 & 6 \\ 2 & -1 & 0 & -5 \\ -3 & 0 & 9 & 8 \end{pmatrix}$$

is a sum of 24 terms (one for each permutation of the numbers 1 through 4):

$$\begin{vmatrix} 1 & 0 & -2 & 3 \\ 5 & -7 & 4 & 6 \\ 2 & -1 & 0 & -5 \\ -3 & 0 & 9 & 8 \end{vmatrix}$$

$$\begin{aligned}
=\;& (1)(-7)(0)(8) - (1)(-7)(-5)(9) - (1)(4)(-1)(8) \\
& + (1)(4)(-5)(0) + (1)(6)(-1)(9) - (1)(6)(0)(0) \\
& - (0)(5)(0)(8) + (0)(5)(-5)(9) + (0)(4)(2)(8) \\
& - (0)(4)(-5)(-3) - (0)(6)(2)(9) + (0)(6)(0)(-3) \\
& + (-2)(5)(-1)(8) - (-2)(5)(-5)(0) - (-2)(-7)(2)(8) \\
& + (-2)(-7)(-5)(-3) + (-2)(6)(2)(0) - (-2)(6)(-1)(-3) \\
& - (3)(5)(-1)(9) + (3)(5)(0)(0) + (3)(-7)(2)(9) \\
& - (3)(-7)(0)(-3) - (3)(4)(2)(0) + (3)(4)(-1)(-3)
\end{aligned}$$

$$= -442.$$

Consequences for Differentials

Let $d\mathbf{x}$ and $d\mathbf{y}$ be $m \times 1$ matrices of differentials:

$$d\mathbf{x} = \begin{pmatrix} dx_1 \\ dx_2 \\ \vdots \\ dx_m \end{pmatrix}, \quad d\mathbf{y} = \begin{pmatrix} dy_1 \\ dy_2 \\ \vdots \\ dy_m \end{pmatrix}.$$

Our definition of the function D and Theorem 6.4 imply the following corollary.

Corollary 6.2 *If $d\mathbf{x}$ and $d\mathbf{y}$ are related by the linear relationship*

$$d\mathbf{y} = \mathbf{L}_c\, d\mathbf{x},$$

then we also have the linear relationship

$$dy_1\, dy_2 \cdots dy_m = \det(\mathbf{L}_c)\, dx_1\, dx_2 \cdots dx_m.$$

Furthermore, the relationship between $dx_1\, dx_2 \cdots dx_m$ and $dy_1\, dy_2 \cdots dy_m$ can be found by multiplication, using the special rules for the products of differentials and scalars given in Section 4.4.

6.3   History and Comments

Ironically, the history of determinants goes back much further than that of matrices. Linear algebra and the theory of matrices was essentially created by Arthur Cayley (1821–1895) and others in the mid-nineteenth century in order to provide a framework in which to place the burgeoning results on systems of linear equations and determinants. As Morris Kline put it in *Mathematical Thought from Ancient to Modern Times*, "One could say that the subject of matrices was well developed before it was created."

Leibniz, in a letter written in 1693, revealed familiarity with 3×3 determinants as a test for linear independence. Cramer's rule, a technique that uses determinants to solve systems of linear equations and will be explained in Section 6.5, is named for Gabriel Cramer (1704–1752), who published it in 1750. Cramer was not the first to discover it; this rule had been published two years earlier in the posthumous *Treatise of Algebra* by Colin MacLaurin (1698–1746). The second half of the eighteenth century saw a great deal of work on determinants by many of the prominent mathematicians of the time. Among them, Joseph-Louis Lagrange (1736–1813) demonstrated the connection between determinants and volume in 1773. We owe the term "determinant" to Carl Friedrich Gauss (1777–1855), who actually used it to refer to what we today call the "discriminant" of a quadratic polynomial. As we shall see, the 2×2 determinant is very closely related to the discriminant. In 1812, Augustin-Louis Cauchy (1789–1857) defined the term "determinant" in its modern sense.

As we have seen in Corollary 6.2, if $\vec{y} = \vec{F}(\vec{x})$ and if $\vec{L}_c$ is the linear transformation of Equation (6.9), then the determinant of the matrix representation of $\vec{L}_c$ plays a crucial role in computing the pullback of $dy_1 \, dy_2 \cdots dy_m$. This was recognized and exploited by Lagrange, Cauchy, and others in the late eighteenth and early nineteenth centuries as they studied wave propagation and other physical phenomena. The use of this determinant was popularized in 1841 with the exposition "De determinantibus functionalibus" of Carl Gustav Jacob Jacobi (1804–1851). This determinant, $\det(\mathbf{L}_c)$, now carries his name; it is the *Jacobian*. The matrix itself is usually referred to as the *Jacobian matrix*.

The development of the determinant given in Section 6.2 may have seemed unnecessarily circuitous. Indeed, many authors cut directly to a definition of the determinant as the function $D(\vec{a}_1, \ldots, \vec{a}_m)$. While this is correct, it obscures the physical significance of the determinant that I have tried to emphasize: that $\det(\vec{a}_1, \ldots, \vec{a}_m)$ is the signed hypervolume of the parallelepiped spanned by $\vec{a}_1, \vec{a}_2, \ldots, \vec{a}_m$. It is this geometric interpretation of the determinant that will be critical in our future investigations.

The second point about determinants, one which cannot be emphasized too strongly, is stated in Corollary 6.2. Our rules for the multiplication of differential forms carry geometric meaning in any number of dimensions. If $d\mathbf{y}$ and $d\mathbf{x}$ have the same dimension and are related by

$$d\vec{y} = \vec{L}(d\vec{x}),$$

then the relationship between the volume in $\vec{x}$ space and the corresponding volume in $\vec{y}$ space is to be found simply by invoking these rules of multiplication. This is the key to the power of the language of differential forms.

Finally, the material in this chapter barely scratches the surface of the theory of linear transformations. This chapter deals only with those basic concepts that will be needed in the remainder of this book.

6.4 Exercises

1. Show that if $\vec{L}$ is an invertible linear transformation, then $\vec{L}^{-1}$ must be linear. Hint: since $\vec{L}$ is 1 to 1,

$$\vec{L}(\vec{x}_1) = \vec{L}(\vec{x}_2) \text{ implies that } \vec{x}_1 = \vec{x}_2.$$

Show that

$$\vec{L}\left(\vec{L}^{-1}(\vec{y}_1 + \vec{y}_2)\right) = \vec{L}\left(\vec{L}^{-1}(\vec{y}_1) + \vec{L}^{-1}(\vec{y}_2)\right),$$
$$\vec{L}\left(\vec{L}^{-1}(c\vec{y}_1)\right) = \vec{L}\left(c\vec{L}^{-1}(\vec{y}_1)\right).$$

2. Which of the following are linear transformations from $\mathbf{R}^2$ to $\mathbf{R}^2$?

   (a) *Shear:* $\vec{T}(x, y) = (x + cy, y)$ for some constant c.

   (b) *Translation:* $\vec{T}(x, y) = (x + a, y + b)$ for some constants a and b.

   (c) *Blow-up:* $\vec{T}(x, y) = (ax, by)$ for some constants a and b.

   (d) *Rotation:* if $x = r \cos\theta$, $y = r \sin\theta$, then

$$\vec{T}(x, y) = (r \cos(\theta + \phi), r \sin(\theta + \phi)),$$

   for some constant angle ϕ.

   (e) *Projection:* given a fixed vector $\vec{r} = (a, b)$, $\vec{T}$ maps each point to the closest point on the line spanned by $\vec{r}$:

$$\vec{T}(\vec{x}) = \vec{x}_r = \frac{\vec{x} \cdot \vec{r}}{|\vec{r}|^2}\vec{r}.$$

   (f) *Reflection:* given a fixed vector $\vec{r} = (a, b)$, $\vec{T}$ maps each point to its reflection with respect to $\vec{r}$:

$$\vec{T}(\vec{x}) = \vec{x} - 2\vec{x}_{r\perp}$$
$$= 2\vec{x}_r - \vec{x}.$$

3. For each transformation in Exercise 2 that is linear, find the 2×2 matrix that corresponds to $\vec{T}$ by finding the images of $(1,0)$ and $(0,1)$. Find the determinant of $\mathbf{T}$.

4. Show that every linear transformation from $\mathbf{R}^2$ to $\mathbf{R}^2$ can be written as a composition of shears, blow-ups, and $90°$ rotations ($\vec{T}(x, y) = (-y, x)$).

5. Define transformations $\vec{L}$ and $\vec{M}$ by

$$\begin{array}{ll} \vec{L}(\vec{\imath}) = \vec{\imath} - 2\vec{k}, & \vec{M}(\vec{\imath}) = -\vec{\jmath} + 3\vec{k}, \\ \vec{L}(\vec{\jmath}) = 3\vec{\jmath} + \vec{k}, & \vec{M}(\vec{\jmath}) = 5\vec{\imath} + \vec{\jmath} - \vec{k}, \\ \vec{L}(\vec{k}) = 2\vec{\imath} - \vec{\jmath} + \vec{k}, & \vec{M}(\vec{k}) = \vec{\imath} + 3\vec{\jmath} + \vec{k}. \end{array}$$

Find the matrix representations for $\vec{L}$, $\vec{M}$, $\vec{L} \circ \vec{M}$, and $\vec{M} \circ \vec{L}$.

6. Among 2×2 matrices, the identity matrix, **I**, and the zero matrix, **0** (all of whose entries are 0), commute with all 2×2 matrices:

$$\mathbf{IM} = \mathbf{MI}, \quad \mathbf{0M} = \mathbf{M0}.$$

Are there any other 2×2 matrices that commute with all 2×2 matrices? Describe them.

7. Find two 3×3 matrices, **A** and **B**, such that

$$\mathbf{AB} = \mathbf{0} \text{ and } \mathbf{BA} = \mathbf{0},$$

but neither **A** nor **B** is the zero matrix.

8. Compute each of the following determinants:

(a) $\begin{vmatrix} 1 & -1 & 0 \\ 2 & 4 & 1 \\ -1 & 0 & 2 \end{vmatrix}$,

(b) $\begin{vmatrix} 1 & 2 & 3 \\ 2 & 3 & 4 \\ 3 & 4 & 5 \end{vmatrix}$,

(c) $\begin{vmatrix} 1 & 1 & 1 \\ 1 & 2 & 3 \\ 1 & 3 & 6 \end{vmatrix}$,

(d) $\begin{vmatrix} 1 & 0 & 2 & 0 \\ 3 & 0 & -1 & 1 \\ 0 & 5 & 0 & -2 \\ 1 & 2 & -3 & 1 \end{vmatrix}$,

(e) $\begin{vmatrix} 2 & -1 & 3 & 1 \\ 0 & 2 & -2 & -1 \\ 4 & -2 & 1 & 0 \\ 0 & 2 & -7 & -3 \end{vmatrix}$,

(f) $\begin{vmatrix} 1 & 1 & 1 & 1 \\ 1 & 2 & 3 & 4 \\ 1 & 3 & 6 & 10 \\ 1 & 4 & 10 & 20 \end{vmatrix}$.

9. Find the signed volume or hypervolume of the parallelepiped that is spanned by

(a) $(2, 0, -1), (0, 1, 3), (-2, 5, 1)$;

(b) $(6, -1, 2), (2, 2, 0), (2, -5, 2)$;

(c) $(1, 3, 0, -2), (2, -5, 1, 0), (0, -1, 0, 3), (-1, 0, 2, -2)$.

10. Give a physical interpretation for the hypervolume of a parallelepiped in 4-dimensional space–time.

11. The *transpose* of a matrix is the matrix obtained by changing the rows into columns, the columns into rows. The transpose of

$$\begin{pmatrix} 1 & -1 & 0 \\ 2 & 4 & 1 \\ -1 & 0 & 2 \end{pmatrix}$$

is

$$\begin{pmatrix} 1 & 2 & -1 \\ -1 & 4 & 0 \\ 0 & 1 & 2 \end{pmatrix}.$$

In general, if

$$\mathbf{L} = (a_{ij})_{i,j=1}^{m},$$

then the transpose of $\mathbf{L}$, written $\mathbf{L}^t$, is

$$\mathbf{L}^t = (a_{ji})_{i,j=1}^{m}.$$

Compute the determinants of the transposes of the matrices in Exercise 8.

12. Prove that $\det(\mathbf{L}) = \det(\mathbf{L}^t)$ for any square matrix.

13. Prove that

$$\begin{vmatrix} 1 & 1 & 1 \\ a & b & c \\ a^2 & b^2 & c^2 \end{vmatrix} = (b-a)(c-a)(c-b).$$

14. Prove that

(a) $$\begin{vmatrix} 1 & 1 & 1 \\ a^2 & b^2 & c^2 \\ a^3 & b^3 & c^3 \end{vmatrix} = (b-a)(c-a)(c-b)(ab+ac+bc),$$

(b) $$\begin{vmatrix} 1 & 1 & 1 \\ a & b & c \\ a^4 & b^4 & c^4 \end{vmatrix} = (b-a)(c-a)(c-b)(a^2+b^2+c^2+ab+ac+bc),$$

(c) $$\begin{vmatrix} 1 & 1 & 1 & 1 \\ a & b & c & d \\ a^2 & b^2 & c^2 & d^2 \\ a^3 & b^3 & c^3 & d^3 \end{vmatrix} = (b-a)(c-a)(d-a)(c-b)(d-b)(d-c).$$

15. Show that the equation of the plane containing the three points (x_0, y_0, z_0), (x_1, y_1, z_1), and (x_2, y_2, z_2) is given by the equation

$$\begin{vmatrix} 1 & 1 & 1 & 1 \\ x & x_0 & x_1 & x_2 \\ y & y_0 & y_1 & y_2 \\ z & z_0 & z_1 & z_2 \end{vmatrix} = 0. \tag{6.27}$$

Explain why this works by using the definition of the determinant as signed hypervolume.

16. Prove that

$$\det(\vec{a}_1, \ldots, \vec{a}_i, \ldots, \vec{a}_k, \ldots, \vec{a}_m) = \det(\vec{a}_1, \ldots, \vec{a}_i, \ldots, \vec{a}_k + c\vec{a}_i, \ldots, \vec{a}_m). \tag{6.28}$$

In other words, you can add a constant multiple of any column to any other column without changing the value of the determinant.

17. For the determinant

$$\begin{vmatrix} 1 & -1 & 0 \\ 2 & 4 & 1 \\ -1 & 0 & 2 \end{vmatrix},$$

note that if we add $1/2$ times the third column to the first, then the third row of the first column becomes 0. Find an appropriate multiple of the second column to add to the new first column so that the second row of the first column becomes 0. Now, compute the determinant.

18. For the determinant

$$\begin{vmatrix} 1 & 0 & 2 & 0 \\ 3 & 0 & -1 & 1 \\ 0 & 5 & 0 & -2 \\ 1 & 2 & -3 & 1 \end{vmatrix},$$

add appropriate multiples of the fourth column to each of the first three columns so that they each have a 0 in the fourth row. Now, add appropriate multiples of the third column to each of the first two columns so that they each have a 0 in the third row. Finally, add an appropriate multiple of the second column to the first so that it now has a 0 in the second row. Now, compute the determinant.

19. The method illustrated in Exercises 17 and 18 is called *Gaussian elimination* (after Carl Friedrich Gauss). Prove that once we have nothing but 0's below the *main diagonal:* $a_{11}, a_{22}, \ldots, a_{mm}$, then the determinant is the product of the terms on the main diagonal.

6.5   Invertibility

Our goal in this section is to prove that a linear transformation,

$$\vec{L} : \mathbf{R}^m \longrightarrow \mathbf{R}^m,$$

is invertible if and only if the determinant of its matrix representation is not zero. The first step is to observe the not very surprising fact that the magnification produced by a composition of two transformations is the product of the individual magnifications.

Theorem 6.5 *Let* **L** *and* **M** *be* $m \times m$ *matrices, then*

$$\det(\mathbf{LM}) = \det(\mathbf{L}) \det(\mathbf{M}).$$

Proof: Let $\vec{b}_1, \vec{b}_2, \ldots, \vec{b}_m$ be the images of the unit basis vectors under $\vec{M}$:

$$\vec{b}_1 = \vec{M}(\vec{\imath}_1), \ldots, \vec{b}_m = \vec{M}(\vec{\imath}_m).$$

The image of $\vec{\imath}_j$ under $\vec{L} \circ \vec{M}$ is therefore

$$\vec{L}\left(\vec{M}(\vec{\imath}_j)\right) = \vec{L}(\vec{b}_j),$$

and so

$$\det(\mathbf{LM}) = \det\left(\vec{L}(\vec{b}_1), \vec{L}(\vec{b}_2), \ldots, \vec{L}(\vec{b}_m)\right).$$

As a function of $\vec{b}_1, \vec{b}_2, \ldots, \vec{b}_m$, this satisfies Equations (6.17) through (6.19), and so by Theorem 6.3,

$$
\begin{aligned}
\det\left(\vec{L}(\vec{b}_1), \ldots, \vec{L}(\vec{b}_m)\right) &= \det(\vec{b}_1, \ldots, \vec{b}_m) \det\left(\vec{L}(\vec{\imath}_1), \ldots, \vec{L}(\vec{\imath}_m)\right) \\
&= \det(\mathbf{M}) \det(\mathbf{L}).
\end{aligned}
$$

Q.E.D.

Corollary 6.3 *If* $\vec{L} : \mathbf{R}^m \longrightarrow \mathbf{R}^m$ *is invertible, then*

$$\det(\mathbf{L}) \neq 0.$$

Proof: Let $\vec{L}^{-1}$ be the inverse transformation, then

$$\mathbf{L}^{-1}\mathbf{L} = \mathbf{I},$$

and so

$$\det(\mathbf{L}^{-1}) \det(\mathbf{L}) = \det(\mathbf{I}) = 1.$$

Q.E.D.

Cofactors

As in the proof of Theorem 5.1, we shall prove that $\vec{L}$ is invertible when $\det(\mathbf{L}) \neq 0$ by actually constructing the inverse. Let $\mathbf{L}$ be an $m \times m$ matrix with entry a_{ij} in the ith row and jth column. It is common to write this as

$$\mathbf{L} = (a_{ij})_{i,j=1}^{m}.$$

As we proved in the previous section,

$$\det(\mathbf{L}) = \sum_{\sigma \in S_m} \operatorname{sgn}(\sigma) \prod_{j=1}^{m} a_{j\sigma(j)}.$$

This summation can be decomposed by separating our set S_m into subsets on which $\sigma^{-1}(1)$ is the same:

$$S_m = \{\sigma \in S_m \,|\, \sigma(1) = 1\} \cup \{\sigma \in S_m \,|\, \sigma(2) = 1\} \cup \cdots \cup \{\sigma \in S_m \,|\, \sigma(m) = 1\}.$$

We shall denote the set of permutations in which $\sigma(i) = 1$ by $S_{m\backslash i}$ so that our summation can be rewritten as

$$\det(\mathbf{L}) = \sum_{i=1}^{m} a_{i1} \sum_{\substack{\sigma \in S_{m\backslash i}}} \operatorname{sgn}(\sigma) \prod_{\substack{j=1 \\ j \neq i}}^{m} a_{j\sigma(j)}. \qquad (6.29)$$

There is a natural correspondence between the permutations $\sigma \in S_{m\backslash i}$ and $\tau \in S_{m-1}$ as follows:

$$\tau(j) = \begin{cases} \sigma(j) - 1, & \text{if } j < i, \\ \sigma(j+1) - 1, & \text{if } j \geq i. \end{cases}$$

$$\sigma(j) = \begin{cases} \tau(j) + 1, & \text{if } j < i, \\ 1, & \text{if } j = i, \\ \tau(j-1) + 1, & \text{if } j > i. \end{cases}$$

For example, we have the following correspondence when $i = 3$:

$$\left. \begin{array}{rcl} \sigma(1) &=& 4 \\ \sigma(2) &=& 3 \\ \sigma(3) &=& 1 \\ \sigma(4) &=& 5 \\ \sigma(5) &=& 2 \end{array} \right\} \longleftrightarrow \left\{ \begin{array}{rcl} \tau(1) &=& 3 \\ \tau(2) &=& 2 \\ \\ \tau(3) &=& 4 \\ \tau(4) &=& 1. \end{array} \right.$$

Since we have

$$\begin{aligned} dx_1 \cdots dx_m &= \operatorname{sgn}(\sigma)\, dx_{\sigma(1)} \cdots dx_{\sigma(m)} \\ &= (-1)^{i-1} \operatorname{sgn}(\sigma)\, dx_1\, dx_{\sigma(1)} \cdots dx_{\sigma(i-1)}\, dx_{\sigma(i+1)} \cdots dx_{\sigma(m)}, \end{aligned}$$

and

$$dx_1 \cdots dx_{m-1} = \operatorname{sgn}(\tau) \, dx_{\tau(1)} \cdots dx_{\tau(m-1)},$$

we see that

$$(-1)^{i-1} \operatorname{sgn}(\sigma) = \operatorname{sgn}(\tau). \tag{6.30}$$

Equation (6.29) can thus be rewritten as

$$\det(\mathbf{L}) = \sum_{i=1}^{m} (-1)^{i-1} a_{i1} \sum_{\tau \in S_{m-1}} \operatorname{sgn}(\tau) \prod_{\substack{j=1 \\ j \neq i}}^{m} a_{j\sigma(j)}. \tag{6.31}$$

The sum over $\tau \in S_{m-1}$ is the determinant of the $(m-1) \times (m-1)$ matrix obtained by deleting the ith row and 1st column from $\mathbf{L}$. We call this smaller matrix the $i1$ *minor of* $\mathbf{L}$ and denote it by $\mathbf{L}_{i1}$. In general, the ik *minor of* $\mathbf{L}$, denoted $\mathbf{L}_{ik}$, is the matrix obtained from $\mathbf{L}$ by deleting the ith row and the kth column. In this notation, Equation (6.29) looks like

$$\det(\mathbf{L}) = \sum_{i=1}^{m} (-1)^{i-1} a_{i1} \det(\mathbf{L}_{i1}). \tag{6.32}$$

In general, if we take the kth column instead of the first, we have the relationship

$$\det(\mathbf{L}) = \sum_{i=1}^{m} (-1)^{i-k} a_{ik} \det(\mathbf{L}_{ik}). \tag{6.33}$$

As an example, taking $k = 2$, we have

$$\begin{vmatrix} 2 & 0 & -1 \\ 3 & -2 & 1 \\ 5 & -3 & 4 \end{vmatrix}$$

$$= -0 \begin{vmatrix} 3 & 1 \\ 5 & 4 \end{vmatrix} + (-2) \begin{vmatrix} 2 & -1 \\ 5 & 4 \end{vmatrix} - (-3) \begin{vmatrix} 2 & -1 \\ 3 & 1 \end{vmatrix}$$

$$= 0(7) - 2(13) + 3(5) = -11.$$

We define the ik *cofactor of* $\mathbf{L}$, denoted A_{ki}, to be

$$A_{ki} = (-1)^{i-k} \det(\mathbf{L}_{ik}). \tag{6.34}$$

Note that we have switched the order of the subscripts in defining A_{ki}. Our Equation (6.33) now looks even simpler:

$$\det(\mathbf{L}) = \sum_{i=1}^{m} A_{ki} a_{ik}. \tag{6.35}$$

If we let

$$\vec{a}_k = (a_{1k}, a_{2k}, \ldots, a_{mk})$$

denote the kth column of $\mathbf{L}$ and

$$\vec{A}_k = (A_{k1}, A_{k2}, \ldots, A_{km}),$$

then Equation (6.33) can be stated even more simply:

$$\det(\mathbf{L}) = \vec{A}_k \cdot \vec{a}_k. \tag{6.36}$$

Note that if

$$\mathbf{A} = (A_{ki})_{k,i=1}^m,$$

then $\vec{A}_k$ is the kth *row* of $\mathbf{A}$.

What if we tried taking the dot product of $\vec{a}_j$ with $\vec{A}_k$ where $j \neq k$? The vector $\vec{A}_k$ depends on everything in $\mathbf{L}$ *except* the kth column: to compute A_{k1}, we delete the 1st row and the kth column; to compute A_{k2}, we delete the 2nd row and the kth column; to compute A_{km}, we delete the mth row and the kth column. The effect of multiplying $\vec{A}_k \cdot \vec{a}_j$ is therefore the same as computing the determinant of the matrix obtained from $\mathbf{A}$ by replacing the kth column with a second copy of the jth column. This means that we are taking the determinant of a matrix in which one of the columns is repeated. By Equation (6.19), this determinant is 0:

$$\vec{A}_k \cdot \vec{a}_j = 0, \quad j \neq k. \tag{6.37}$$

The Inverse Matrix

Theorem 6.6 *If $\mathbf{L}$ is an $m \times m$ matrix and*

$$\delta = \det(\mathbf{L}) \neq 0,$$

then $\mathbf{L}$ has an inverse:

$$\mathbf{L}^{-1} = (A_{ki}/\delta)_{k,i=1}^m. \tag{6.38}$$

Proof: The entry in the kth row and lth column of $\mathbf{L}^{-1}\mathbf{L}$ is the dot product of the kth row of $\mathbf{L}^{-1}$ with the lth column of $\mathbf{L}$:

$$\frac{1}{\delta}\vec{A}_k \cdot \vec{a}_l = \begin{cases} \delta/\delta = 1, & \text{if } k = l, \\ 0/\delta = 0, & \text{if } k \neq l. \end{cases}$$

Thus, we have

$$\mathbf{L}^{-1}\mathbf{L} = \mathbf{I}.$$

Q.E.D.

Example

Taking **L** to be

$$\begin{pmatrix} 2 & 0 & -1 \\ 3 & -2 & 1 \\ 5 & -3 & 4 \end{pmatrix},$$

the matrix of cofactors is

$$\begin{pmatrix} \begin{vmatrix} -2 & 1 \\ -3 & 4 \end{vmatrix} & -\begin{vmatrix} 0 & -1 \\ -3 & 4 \end{vmatrix} & \begin{vmatrix} 0 & -1 \\ -2 & 1 \end{vmatrix} \\[2mm] -\begin{vmatrix} 3 & 1 \\ 5 & 4 \end{vmatrix} & \begin{vmatrix} 2 & -1 \\ 5 & 4 \end{vmatrix} & -\begin{vmatrix} 2 & -1 \\ 3 & 1 \end{vmatrix} \\[2mm] \begin{vmatrix} 3 & -2 \\ 5 & -3 \end{vmatrix} & -\begin{vmatrix} 2 & 0 \\ 5 & -3 \end{vmatrix} & \begin{vmatrix} 2 & 0 \\ 3 & -2 \end{vmatrix} \end{pmatrix}$$

$$= \begin{pmatrix} -5 & 3 & -2 \\ -7 & 13 & -5 \\ 1 & 6 & -4 \end{pmatrix}.$$

If we take any row of this cofactor matrix and take its dot product with the corresponding column of the original matrix, we always get the determinant:

$$\begin{aligned} (-5,3,-2) \cdot (2,3,5) &= -10 + 9 - 10 = -11, \\ (-7,13,-5) \cdot (0,-2,-3) &= 0 - 26 + 15 = -11, \\ (1,6,-4) \cdot (-1,1,4) &= -1 + 6 - 16 = -11. \end{aligned}$$

Since the determinant of **L** is -11, the inverse of **L** is given by

$$\mathbf{L}^{-1} = \begin{pmatrix} 5/11 & -3/11 & 2/11 \\ 7/11 & -13/11 & 5/11 \\ -1/11 & -6/11 & 4/11 \end{pmatrix}.$$

Cramer's Rule

Cramer's rule arises from our treatment of inverses. It yields the solution of a system of m equations in m unknowns:

$$\begin{aligned} a_{11}x_1 + a_{12}x_2 + \ldots + a_{1m}x_m &= b_1, \\ a_{21}x_1 + a_{22}x_2 + \ldots + a_{2m}x_m &= b_2, \\ &\vdots \\ a_{m1}x_1 + a_{m2}x_2 + \ldots + a_{mm}x_m &= b_m, \end{aligned} \qquad (6.39)$$

when $\det(a_{ij})_{i,j=1}^{m}$ is not zero.

If we let $\mathbf{L}$ be the matrix of coefficients and

$$\mathbf{x} = \begin{pmatrix} x_1 \\ x_2 \\ \vdots \\ x_m \end{pmatrix}, \quad \mathbf{b} = \begin{pmatrix} b_1 \\ b_2 \\ \vdots \\ b_m \end{pmatrix},$$

then our system of equations (6.39) can be written as

$$\mathbf{b} = \mathbf{Lx}.$$

If

$$\delta = \det(\mathbf{L}) \neq 0$$

and $\mathbf{A}$ is the matrix of cofactors,

$$\mathbf{A} = \begin{pmatrix} A_{11} & A_{12} & \cdots & A_{1m} \\ A_{21} & A_{22} & \cdots & A_{2m} \\ \vdots & & & \\ A_{m1} & A_{m2} & \cdots & A_{mm} \end{pmatrix},$$

so that the inverse of $\mathbf{L}$ can be written as

$$\mathbf{L}^{-1} = \begin{pmatrix} A_{11}/\delta & A_{12}/\delta & \cdots & A_{1m}/\delta \\ A_{21}/\delta & A_{22}/\delta & \cdots & A_{2m}/\delta \\ \vdots & & & \\ A_{m1}/\delta & A_{m2}/\delta & \cdots & A_{mm}/\delta \end{pmatrix},$$

then we have

$$\begin{pmatrix} x_1 \\ x_2 \\ \vdots \\ x_m \end{pmatrix} = \mathbf{L}^{-1}\mathbf{b} = \begin{pmatrix} A_{11}/\delta & A_{12}/\delta & \cdots & A_{1m}/\delta \\ A_{21}/\delta & A_{22}/\delta & \cdots & A_{2m}/\delta \\ \vdots & & & \\ A_{m1}/\delta & A_{m2}/\delta & \cdots & A_{mm}/\delta \end{pmatrix} \begin{pmatrix} b_1 \\ b_2 \\ \vdots \\ b_m \end{pmatrix}.$$

From this, it follows that

$$\begin{aligned} x_k &= \frac{1}{\delta} \left(A_{k1}b_1 + A_{k2}b_2 + \cdots + A_{km}b_m \right) \\ &= \frac{1}{\delta} \vec{A}_k \cdot \vec{b}. \end{aligned}$$

But, $\vec{A}_k \cdot \vec{b}$ is the determinant of the matrix obtained by replacing the kth column of $\mathbf{L}$ by $\mathbf{b}$.

Cramer's rule states what we have just proven, namely, if we have a system of m equations in m unknowns for which the matrix of coefficients has a nonzero determinant, then there is a unique solution and the kth variable is found by taking the ratio whose numerator is the determinant of the matrix of coefficients with the kth column replaced by the column of constants and whose denominator is the determinant of the matrix of coefficients.

Example

To solve the following system of equations for x_2:

$$
\begin{array}{rrrrrr}
3x_1 & - & 5x_2 & & + & x_4 & = & 1, \\
x_1 & & & - 2x_3 & + & 4x_4 & = & 0, \\
2x_1 & + & x_2 & - x_3 & + & 3x_4 & = & -1, \\
x_1 & - & 2x_2 & & - & x_4 & = & 1,
\end{array}
$$

we compute the ratio of determinants:

$$
x_2 = \frac{\begin{vmatrix} 3 & 1 & 0 & 1 \\ 1 & 0 & -2 & 4 \\ 2 & -1 & -1 & 3 \\ 1 & 1 & 0 & -1 \end{vmatrix}}{\begin{vmatrix} 3 & -5 & 0 & 1 \\ 1 & 0 & -2 & 4 \\ 2 & 1 & -1 & 3 \\ 1 & -2 & 0 & -1 \end{vmatrix}} = \frac{10}{-27}.
$$

6.6 Exercises

1. Use the fact that multiplication of matrices is associative to prove that if $\mathbf{A}$, $\mathbf{B}$, and $\mathbf{C}$ are square matrices and $\mathbf{AB} = \mathbf{I} = \mathbf{CA}$, then $\mathbf{B} = \mathbf{C}$. This implies that once we have found an inverse that works on one side of $\mathbf{A}$ it must also work on the other side.

2. If the determinant of an $n \times n$ matrix $\mathbf{L}$ is $\delta \neq 0$, what is the determinant of $\mathbf{A}$, the matrix of its cofactors?

3. Let $\mathbf{L}$ be a matrix whose determinant is zero and let $\mathbf{A}$ be its matrix of cofactors. What is the product $\mathbf{AL}$?

4. Find the matrix of cofactors for each of the following matrices:

   (a) $\begin{pmatrix} 2 & 3 \\ 1 & -2 \end{pmatrix}$;

   (b) $\begin{pmatrix} 3 & 0 & 1 \\ -2 & 1 & 4 \\ 2 & 5 & 0 \end{pmatrix}$;

(c) $\begin{pmatrix} 1 & 2 & 3 \\ 2 & 3 & 4 \\ 3 & 4 & 5 \end{pmatrix}$;

(d) $\begin{pmatrix} 1 & 1 & 1 \\ 1 & 2 & 3 \\ 1 & 3 & 6 \end{pmatrix}$;

(e) $\begin{pmatrix} 1 & 0 & 1 & 1 \\ 1 & 1 & 0 & 1 \\ 0 & 1 & 1 & 1 \\ 1 & 1 & 1 & 0 \end{pmatrix}$.

5. Find the inverse of each matrix of Exercise 4 that is invertible.

6. Use Cramer's rule to solve each of the following systems of equations for x, y, and z.

(a)

$$\begin{aligned} x + 3y - 5z &= 1, \\ 2x - y + 3z &= 0, \\ 5x + 2y - z &= 2. \end{aligned}$$

(b)

$$\begin{aligned} 3x - y + z &= 3, \\ x + 2y - z &= -2, \\ 4x - 3y + 3z &= 1. \end{aligned}$$

7

Differential Calculus

7.1 Limits

We recall that a vector field, $\vec{F}$, is differentiable at $\vec{c}$ if and only if there exists a linear transformation, $\vec{L}_c$, such that

$$\vec{F}(\vec{x}) - \vec{F}(\vec{c}) \;=\; \vec{L}_c(\vec{x} - \vec{c}) \;+\; |\vec{x} - \vec{c}|\, \vec{E}(\vec{c}, \vec{x} - \vec{c}), \qquad (7.1)$$

where

$$\lim_{\vec{x} \to \vec{c}} \vec{E}(\vec{c}, \vec{x} - \vec{c}) = \vec{0}.$$

We first need to define what we mean by this limit.

The notion of a limit in one dimension is fairly intuitive:

$$\lim_{x \to c} f(x) = l$$

means that as x gets closer to c, $f(x)$ gets closer to l. There are no problems because we know what it means for x to be "close" to c: it is either just a little bit larger or just a little bit smaller. The problems arise in higher dimensions, where we have infinitely many different paths by which $\vec{x}$ can approach $\vec{c}$. It could come in along any of the straight lines passing through $\vec{c}$, or it could approach along some curving trajectory, even spiraling in to $\vec{c}$. We want a definition of limit so that

$$\lim_{\vec{x} \to \vec{c}} \vec{F}(\vec{x}) = \vec{l}$$

means that along *any* path approaching $\vec{c}$, as $\vec{x}$ gets closer to $\vec{c}$, $\vec{F}(\vec{x})$ gets closer to $\vec{l}$.

We define the multidimensional limit in terms of the 1-dimensional limit:

$$\lim_{\vec{x} \to \vec{c}} \vec{F}(\vec{x}) = \vec{l}$$

means that

$$\lim_{|\vec{x} - \vec{c}| \to 0} |\vec{F}(\vec{x}) - \vec{l}| = 0,$$

where $|\vec{x} - \vec{c}|$ is the distance from $\vec{x}$ to $\vec{c}$:

$$|\vec{x} - \vec{c}|^2 = (\vec{x} - \vec{c}) \cdot (\vec{x} - \vec{c}) = (x_1 - c_1)^2 + (x_2 - c_2)^2 + \cdots + (x_n - c_n)^2,$$

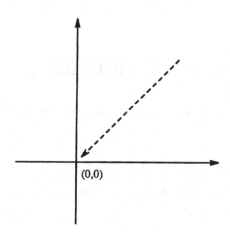

FIGURE 7.1. Approaching (0,0) from the northeast.

$$|\vec{x} - \vec{c}| = \sqrt{(x_1 - c_1)^2 + (x_2 - c_2)^2 + \cdots + (x_n - c_n)^2}.$$

Equivalently, this says that we can force $\vec{F}(\vec{x})$ to be as close as we want to $\vec{l}$ simply by keeping $\vec{x}$ sufficiently close to $\vec{c}$. In the language of ϵ and δ: given any $\epsilon > 0$ (how close we want $\vec{F}(\vec{x})$ to be to $\vec{l}$), there exists a δ (how close $\vec{x}$ has to be to $\vec{c}$) such that $0 < |\vec{x} - \vec{c}| < \delta$ implies that $|\vec{F}(\vec{x}) - \vec{l}| < \epsilon$.

Multidimensional limits can do unexpected things. Consider the scalar field defined on all of $\mathbf{R}^2$ except the origin, $(0,0)$:

$$F(x, y) = \frac{xy}{x^2 + y^2}. \tag{7.2}$$

If we approach $(0,0)$ along either the x or y axis, F is 0 and so the limit is also 0. But, if we come down from the northeast along the line $y = x$ (Figure 7.1), then

$$F(x, x) = \frac{x^2}{x^2 + x^2} = \frac{1}{2},$$

and so the limit is 1/2. No matter how tight a circle we draw around the origin, we can find points inside that circle where F is 0, and also points where F is 1/2. In fact, on the line $y = mx$,

$$F(x, mx) = \frac{mx^2}{x^2 + m^2 x^2} = \frac{m}{1 + m^2},$$

and so no matter how close we are to $(0,0)$, F takes on all possible values in $[-1/2, 1/2]$. This implies that F cannot have a limit as (x, y) approaches $(0,0)$.

As a more rigorous construction: both $(\delta/2, 0)$ and $(\delta/2, \delta/2)$ lie inside the circle with its center at the origin and radius δ. But, as we have shown,

$$F\left(\frac{\delta}{2}, 0\right) = 0, \quad F\left(\frac{\delta}{2}, \frac{\delta}{2}\right) = \frac{1}{2}.$$

If we are given $\epsilon = 1/4$, there is no limit l and positive δ such that

$$\left| F\left(\frac{\delta}{2}, 0\right) - l \right| < \frac{1}{4} \quad \text{and} \quad \left| F\left(\frac{\delta}{2}, \frac{\delta}{2}\right) - l \right| < \frac{1}{4}.$$

This example also provides an illustration of the fact that a vector limit is *not* the same as an iterated limit:

$$\lim_{x \to 0} \left(\lim_{y \to 0} \frac{xy}{x^2 + y^2} \right) = \lim_{x \to 0} \frac{0}{x^2} = 0,$$

$$\lim_{y \to 0} \left(\lim_{x \to 0} \frac{xy}{x^2 + y^2} \right) = \lim_{y \to 0} \frac{0}{y^2} = 0,$$

but

$$\lim_{(x,y) \to (0,0)} \frac{xy}{x^2 + y^2}$$

does not exist!

It gets even less intuitive. Consider next the scalar field defined on all of $\mathbf{R}^2$ except $(0,0)$:

$$G(x, y) = \frac{x^2 y}{x^4 + y^2}. \tag{7.3}$$

Again, G is 0 everywhere on the x and y axes. If we take any other line through the origin, $y = mx, m \neq 0$, then

$$\lim_{x \to 0} G(x, mx) = \lim_{x \to 0} \frac{mx^3}{x^4 + m^2 x^2}$$

$$= \lim_{x \to 0} \frac{mx}{x^2 + m^2} = 0.$$

It appears that surely in this instance, the limit must exist (and be 0). However, if we approach the origin along the parabola $y = x^2$, then

$$G(x, x^2) = \frac{x^4}{x^4 + x^4} = \frac{1}{2},$$

and so we can find points arbitrarily close to the origin for which G is 1/2. Again,

$$\lim_{(x,y) \to (0,0)} G(x, y)$$

does not exist.

Level Curves

To understand what was happening in the last two examples, it helps to be able to visualize the functions. For a scalar field $F : \mathbf{R}^2 \longrightarrow \mathbf{R}$, this can be done very effectively by using *level curves*. We picture the points of the domain, (x, y), as position in a horizontal plane. The function $F(x, y)$ can be thought of as height above or below this plane, giving rise to a 3-dimensional surface of hills and valleys. For the graph of this function, we create a topographic map, tracing out the lines of constant height. That is, we choose several different values of c from the range of F, and for each one we graph the equation

$$F(x, y) = c.$$

For the function F defined in Equation (7.2), we see that

$$\frac{xy}{x^2 + y^2} = 0$$

if and only if x or y is 0. Thus the *level curve of height* 0 consists of the x and y axes. In general, if

$$\frac{xy}{x^2 + y^2} = c,$$

then we can solve for y in terms of x:

$$cy^2 - xy + cx^2 = 0,$$
$$y = \frac{1 \pm \sqrt{1 - 4c^2}}{2c} x.$$

The level curves are pairs of straight lines passing through the origin (Figure 7.2). A 3-dimensional representation of this function is given in Figure 7.3.

For the function G defined in Equation (7.3), $G(x, y) = c$, $c \neq 0$, is equivalent to the equation

$$y = \frac{1 \pm \sqrt{1 - 4c^2}}{2c} x^2.$$

The level curves are pairs of parabolas passing through the origin (Figure 7.4). A 3-dimensional representation of this function is given in Figure 7.5.

The problem with these functions is now apparent. All of our level curves are converging at the origin. Topographically, the surface described by $G(x, y)$ has a ridge of constant height following the curve $y = x^2$. It has a valley of uniform depth along the curve $y = -x^2$. As we get closer to the origin, the slope between the two is progressively steeper until at the origin itself we have a vertical cliff running all the way from the lowest to the highest points of our terrain.

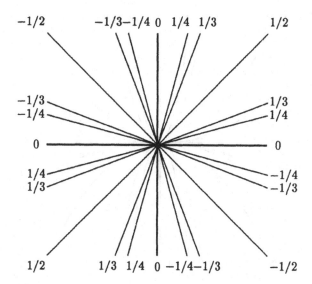

FIGURE 7.2. The level curves of $F(x, y) = xy/(x^2 + y^2)$.

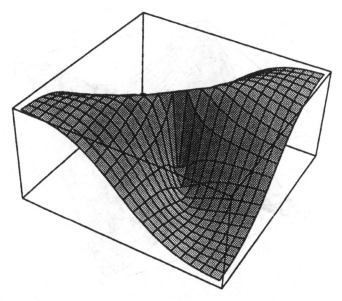

FIGURE 7.3. A 3-dimensional representation of $F(x, y) = xy/(x^2 + y^2)$.

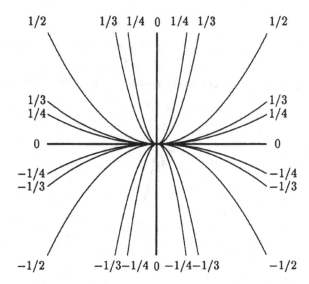

FIGURE 7.4. The level curves of $G(x,y) = x^2y/(x^4 + y^2)$.

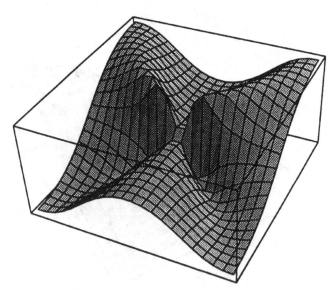

FIGURE 7.5. A 3-dimensional representation of $F(x,y) = x^2y/(x^4 + y^2)$.

Continuity

If a scalar field is to be called *continuous*, it must not possess such cliffs. More precisely, a function $\vec{F}(\vec{x})$, which can be either a scalar or a vector field, is continuous at $\vec{c}$ if and only if

$$\lim_{\vec{x} \to \vec{c}} \vec{F}(\vec{x}) = \vec{F}(\vec{c}).$$

For functions of a single variable, differentiability implies continuity. The same is true in higher dimensions.

Theorem 7.1 *If $\vec{F}(\vec{x})$ is differentiable at $\vec{x} = \vec{c}$, then it is also continuous at $\vec{c}$.*

Proof: From the definition of differentiability, we have a linear transformation $\vec{L}_c$ such that

$$\lim_{\vec{x} \to \vec{c}} \left(\vec{F}(\vec{x}) - \vec{F}(\vec{c}) \right) = \lim_{\vec{x} \to \vec{c}} \left(\vec{L}_c(\vec{x} - \vec{c}) + |\vec{x} - \vec{c}| \, \vec{E}(\vec{c}, \vec{x} - \vec{c}) \right)$$
$$= \vec{L}_c(\vec{0}) + \vec{0} = \vec{0}.$$

Q.E.D.

As a corollary to this theorem, there is no possible value at the origin for the functions defined in Equations (7.2) and (7.3) that will make them differentiable at the origin.

The Good News

If we have spent this much effort on scalar fields from $\mathbf{R}^2$ to $\mathbf{R}$, what must the complications be like as we move to higher dimensions? The good news is that for the material to be covered in this book, it does not get any worse. First of all, a vector field is just an ordered set of scalar fields. If $\vec{F} : \mathbf{R}^n \longrightarrow \mathbf{R}^m$, then there are m scalar fields such that

$$\vec{F}(\vec{x}) = (f_1(\vec{x}), f_2(\vec{x}), \ldots, f_m(\vec{x})).$$

For this reason, most of our attention in this chapter will be directed toward scalar fields. Once we understand their derivatives, we can piece them together to form the derivative of the vector field.

Second, aside from problems of visualization, a scalar field in n variables is no more complicated than a scalar field in 2 variables. The problems that we shall be confronting arise as we move from 1 to 2 variables. Once we have found how to deal with them in $\mathbf{R}^2$, the solutions will carry into higher dimensions without much difficulty.

One way of visualizing a scalar field defined on $\mathbf{R}^3$ is to consider the *level surfaces*:

$$F(x, y, z) = c.$$

If we think of the scalar field as describing temperature at the point (x, y, z), then the level surfaces correspond to *isotherms*, sheets of points sharing the same temperature.

7.2 Exercises

1. For the scalar field $G(x, y)$ defined in Equation (7.3), prove that there is no limit l and no $\delta > 0$ such that $|(x, y)| < \delta$ implies that $|G(x, y) - l| < 1/4$.

2. Let f and g be scalar fields for which

$$\lim_{\vec{x} \to \vec{c}} f(\vec{x}) = l, \quad \lim_{\vec{x} \to \vec{c}} g(\vec{x}) = m,$$

and let λ be a real number. Using the definition of multidimensional limits in terms of 1-dimensional limits, prove that

(a) $\lim_{\vec{x} \to \vec{c}} \lambda f(\vec{x}) = \lambda l$,

(b) $\lim_{\vec{x} \to \vec{c}} f(\vec{x}) + g(\vec{x}) = l + m$,

(c) $\lim_{\vec{x} \to \vec{c}} f(\vec{x}) g(\vec{x}) = lm$.

Hint: for parts b and c, use the triangle inequality:

$$|f(\vec{x}) + g(\vec{x}) - l - m| \leq |f(\vec{x}) - l| + |g(\vec{x}) - m|,$$
$$|f(\vec{x}) g(\vec{x}) - lm| \leq |f(\vec{x}) g(\vec{x}) - lg(\vec{x})| + |lg(\vec{x}) - lm|.$$

3. Prove that if $|f(\vec{x})|$ is bounded for all $\vec{x}$ sufficiently close to the origin (there is some bound, M, and some $\delta > 0$ such that if $|\vec{x}| < \delta$, then $|f(\vec{x})| < M$) and if

$$\lim_{\vec{x} \to \vec{0}} g(\vec{x}) = 0,$$

then

$$\lim_{\vec{x} \to \vec{0}} f(\vec{x}) g(\vec{x}) = 0.$$

4. Prove that

$$\lim_{(x,y) \to (1,0)} \frac{xy}{x^2 + y^2} = 0,$$

by showing that given some $\epsilon > 0$ you can find an appropriate δ (δ as a function of ϵ) so that

$$\sqrt{(x-1)^2 + y^2} < \delta$$

implies

$$\left|\frac{xy}{x^2 + y^2}\right| < \epsilon.$$

(a) Show that if $|(x, y) - (1, 0)| < \delta$, then $-\delta < y < \delta$, $1 - \delta < x < 1 + \delta$.

(b) Using these inequalities, show that if $0 < \delta < 1$, then

$$\left|\frac{xy}{x^2 + y^2}\right| < \frac{\delta(1 + \delta)}{(1 - \delta)^2}.$$

(c) Use part b to find δ as a function of ϵ.

5. Prove that the order of iteration of limits can be significant by showing that

$$\lim_{x \to 0}\left(\lim_{y \to 0}\frac{x^2}{x^2 + y^2}\right) \neq \lim_{y \to 0}\left(\lim_{x \to 0}\frac{x^2}{x^2 + y^2}\right).$$

6. Show that if the iterated limits exist and are not equal:

$$\lim_{x \to 0}\left(\lim_{y \to 0} f(x, y)\right) \neq \lim_{y \to 0}\left(\lim_{x \to 0} f(x, y)\right),$$

then

$$\lim_{(x,y) \to (0,0)} f(x, y)$$

cannot exist.

7. Let

$$f(x, y) = \begin{cases} x \sin \frac{1}{y} & \text{if} \quad y \neq 0, \\ 0 & \text{if} \quad y = 0. \end{cases}$$

Show that

$$\lim_{(x,y) \to (0,0)} f(x, y) = 0,$$

but that

$$\lim_{x \to 0}\left(\lim_{y \to 0} f(x, y)\right) \neq \lim_{y \to 0}\left(\lim_{x \to 0} f(x, y)\right).$$

Explain what is happening. Why does this not contradict Exercise 6?

8. Graph enough level curves for each of the following scalar fields so that you have a picture of the surface:

(a) $f(x, y) = xy$,

(b) $f(x, y) = x^2 - y^2$,

(c) $f(x, y) = 3x - 2y$,

    (d) $f(x,y) = y - \sin x$,
    (e) $f(x,y) = (x-y)/(x+y)$,
    (f) $f(x,y) = (x^2 - y^2)/(x^2 + y^2)$,
    (g) $f(x,y) = 1/(x^2 + y^2 + 1)$,
    (h) $f(x,y) = \sin(x^2 + y^2)/(x^2 + y^2)$.

9. Describe the level surfaces at $c = 0, 1, 4, 9$ for each of the following
   scalar fields:

    (a) $f(x,y,z) = x^2 + y^2 + z^2$,
    (b) $f(x,y,z) = x^2 + y^2 - z$,
    (c) $f(x,y,z) = z/(x^2 + y^2)$, $(x,y) \neq (0,0)$.

10. Let
$$f(x,y) = \frac{\sin(x^2 + y^2)}{x^2 + y^2}, \quad (x,y) \neq (0,0).$$
Define $f(0,0)$ so that $f(x,y)$ is continuous at (0,0).

11. Show that
$$f(x,y) = \begin{cases} \frac{x^2 - y^2}{x^2 + y^2}, & (x,y) \neq (0,0) \\ 0, & (x,y) = (0,0) \end{cases}$$
is not continuous at (0,0).

12. Let
$$f(x,y) = \frac{x-y}{x+y}, \quad x+y \neq 0.$$
Can this function be defined on the line $x + y = 0$ so that it is
continuous at some point on this line?

13. Is
$$f(x,y) = \begin{cases} \frac{x^2 y^2}{x^2 + y^2}, & (x,y) \neq (0,0) \\ 0, & (x,y) = (0,0) \end{cases}$$
continuous at (0,0)? Justify your answer.

14. Show that if f is constant, then $f(\vec{x}) = c$ is differentiable.

15. Show that $f(x_1, x_2, \ldots, x_n) = x_1$ and $f(\vec{x}) = |\vec{x}|$ are both differen-
    tiable functions, provided $|\vec{x}| \neq 0$.

16. Show that the sum of differentiable functions is always a differentiable
    function.

17. Show that the product of differentiable functions is always a differ-
    entiable function.

18. Show that the quotient of differentiable functions is always a differ-
    entiable function provided the denominator is not zero.

7.3 Directional Derivatives

Having cleared up our definition of differentiability, we are now ready to turn our attention to the key element of that definition, the linear transformation $\vec{L}_c$. How do we know if it exists? How do we find it?

Our starting point is the concept of the derivative as the rate of change. For real-valued functions of a single variable, the derivative tracks the rate at which the value of the function is changing as we move along the x axis from left to right. For a scalar field, $f : \mathbf{R}^2 \longrightarrow \mathbf{R}$, which we model as a surface in three dimensions, we can ask for a description of the rate at which the height of the surface changes. But, a complication arises. When the domain was $\mathbf{R}$, we had only two directions in which we could change the independent variable: left to right or right to left. Since reversing direction reverses the sign on the rate of change, it is common to consider only what happens as we move from left to right.

But now, as we stand on the surface describing our scalar field f, we have infinitely many directions in which we may move. Imagine we are standing on the side of a hill. We can take a step toward or away from the summit, but we could equally well walk around the hill staying at a constant height or head obliquely up the slope. The rate at which the height is changing depends on the direction in which we choose to move. This leads to the following definition.

Definition: Let $\vec{c}$ be a point in the domain of the scalar field $f : \mathbf{R}^n \longrightarrow \mathbf{R}$ and let $\vec{u}$ be a unit vector ($|\vec{u}| = 1$) specifying a direction. The *directional derivative of f at $\vec{c}$ in the direction $\vec{u}$*, denoted $f'(\vec{c}; \vec{u})$, is defined to be the limit

$$f'(\vec{c}; \vec{u}) = \lim_{h \to 0} \frac{f(\vec{c} + h\vec{u}) - f(\vec{c})}{h}. \tag{7.4}$$

Example

As an example, consider

$$f(x, y) = x^2 - y^2,$$

and let $\vec{c} = (1, 0), \vec{u}_1 = (1, 0), \vec{u}_2 = (\sqrt{2}/2, \sqrt{2}/2), \vec{u}_3 = (0, 1), \vec{u}_4 = (-1, 0)$ (see Figure 7.6):

$$f'(\vec{c}; \vec{u}_1) = \lim_{h \to 0} \frac{f(1 + h, 0) - f(1, 0)}{h} = \lim_{h \to 0} \frac{2h + h^2}{h} = 2,$$

$$f'(\vec{c}; \vec{u}_2) = \lim_{h \to 0} \frac{f(1 + h\sqrt{2}/2, h\sqrt{2}/2) - f(1, 0)}{h} = \lim_{h \to 0} \frac{h\sqrt{2}}{h} = \sqrt{2},$$

$$f'(\vec{c}; \vec{u}_3) = \lim_{h \to 0} \frac{f(1, h) - f(1, 0)}{h} = \lim_{h \to 0} \frac{-h^2}{h} = 0,$$

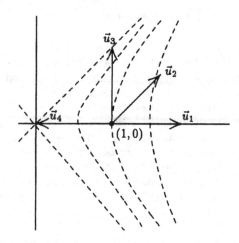

FIGURE 7.6. Directional derivatives on $f(x, y) = x^2 - y^2$.

$$f'(\vec{c}; \vec{u}_4) = \lim_{h \to 0} \frac{f(1 - h, 0) - f(1, 0)}{h} = \lim_{h \to 0} \frac{-2h + h^2}{h} = -2.$$

Note the importance of using the *unit vector* in the desired direction. If we replace $\vec{u}_2$ by the more convenient vector $(1,1)$, which points in the same direction, we get a different result:

$$f'(\vec{c}; (1,1)) = \lim_{h \to 0} \frac{f(1 + h, h) - f(1, 0)}{h} = \lim_{h \to 0} \frac{2h}{h} = 2.$$

Reversing Direction

Proposition 7.1 *Reversing direction changes the sign of the directional derivative:*

$$f'(\vec{c}; -\vec{u}) = -f'(\vec{c}; \vec{u}).$$

Proof: We substitute $-k$ for h in the definition of the directional derivative:

$$
\begin{aligned}
f'(\vec{c}; -\vec{u}) &= \lim_{h \to 0} \frac{f(\vec{c} - h\vec{u}) - f(\vec{c})}{h} \\
&= \lim_{k \to 0} \frac{f(\vec{c} + k\vec{u}) - f(\vec{c})}{-k} \\
&= -\lim_{k \to 0} \frac{f(\vec{c} + k\vec{u}) - f(\vec{c})}{k} = -f'(\vec{c}; \vec{u}).
\end{aligned}
$$

Q.E.D.

Partial Derivatives

The *partial derivatives* of the scalar field f are the directional derivatives in the directions of the unit basis vectors. For example, if $f : \mathbf{R}^3 \longrightarrow \mathbf{R}$, then the *partial of f with respect to x* (the first variable) or *with respect to $\vec{\imath}$* (the first basis vector) is

$$f'(x, y, z; \vec{\imath}) = \lim_{h \to 0} \frac{f(x + h, y, z) - f(x, y, z)}{h}.$$

This is denoted by

$$\frac{\partial f}{\partial x}, \quad D_1 f, \quad \text{or} \quad f_x.$$

Similarly, we have

$$\frac{\partial f}{\partial y} = D_2 f = f'(x, y, z; \vec{\jmath}) = \lim_{h \to 0} \frac{f(x, y + h, z) - f(x, y, z)}{h},$$

$$\frac{\partial f}{\partial z} = D_3 f = f'(x, y, z; \vec{k}) = \lim_{h \to 0} \frac{f(x, y, z + h) - f(x, y, z)}{h}.$$

In general, if $f : \mathbf{R}^n \longrightarrow \mathbf{R}$, then

$$\frac{\partial f}{\partial x_j} = D_j f = f'(\vec{x}; \vec{\imath}_j) = \lim_{h \to 0} \frac{f(\vec{x} + h\vec{\imath}_j) - f(\vec{x})}{h}.$$

In practice, $D_j f$ is computed by performing an ordinary differentiation with respect to x_j, treating all other variables as constants, which is what they are. The other variables do not vary; they remain constant when taking this partial derivative.

Example

$$f(x, y, z) = \frac{x^2 + y}{y^2 + z^2}, \quad (y, z) \neq (0, 0),$$

$$\frac{\partial f}{\partial x} = \frac{2x}{y^2 + z^2},$$

$$\frac{\partial f}{\partial y} = \frac{(y^2 + z^2) - (x^2 + y)(2y)}{(y^2 + z^2)^2} = \frac{z^2 - 2x^2 y - y^2}{(y^2 + z^2)^2},$$

$$\frac{\partial f}{\partial z} = \frac{-(x^2 + y)(2z)}{(y^2 + z^2)^2} = \frac{-2x^2 z - 2yz}{(y^2 + z^2)^2}.$$

Warning

Although partial derivatives will play an important role in defining $\vec{L}_c$, their existence means less than differentiability. Consider the scalar field of Equation (7.2) which we shall now define to be 0 at $(0,0)$:

$$F(x,y) = \begin{cases} \frac{xy}{x^2+y^2} & \text{if } (x,y) \neq (0,0), \\ 0 & \text{if } (x,y) = (0,0). \end{cases}$$

Both partial derivatives exist at (0,0):

$$\frac{\partial F}{\partial x}(0,0) = \lim_{h \to 0} \frac{F(h,0) - F(0,0)}{h} = \lim_{h \to 0} \frac{0-0}{h} = 0,$$

$$\frac{\partial F}{\partial y}(0,0) = \lim_{h \to 0} \frac{F(0,h) - F(0,0)}{h} = \lim_{h \to 0} \frac{0-0}{h} = 0.$$

Yet, as we have seen, $F(x,y)$ is not continuous at $(0,0)$, and so it is not differentiable at $(0,0)$.

In fact, it is possible for all directional derivatives to exist without the scalar field being differentiable. Consider the scalar field

$$G(x,y) = \begin{cases} \frac{x^2 y}{x^4+y^2} & \text{if } (x,y) \neq (0,0), \\ 0 & \text{if } (x,y) = (0,0). \end{cases}$$

Again, both of its partial derivatives exist and are 0 at $(0,0)$. Let $\vec{u}$ be any other unit vector, $\vec{u} = (u_1, u_2), u_1^2 + u_2^2 = 1, u_1 u_2 \neq 0$, then

$$
\begin{aligned}
G'(0,0;\vec{u}) &= \lim_{h \to 0} \frac{G(hu_1, hu_2) - G(0,0)}{h} \\
&= \lim_{h \to 0} \frac{h^3 u_1^2 u_2}{h^4 u_1^4 + h^2 u_2^2} \\
&= \lim_{h \to 0} \frac{h u_1^2 u_2}{h^2 u_1^4 + u_2^2} \\
&= 0.
\end{aligned}
$$

But $G(x,y)$ is not continuous at $(0,0)$ and so it is not differentiable at $(0,0)$.

The existence of the partial derivatives is not sufficient to guarantee differentiability. It does not even guarantee continuity.

Higher Order Partials

We can take the partial derivatives of partial derivatives. For example, let

$$f(x,y,z) = x^2 z - 3xy + y^3 z^2.$$

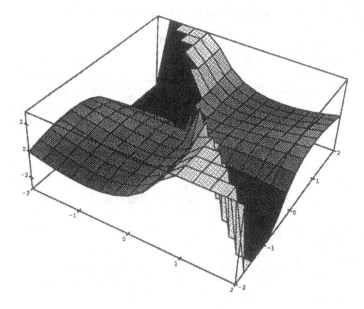

FIGURE 7.7. $f(x, y) = (x^2 y - xy^2)/(x + y)$.

The *second partial derivative of f with respect to x* is

$$\frac{\partial^2 f}{\partial x^2} = \frac{\partial}{\partial x} \left(\frac{\partial f}{\partial x} \right) = \frac{\partial}{\partial x} (2xz - 3y) = 2z.$$

This is also written $D_{11} f = D_1(D_1 f)$.

We can mix the partial derivatives:

$$\frac{\partial^2 f}{\partial y \, \partial x} = \frac{\partial}{\partial y} \left(\frac{\partial f}{\partial x} \right) = \frac{\partial}{\partial y} (2xz - 3y) = -3.$$

This is also written $D_{21} f = D_2(D_1 f)$. Note that

$$\frac{\partial^2 f}{\partial x \, \partial y} = \frac{\partial}{\partial x} \left(-3x + 3y^2 z^2 \right) = -3.$$

In this case, we have

$$\frac{\partial^2 f}{\partial x \, \partial y} = \frac{\partial^2 f}{\partial y \, \partial x}.$$

This is often the case *but not always*. Consider, for example (Figure 7.7),

$$f(x, y) = \begin{cases} \frac{x^2 y - xy^2}{x + y} & \text{if } x + y \neq 0, \\ 0 & \text{if } x + y = 0. \end{cases}$$

Using the definition, we compute one of the mixed partials at $(0,0)$:

$$
\begin{aligned}
\frac{\partial^2 f}{\partial x\,\partial y}(0,0) &= \lim_{h\to 0} \frac{\frac{\partial f}{\partial y}(h,0) - \frac{\partial f}{\partial y}(0,0)}{h} \\[2mm]
&= \lim_{h\to 0} \frac{\lim_{k\to 0}\frac{f(h,k)-f(h,0)}{k} - \lim_{k\to 0}\frac{f(0,k)-f(0,0)}{k}}{h} \\[2mm]
&= \lim_{h\to 0}\left(\lim_{k\to 0}\frac{f(h,k)-f(h,0)-f(0,k)+f(0,0)}{hk}\right) \quad (7.5) \\[2mm]
&= \lim_{h\to 0}\left(\lim_{k\to 0}\frac{\frac{h^2 k - hk^2}{h+k}-0-0+0}{hk}\right) \\[2mm]
&= \lim_{h\to 0}\left(\lim_{k\to 0}\frac{h-k}{h+k}\right) = \lim_{h\to 0} 1 = 1.
\end{aligned}
$$

Similarly, we can show that

$$
\frac{\partial^2 f}{\partial y\,\partial x}(0,0) = \lim_{k\to 0}\left(\lim_{h\to 0}\frac{h-k}{h+k}\right) = \lim_{k\to 0}-1 = -1.
$$

The Mean Value Theorem

After showing some of the things that the mere existence of directional derivatives does not imply, we shall finish this section by demonstrating that the existence of the directional derivative is, nevertheless, a very powerful notion.

Theorem 7.2 (The Mean Value Theorem) *Let f be a scalar field and $\vec{a}$ and $\vec{b}$ two points in its domain. Let $m = |\vec{b}-\vec{a}|$ and $\vec{u} = (\vec{b}-\vec{a})/m$, the unit vector in the direction from $\vec{a}$ to $\vec{b}$, and assume that $f'(\vec{x};\vec{u})$ exists for every $\vec{x}$ on the line segment from $\vec{a}$ to $\vec{b}$:*

$$
L = \{\vec{a}+t\vec{u} \mid 0 \le t \le m\}.
$$

Then, the average rate of change of f from $\vec{a}$ to $\vec{b}$ is equal to the directional derivative at some point between $\vec{a}$ and $\vec{b}$. That is, there exists a point $\vec{c} \in L, \vec{c} \ne \vec{a}$ or $\vec{b}$, such that

$$
\frac{f(\vec{b})-f(\vec{a})}{|\vec{b}-\vec{a}|} = f'(\vec{c};\vec{u}).
$$

Proof: Let $g(t) = f(\vec{a}+t\vec{u})$. Then $g(0) = f(\vec{a})$ and $g(m) = f(\vec{b})$. We also have

$$g'(t) = \lim_{h \to 0} \frac{g(t+h) - g(t)}{h}$$

$$= \lim_{h \to 0} \frac{f(\vec{a} + t\vec{u} + h\vec{u}) - f(\vec{a} + t\vec{u})}{h}$$

$$= f'(\vec{a} + t\vec{u}; \vec{u}),$$

and therefore $g(t)$ is differentiable for all t in $[0, m]$. By the mean value theorem for real-valued functions of a single variable, there is a t_0, $0 < t_0 < m$, such that

$$\frac{g(m) - g(0)}{m} = g'(t_0).$$

Let $\vec{c} = \vec{a} + t_0 \vec{u}$, then

$$f'(\vec{c}; \vec{u}) = g'(t_0)$$

$$= \frac{g(m) - g(0)}{m}$$

$$= \frac{f(\vec{b}) - f(\vec{a})}{|\vec{b} - \vec{a}|}.$$

Q.E.D.

7.4   The Derivative

We begin by assuming that the scalar field $f : \mathbf{R}^n \longrightarrow \mathbf{R}$ is differentiable at $\vec{c}$ and observe what we can say about the linear transformation $L_c : \mathbf{R}^n \longrightarrow \mathbf{R}$ satisfying

$$f(\vec{x}) - f(\vec{c}) = L_c(\vec{x} - \vec{c}) + |\vec{x} - \vec{c}| E(\vec{c}, \vec{x} - \vec{c}),$$

where

$$\lim_{\vec{x} \to \vec{c}} E(\vec{c}, \vec{x} - \vec{c}) = 0.$$

We can use this equality to express our directional derivatives in terms of L_c:

$$f'(\vec{c}; \vec{u}) = \lim_{h \to 0} \frac{f(\vec{c} + h\vec{u}) - f(\vec{c})}{h}$$

$$= \lim_{h \to 0} \frac{1}{h} [L_c(\vec{c} + h\vec{u} - \vec{c}) + |h\vec{u}| E(\vec{c}, h\vec{u})]$$

$$= \lim_{h \to 0} \frac{1}{h} [L_c(h\vec{u}) + |h| E(\vec{c}, h\vec{u})]$$

$$= \lim_{h \to 0} \left[\frac{h}{h} L_c(\vec{u}) + \frac{|h|}{h} E(\vec{c}, h\vec{u}) \right]$$

$$= L_c(\vec{u}). \tag{7.6}$$

Since L_c is a linear transformation, it is uniquely determined by its values at $\vec{\imath}_1, \vec{\imath}_2, \ldots, \vec{\imath}_n$. What Equation (7.6) demonstrates is that if f is differentiable at $\vec{c}$, then these values are given by the partial derivatives:

$$L_c(\vec{\imath}_j) = \frac{\partial f}{\partial x_j}(\vec{c}).\qquad(7.7)$$

This fact is expressed in Theorem 7.3. It is convenient to establish a special notation for the vector of partial derivatives at $\vec{c}$:

$$\nabla f(\vec{c}) = \left(\frac{\partial f}{\partial x_1}(\vec{c}), \frac{\partial f}{\partial x_2}(\vec{c}), \ldots, \frac{\partial f}{\partial x_n}(\vec{c})\right).\qquad(7.8)$$

This vector is called the *gradient of f at $\vec{c}$*. The symbol ∇, read "del" or "nabla", was introduced by Hamilton in his work on quaternions.

Theorem 7.3 *If $f : \mathbf{R}^n \longrightarrow \mathbf{R}$ is differentiable at $\vec{c}$, then*

$$
\begin{aligned}
L_c(\Delta x_1, \Delta x_2, \ldots, \Delta x_n) &= \frac{\partial f}{\partial x_1}(\vec{c})\,\Delta x_1 + \frac{\partial f}{\partial x_2}(\vec{c})\,\Delta x_2 + \ldots \\
&\quad + \frac{\partial f}{\partial x_n}(\vec{c})\,\Delta x_n \\
&= \nabla f(\vec{c}) \cdot \Delta \vec{x}.\qquad(7.9)
\end{aligned}
$$

Similarly, if $y = f(x_1,\ldots,x_n)$, then the differentials of y and the x_j's are related by

$$dy = \nabla f(\vec{c}) \cdot d\vec{x},\qquad(7.10)$$

or equivalently,

$$dy = \frac{\partial f}{\partial x_1}dx_1 + \frac{\partial f}{\partial x_2}dx_2 + \ldots + \frac{\partial f}{\partial x_n}dx_n,\qquad(7.11)$$

where all partial derivatives are evaluated at $\vec{c}$.

Corollary 7.1 *If f is differentiable at $\vec{c}$, then the directional derivative at $\vec{c}$ is given by*

$$f'(\vec{c}; \vec{u}) = \nabla f(\vec{c}) \cdot \vec{u}.\qquad(7.12)$$

Warning

This is an appropriate place to repeat the warning on page 184. If f is differentiable, then L_c is defined in terms of the partial derivatives. But, as we have seen, the partial derivatives can exist when f is not differentiable.

Examples

Let
$$f(x, y) = x^2 y - 3xy^2,$$
so that
$$\nabla f = (2xy - 3y^2, x^2 - 6xy).$$

As we shall see at the end of this section, f is differentiable whenever the partial derivatives exist and are continuous, and so our function is differentiable everywhere. At $(1, 2)$, the value of ∇f is $(4 - 12, 1 - 12) = (-8, -11)$, and so
$$L_{(1,2)}(\Delta x, \Delta y) = -8\,\Delta x - 11\,\Delta y.$$

The directional derivative at $(1,2)$ in the direction $(3/5, -4/5)$ is
$$L_{(1,2)}\left(\frac{3}{5}, \frac{-4}{5}\right) = -8 \cdot \frac{3}{5} - 11 \cdot \frac{-4}{5} = 4.$$

The differentials of f, x, and y are related by
$$df = -8\,dx - 11\,dy.$$

Let
$$f(x, y, z) = x^2 z - 2xyz + y^3,$$
so that
$$\nabla f = (2xz - 2yz, -2xz + 3y^2, x^2 - 2xy).$$

Again, f is differentiable everywhere. Setting $\vec{c} = (1, 2, 3)$, we see that
$$\nabla f(\vec{c}) = (-6, 6, -3),$$
and so
$$L_{(1,2,3)}(\Delta x, \Delta y, \Delta z) = -6\,\Delta x + 6\,\Delta y - 3\,\Delta z.$$

The directional derivative at $(1, 2, 3)$ in the direction of $(1/3, -2/3, 2/3)$ is given by
$$f'\left(1, 2, 3; \tfrac{1}{3}, -\tfrac{2}{3}, \tfrac{2}{3}\right) = (-6, 6, -3) \cdot \left(\tfrac{1}{3}, -\tfrac{2}{3}, \tfrac{2}{3}\right) = -8.$$

The differentials of f, x, y, and z are related by
$$df = -6\,dx + 6\,dy - 3\,dz.$$

Estimating Error

From the definition of the derivative, we also see that we can approximate the amount

$$f(x, y, z) = x^2 z - 2xyz + y^3$$

changes when x, y, and z change. For example, if $\Delta x = 0.1$, $\Delta y = -0.3$, and $\Delta z = 0.2$, then Δf is approximately

$$
\begin{aligned}
\Delta f &\simeq \nabla f(1, 2, 3) \cdot (\Delta x, \Delta y, \Delta z) \\
&= (-6, 6, -3) \cdot (0.1, -0.3, 0.2) \\
&= -0.6 - 1.8 - 0.6 = -3.
\end{aligned}
$$

If we are given the possible percentage error in the independent variables x, y and z, these percentages represent the ratios

$$\frac{\Delta x}{x}, \frac{\Delta y}{y}, \text{ and } \frac{\Delta z}{z}.$$

To approximate the resulting percentage error in f, we divide our approximate equality by f:

$$\frac{\Delta f}{f} \simeq \frac{1}{f} \nabla f \cdot \left(\frac{\Delta x}{x} x, \frac{\Delta y}{y} y, \frac{\Delta z}{z} z\right).$$

Let

$$\frac{\Delta x}{x} = 5\% = 0.05, \quad \frac{\Delta y}{y} = 8\% = 0.08, \quad \frac{\Delta z}{z} = 3\% = 0.03,$$

then at $(1, 2, 3)$ we have

$$
\begin{aligned}
\frac{\Delta f}{f} &\simeq \frac{(-6, 6, -3) \cdot (0.05 \times 1, 0.08 \times 2, 0.03 \times 3)}{f(1, 2, 3)} \\
&= \frac{-0.3 + 0.96 - 0.27}{-1} = -0.39 = -39\%.
\end{aligned}
$$

Be aware that this technique is useful when a quick and rough idea of the error is desired, but it does not constitute a careful error analysis. All we know is that the true error is somewhere in the vicinity of 39%. What do we mean by "in the vicinity"? Chapter 9 will take a closer look at determining the accuracy of our approximations.

The Jacobian Matrix

If $\vec{F} : \mathbf{R}^n \longrightarrow \mathbf{R}^m$ is a vector field differentiable at $\vec{c}$ and $\vec{y} = \vec{F}(\vec{x}) = (f_1(\vec{x}), f_2(\vec{x}), \ldots, f_m(\vec{x}))$, then we have a system of scalar equations:

$$y_1 = f_1(x_1, \ldots, x_n),$$
$$y_2 = f_2(x_1, \ldots, x_n),$$
$$\vdots$$
$$y_m = f_m(x_1, \ldots, x_n),$$

giving us a system of differential equations:

$$dy_1 = \frac{\partial f_1}{\partial x_1} dx_1 + \frac{\partial f_1}{\partial x_2} dx_2 + \ldots + \frac{\partial f_1}{\partial x_n} dx_n,$$

$$dy_2 = \frac{\partial f_2}{\partial x_1} dx_1 + \frac{\partial f_2}{\partial x_2} dx_2 + \ldots + \frac{\partial f_2}{\partial x_n} dx_n,$$

$$\vdots$$

$$dy_m = \frac{\partial f_m}{\partial x_1} dx_1 + \frac{\partial f_m}{\partial x_2} dx_2 + \ldots + \frac{\partial f_m}{\partial x_n} dx_n,$$

where the partial derivatives are to be evaluated at $\vec{c}$. We define the $m \times n$ matrix

$$\mathbf{F}'_c = \begin{pmatrix} \frac{\partial f_1}{\partial x_1}(\vec{c}) & \frac{\partial f_1}{\partial x_2}(\vec{c}) & \cdots & \frac{\partial f_1}{\partial x_n}(\vec{c}) \\ \frac{\partial f_2}{\partial x_1}(\vec{c}) & \frac{\partial f_2}{\partial x_2}(\vec{c}) & \cdots & \frac{\partial f_2}{\partial x_n}(\vec{c}) \\ \vdots & & & \\ \frac{\partial f_m}{\partial x_1}(\vec{c}) & \frac{\partial f_m}{\partial x_2}(\vec{c}) & \cdots & \frac{\partial f_m}{\partial x_n}(\vec{c}) \end{pmatrix}, \tag{7.13}$$

called the *Jacobian matrix* of $\vec{F}$ at $\vec{c}$. If $m = n$, then the Jacobian matrix has a determinant which is called the *Jacobian* of $\vec{F}$ at $\vec{c}$ and is denoted by

$$\frac{\partial(f_1, f_2, \ldots, f_n)}{\partial(x_1, x_2, \ldots, x_n)} = \det(\mathbf{F}'_c). \tag{7.14}$$

These observations are expressed in the following theorem.

Theorem 7.4 *If the vector field* $\vec{F} : \mathbf{R}^n \longrightarrow \mathbf{R}^m$ *is differentiable at* $\vec{c}$, *then*

$$\Delta \mathbf{y} = \mathbf{F}'_c \Delta \mathbf{x} + |\Delta \vec{x}| \mathbf{E}(\vec{c}, \Delta \vec{x}), \tag{7.15}$$

where $\mathbf{F}'_c$ *is the Jacobian matrix of* $\vec{F}$ *at* $\vec{c}$, $\Delta \mathbf{y}$ *and* $\mathbf{E}(\vec{c}, \Delta \vec{x})$ *are* $m \times 1$ *column matrices, and the entries of* $\mathbf{E}(\vec{c}, \Delta \vec{x})$ *approach 0 as* $\Delta \vec{x}$ *approaches* $\vec{0}$. *We also have*

$$d\mathbf{y} = \mathbf{F}'_c d\mathbf{x}. \tag{7.16}$$

If $m = n$, *then*

$$dy_1 dy_2 \cdots dy_n = \frac{\partial(f_1, f_2, \ldots, f_n)}{\partial(x_1, x_2, \ldots, x_n)} dx_1 dx_2 \cdots dx_n. \tag{7.17}$$

The linear transformation corresponding to the Jacobian matrix at $\vec{c}$, $\mathbf{F}'_c$, is sometimes called the *total derivative* of $\vec{F}$ at $\vec{c}$. Unless there is danger of confusion, we shall simply refer to it as the *derivative* of $\vec{F}$ at $\vec{c}$ and denote it by $\vec{F}'_c$. If $\vec{F} : \mathbf{R}^n \longrightarrow \mathbf{R}^m$ is a vector field differentiable at $\vec{c}$, then $\vec{F}'_c$ is a linear transformation from $\mathbf{R}^n$ to $\mathbf{R}^m$. The derivative, $\vec{F}'_c$, is the linear transformation that best approximates the change in $\vec{F}$ at $\vec{c}$:

$$\Delta \vec{y} = \vec{F}'_c(\Delta \vec{x}) + |\vec{x}| \, \vec{E}(\vec{c}, \Delta \vec{x}), \qquad (7.18)$$

where $\vec{E}$ approaches $\vec{0}$ as $\Delta \vec{x}$ approaches $\vec{0}$.

Examples

Let $\vec{F}$ be the mapping from $\mathbf{R}^3$ to $\mathbf{R}^4$ given by

$$\vec{F}(x, y, z) = (3x - 2z + 1, \, -x + 5y + 3z - 2, \, 4x - 6y + 5, \, x - 5y + z).$$

The Jacobian matrix is a constant matrix:

$$\mathbf{F}'_{(x,y,z)} = \begin{pmatrix} 3 & 0 & -2 \\ -1 & 5 & 3 \\ 4 & -6 & 0 \\ 1 & -5 & 1 \end{pmatrix}.$$

If $\vec{G}$ is given by
$$\vec{G}(r, \theta) = (r \cos \theta, \, r \sin \theta),$$

then

$$\mathbf{G}'_{(r,\theta)} = \begin{pmatrix} \cos \theta & -r \sin \theta \\ \sin \theta & r \cos \theta \end{pmatrix}.$$

If $(x, y) = \vec{G}(r, \theta)$, then

$$\begin{pmatrix} dx \\ dy \end{pmatrix} = \begin{pmatrix} \cos \theta & -r \sin \theta \\ \sin \theta & r \cos \theta \end{pmatrix} \begin{pmatrix} dr \\ d\theta \end{pmatrix}$$

$$= \begin{pmatrix} \cos \theta \, dr - r \sin \theta \, d\theta \\ \sin \theta \, dr + r \cos \theta \, d\theta \end{pmatrix}.$$

It follows that

$$\begin{aligned} dx \, dy &= (\cos \theta \, dr - r \sin \theta \, d\theta)(\sin \theta \, dr + r \cos \theta \, d\theta) \\ &= r \cos^2 \theta \, dr \, d\theta - r \sin^2 \theta \, d\theta \, dr \\ &= r(\cos^2 \theta + \sin^2 \theta) \, dr \, d\theta \\ &= r \, dr \, d\theta. \end{aligned}$$

Equivalently, we have

$$\frac{\partial(x, y)}{\partial(r, \theta)} = \begin{vmatrix} \cos \theta & -r \sin \theta \\ \sin \theta & r \cos \theta \end{vmatrix} = r.$$

As an application of this example, consider an airport radar station that can determine the distance, r, and direction, θ, of an incoming plane to within an accuracy of 1%. The position is to be plotted in rectangular co-ordinates. We want to find the approximate percentage error in x and y when $r = 5.3$ km and $\theta = 37.8° \simeq 0.66$ radians and thus

$$
\begin{aligned}
x &= 5.3 \cos 37.8° \simeq 4.19, \\
y &= 5.3 \sin 37.8° \simeq 3.25, \\
\Delta r &= .01 \times 5.3 = .053, \\
\Delta\theta &= .01 \times 0.66 = .0066.
\end{aligned}
$$

We have

$$
\begin{pmatrix} \Delta x \\ \Delta y \end{pmatrix} \simeq \begin{pmatrix} \cos\theta & -r\sin\theta \\ \sin\theta & r\cos\theta \end{pmatrix} \begin{pmatrix} \Delta r \\ \Delta\theta \end{pmatrix}
$$

$$
\simeq \begin{pmatrix} \cos.66 & -5.3\sin.66 \\ \sin.66 & 5.3\cos.66 \end{pmatrix} \begin{pmatrix} .053 \\ .0066 \end{pmatrix}
$$

$$
\simeq \begin{pmatrix} .79 & -3.25 \\ .613 & 4.19 \end{pmatrix} \begin{pmatrix} .053 \\ .0066 \end{pmatrix}
$$

$$
\simeq \begin{pmatrix} 0.02 \\ 0.06 \end{pmatrix}.
$$

It follows that

$$
\begin{aligned}
\frac{\Delta x}{x} &\simeq \frac{0.02}{4.19} \simeq 0.5\%, \\
\frac{\Delta y}{y} &\simeq \frac{0.06}{3.25} \simeq 1.8\%.
\end{aligned}
$$

Caution

In general, the gradient of a scalar field or the Jacobian matrix of a vector field are expressed, respectively, as a vector or a matrix whose entries are functions of one or more variables. If $f(x, y, z) = x^2 y - y^2 z$, then we write

$$
\nabla f = (2xy, x^2 - 2yz, -y^2),
$$

and if $\vec{F}(x, y, z) = (2xy, x^2 - 2yz, -y^2)$, then

$$
\mathbf{F}' = \begin{pmatrix} 2y & 2x & 0 \\ 2x & -2z & -2y \\ 0 & -2y & 0 \end{pmatrix}.
$$

Note that it is the gradient at f at some specific point such as $(1,2,1)$ that is a linear transformation from $\mathbf{R}^3$ to $\mathbf{R}$:

$$
\begin{aligned}
f'_{(1,2,1)}(\Delta x, \Delta y, \Delta z) &= (4, -3, -4) \cdot (\Delta x, \Delta y, \Delta z) \\
&= 4\,\Delta x - 3\,\Delta y - 4\,\Delta z.
\end{aligned}
$$

Similarly, it is the Jacobian matrix of $\vec{F}$ at some specific point such as $(1,2,1)$ which is a linear transformation from $\mathbf{R}^3$ to $\mathbf{R}^3$:

$$
\begin{aligned}
\vec{F}'_{(1,2,1)}(\Delta x, \Delta y, \Delta z) &= \begin{pmatrix} 4 & 2 & 0 \\ 2 & -2 & -4 \\ 0 & -4 & 0 \end{pmatrix} \begin{pmatrix} \Delta x \\ \Delta y \\ \Delta x \end{pmatrix} \\
&= \begin{pmatrix} 4\,\Delta x - 2\,\Delta y \\ 2\,\Delta x - 2\,\Delta y - 4\,\Delta z \\ -4\,\Delta y \end{pmatrix}.
\end{aligned}
$$

The gradient itself is not a linear transformation from $\mathbf{R}^3$ to $\mathbf{R}$. It is something slightly more complicated. It takes a vector in $\mathbf{R}^3$ and transforms it into a linear transformation. Similarly, the Jacobian matrix is a mapping from vectors to linear transformations.

The gradient can also be considered to be a vector field, in this case from $\mathbf{R}^3$ to $\mathbf{R}^3$, so that it makes sense to speak of the Jacobian matrix of a gradient. This is called the *Hessian*, after Ludwig Otto Hesse (1811-1874). We shall be exploiting the Hessian in Chapter 9. Our Jacobian matrix can be considered to be a vector field from $\mathbf{R}^3$ to $\mathbf{R}^9$. It also has a Jacobian matrix that is a vector field from $\mathbf{R}^3$ to $\mathbf{R}^{27}$, where it is convenient to arrange the 27 functions in a $3 \times 3 \times 3$ cube. Its Jacobian matrix is a vector field from $\mathbf{R}^3$ to $\mathbf{R}^{81}$, and so on indefinitely.

When is f differentiable?

All of this has been predicated on the assumption that f (or $\vec{F}$) *is* differentiable at $\vec{c}$. But, as we have seen, differentiability is a strong condition, stronger than the mere existence of the partial derivatives. Fortunately, we do not need too much more than existence. Existence plus continuity of the partial derivatives is sufficient. Exercise 7 in Section 7.5 illustrates that continuity of the partial derivatives is not a *necessary* condition for differentiability. It gives an example of a function that is differentiable at $(0,0)$, but for which the partial derivatives are not continuous there.

Theorem 7.5 *Given a scalar field $f : \mathbf{R}^n \longrightarrow \mathbf{R}$, if each partial derivative $\frac{\partial f}{\partial x_j}(\vec{x})$ exists for all $\vec{x}$ sufficiently close to $\vec{c}$ (there is some $\delta > 0$ such that $|\vec{x} - \vec{c}| < \delta$ implies that $\frac{\partial f}{\partial x_j}(\vec{x})$ exists) and $\frac{\partial f}{\partial x_j}$ is continuous at $\vec{c}$, then f is differentiable at $\vec{c}$.*

Proof: The only candidate for $L_c(\Delta \vec{x})$ is $\nabla f(\vec{c}) \cdot \Delta \vec{x}$. We must show that this works. If we are to have

$$f(\vec{c} + \Delta \vec{x}) - f(\vec{c}) = \nabla f(\vec{c}) \cdot \Delta \vec{x} + |\Delta \vec{x}| E(\vec{c}, \Delta \vec{x}),$$

then we can solve for $E(\vec{c}, \Delta \vec{x})$ in terms of f and ∇f:

$$E(\vec{c}, \Delta \vec{x}) = \frac{1}{|\Delta \vec{x}|} \left(f(\vec{c} + \Delta \vec{x}) - f(\vec{c}) - \nabla f(\vec{c}) \cdot \Delta \vec{x} \right). \qquad (7.19)$$

We need to show that

$$\lim_{\Delta \vec{x} \to \vec{0}} E(\vec{c}, \Delta \vec{x}) = 0.$$

Let $|\Delta \vec{x}| = m$ and set

$$\vec{u} = \frac{\Delta \vec{x}}{m} = u_1 \vec{i}_1 + u_2 \vec{i}_2 + \ldots + u_n \vec{i}_n,$$

so that $\Delta x_1 = m u_1, \Delta x_2 = m u_2, \ldots, \Delta x_n = m u_n$. Set

$$
\begin{aligned}
\vec{v}_1 &= u_1 \vec{i}_1, \\
\vec{v}_2 &= u_1 \vec{i}_1 + u_2 \vec{i}_2, \\
&\vdots \\
\vec{v}_{n-1} &= u_1 \vec{i}_1 + \cdots + u_{n-1} \vec{i}_{n-1}, \\
\vec{v}_n &= u_1 \vec{i}_1 + \cdots + u_{n-1} \vec{i}_{n-1} + u_n \vec{i}_n = \vec{u}.
\end{aligned}
$$

We can then rewrite

$$
\begin{aligned}
f(\vec{c} + \Delta \vec{x}) - f(\vec{c}) &= f(\vec{c} + m\vec{u}) - f(\vec{c}) \\
&= f(\vec{c} + m\vec{v}_n) - f(\vec{c} + m\vec{v}_{n-1}) \\
&\quad + f(\vec{c} + m\vec{v}_{n-1}) - f(\vec{c} + m\vec{v}_{n-2}) \\
&\quad + \cdots \\
&\quad + f(\vec{c} + m\vec{v}_2) - f(\vec{c} + m\vec{v}_1) \\
&\quad + f(\vec{c} + m\vec{v}_1) - f(\vec{c}).
\end{aligned}
$$

By Theorem 7.2, the mean value theorem, each of these differences can be expressed in terms of a partial derivative. For example, if $u_n \neq 0$, then

$$
\begin{aligned}
\frac{f(\vec{c} + m\vec{v}_n) - f(\vec{c} + m\vec{v}_{n-1})}{|(\vec{c} + m\vec{v}_n) - (\vec{c} + m\vec{v}_{n-1})|} \\
= \frac{f(\vec{c} + m\vec{v}_{n-1} + m u_n \vec{i}_n) - f(\vec{c} + m\vec{v}_{n-1})}{m|u_n|} \\
= f'\left(\vec{c} + \vec{k}_n; \frac{u_n}{|u_n|} \vec{i}_n \right) \\
= \frac{u_n}{|u_n|} \frac{\partial f}{\partial x_n}(\vec{c} + \vec{k}_n),
\end{aligned}
$$

where $\vec{k}_n$ lies between $m\vec{v}_{n-1}$ and $m\vec{v}_n$. This implies

$$f(\vec{c} + m\vec{v}_n) - f(\vec{c} + m\vec{v}_{n-1}) = mu_n \frac{\partial f}{\partial x_n}(\vec{c} + \vec{k}_n). \qquad (7.20)$$

If $u_n = 0$, then $\vec{v}_n = \vec{v}_{n-1}$ and so Equation (7.20) is still valid. In general, we see that

$$\begin{aligned}
f(\vec{c} + \Delta \vec{x}) - f(\vec{c}) &= mu_n \frac{\partial f}{\partial x_n}(\vec{c} + \vec{k}_n) \\
&\quad + mu_{n-1} \frac{\partial f}{\partial x_{n-1}}(\vec{c} + \vec{k}_{n-1}) \\
&\quad + \cdots \\
&\quad + mu_1 \frac{\partial f}{\partial x_1}(\vec{c} + \vec{k}_1),
\end{aligned} \qquad (7.21)$$

where $\vec{k}_j$ lies between $m\vec{v}_j$ and $m\vec{v}_{j-1}$ when j is at least 2 and most n, and $\vec{k}_1$ lies betweem $m\vec{v}_1$ and $\vec{0}$. We also observe that

$$\begin{aligned}
\nabla f(\vec{c}) \cdot \Delta \vec{x} &= mu_n \frac{\partial f}{\partial x_n}(\vec{c}) \\
&\quad + mu_{n-1} \frac{\partial f}{\partial x_{n-1}}(\vec{c}) \\
&\quad + \cdots \\
&\quad + mu_1 \frac{\partial f}{\partial x_1}(\vec{c}).
\end{aligned} \qquad (7.22)$$

Using Equations (7.21) and (7.22) in Equation (7.19), we see that

$$\begin{aligned}
E(\vec{c}, \Delta \vec{x}) &= \frac{1}{m}\left(mu_n \frac{\partial f}{\partial x_n}(\vec{c} + \vec{k}_n) + mu_{n-1}\frac{\partial f}{\partial x_{n-1}}(\vec{c} + \vec{k}_{n-1}) \right. \\
&\quad + \cdots + mu_1 \frac{\partial f}{\partial x_1}(\vec{c} + \vec{k}_1) - mu_1 \frac{\partial f}{\partial x_1}(\vec{c}) \\
&\quad \left. - mu_2 \frac{\partial f}{\partial x_2}(\vec{c}) - \cdots - mu_n \frac{\partial f}{\partial x_n}(\vec{c}) \right) \\
&= u_1\left(\frac{\partial f}{\partial x_1}(\vec{c} + \vec{k}_1) - \frac{\partial f}{\partial x_1}(\vec{c}) \right) \\
&\quad + u_2\left(\frac{\partial f}{\partial x_2}(\vec{c} + \vec{k}_2) - \frac{\partial f}{\partial x_2}(\vec{c}) \right) \\
&\quad + \cdots \\
&\quad + u_n\left(\frac{\partial f}{\partial x_n}(\vec{c} + \vec{k}_n) - \frac{\partial f}{\partial x_n}(\vec{c}) \right).
\end{aligned}$$

As $\Delta \vec{x}$ approaches $\vec{0}$, $m = |\Delta \vec{x}|$ approaches 0, and so the point $\vec{k}_j$ between $m\vec{v}_{j-1}$ and $m\vec{v}_j$ also approaches $\vec{0}$. Since each $\partial f/\partial x_j$ is continuous at $\vec{c}$,

we have

$$\lim_{\Delta \vec{x} \to \vec{0}} \left(\frac{\partial f}{\partial x_j}(\vec{c} + \vec{k}_j) - \frac{\partial f}{\partial x_j}(\vec{c}) \right) = 0,$$

and so

$$\lim_{\Delta \vec{x} \to \vec{0}} E(\vec{c}, \Delta \vec{x}) = 0.$$

Q.E.D.

7.5   Exercises

1. Let $f(x, y) = x^3 y - xy^2$, find $f'(\vec{c}; \vec{u})$ for

   (a) $\vec{c} = (0, 0), \vec{u} = (1, 0)$,
   (b) $\vec{c} = (2, 1), \vec{u} = (1/\sqrt{2}, 1/\sqrt{2})$,
   (c) $\vec{c} = (-2, 1), \vec{u} = (3/5, -4/5)$,
   (d) $\vec{c} = (3, -4), \vec{u} = (-1/2, \sqrt{3}/2)$.

2. Let $f(x, y, z) = xy^2 - xz^2$, find $f'(\vec{c}; \vec{u})$ for

   (a) $\vec{c} = (1, 1, 1), \vec{u} = (0, 1, 0)$,
   (b) $\vec{c} = (2, -1, 3), \vec{u} = (-1/3, 2/3, -2/3)$,
   (c) $\vec{c} = (-1, 2, -1), \vec{u} = (4/5, 0, -3/5)$.

3. Let $f(x, y, z) = x\sqrt{y^2 + z^2}$. Prove that at $\vec{c} = (0, 0, 0)$, $f'(\vec{c}; \vec{u}) = 0$ for any unit vector $\vec{u}$.

4. Is $f(x, y, z) = x\sqrt{y^2 + z^2}$ differentiable at $(0,0,0)$? Prove your assertion.

5. Let $f(x, y, z) = \sqrt{x^2 + y^2 + z^2}, \vec{c} = (0, 0, 0)$. Show that $f'(\vec{c}; \vec{u})$ does not exist for any unit vector $\vec{u}$.

6. Is the scalar field

$$f(x, y) = \begin{cases} \frac{xy}{\sqrt{x^2 + y^2}} & \text{if } (x, y) \neq (0, 0), \\ 0 & \text{if } (x, y) = (0, 0), \end{cases}$$

continuous at the origin? Is it differentiable at the origin?

7. Let

$$f(x, y) = \begin{cases} (x^2 + y^2) \sin\left(\frac{1}{x^2 + y^2}\right), & (x, y) \neq (0, 0), \\ 0, & (x, y) = (0, 0). \end{cases}$$

(a) Show that the partial derivatives of f with respect to x and y are not continuous at $(0,0)$.

(b) Show that the partial derivatives of f with respect to x and y do exist at $(0,0)$ and are equal to 0 there. It follows that if f is differential at $(0,0)$, then

$$L_{(0,0)}(x,y) = 0.$$

(c) Show that f *is* differentiable at $(0,0)$ by showing that if $E\left((0,0),(x,y)\right)$ satisfies

$$f(x,y) - f(0,0) \;=\; 0 \,+\, |(x,y)|\, E\left((0,0),(x,y)\right),$$

then

$$\lim_{(x,y)\to(0,0)} E\left((0,0),(x,y)\right) = 0.$$

8. Show that $f(x,y) = \sqrt{|xy|}$ is continuous at $(0,0)$, but that $\partial f/\partial x$ and $\partial f/\partial y$ are not continuous at $(0,0)$. Is $f(x,y)$ differentiable at $(0,0)$? Justify your answer.

9. Use the mean value theorem, Theorem 7.2, to prove that if $f'(\vec{c};\vec{u}) = 0$ for every $\vec{c}$ in the domain of f and every unit vector $\vec{u}$, then $f(\vec{x})$ is a constant.

10. Suppose there is a unit vector $\vec{u}$ such that $f'(\vec{c};\vec{u}) = 0$ for every $\vec{c}$ in the domain of f. What can you conclude about f?

11. Let $\vec{u}_1 = (1/\sqrt{2}, 1/\sqrt{2}), \vec{u}_2 = (-1/\sqrt{5}, 2/\sqrt{5})$ and suppose that we have a differentiable scalar field that has the directional derivatives

$$f'(\vec{c};\vec{u}_1) = 3\sqrt{2}, \quad f'(\vec{c};\vec{u}_2) = -1/\sqrt{5}$$

at $\vec{c} = (3,5)$. Find the values of $\partial f/\partial x_1$ and $\partial f/\partial x_2$ at $(3,5)$.

12. Prove that if a and b are constants and f and g are scalar fields, then

$$\nabla(af + bg) = a\,\nabla f + b\,\nabla g.$$

13. Prove that if $f : \mathbf{R} \longrightarrow \mathbf{R}$ and $g : \mathbf{R}^n \longrightarrow \mathbf{R}$ and the range of g lies in the domain of f, then

$$\nabla(f \circ g) = f'(g)\,\nabla g.$$

14. Find the gradient of f for each of the following scalar fields:

(a) $f(x,y,z) = \sin xy + \cos yz$,

(b) $f(x,y,z) = xe^{yz}$,

(c) $f(x,y,z) = 1/\sqrt{x^2 + y^2 + z^2}$, $(x,y,z) \neq (0,0,0)$,

(d) $f(x,y,z) = 1/\sqrt{x^2 + y^2}$, $(x,y) \neq (0,0)$.

15. For each scalar field given in Exercise 14, approximate Δf at (x,y,z) $= (1,0,0)$ when $\Delta x = 0.2$, $\Delta y = 0.1$, and $\Delta z = -0.3$.

16. For each scalar field given in Exercise 14, find the approximate percentage error of f at $(1, \pi, 1)$ when the percentage errors in x, y, and z are each 1%.

17. The diameter and height of a cylinder are measured to be 6 and 10 inches, respectively, with a possible error of 1%. What is the approximate possible percentage error in the volume? In the total surface area (top, bottom, and sides)?

18. The period of a simple pendulum is

$$T = 2\pi \sqrt{l/g},$$

where l is the length and g is the gravitational constant. If we compute T by taking $\pi = 3.14$ ($|\text{error}| < 0.002$), $l = 40$ cm ($|\text{error}| < 0.1$), and $g = 980$ cm/s^2 ($|\text{error}| < 0.7$), find the approximate possible error in T.

19. Find the length of a pendulum whose period is one second. How accurately do we need to know π and g and how large an error in length can be tolerated if we are to use this pendulum to regulate a clock accurate to within one second per week?

20. Find all mixed partial derivatives:

$$\frac{\partial^2 f}{\partial x \, \partial y}, \ \frac{\partial^2 f}{\partial y \, \partial x}, \ \frac{\partial^2 f}{\partial x \, \partial z}, \ \frac{\partial^2 f}{\partial z \, \partial x}, \ \frac{\partial^2 f}{\partial y \, \partial z}, \ \frac{\partial^2 f}{\partial z \, \partial y},$$

for the scalar fields given in Exercise 14.

21. Find the Jacobian matrix of each of the following vector fields:

(a) $\vec{F}(x,y) = (x^2 y, xy, xy^2)$,

(b) $\vec{F}(x,y,z) = (x \sin y, z \cos y)$,

(c) $\vec{F}(x,y,z) = (xe^{yz}, ye^{xz}, ze^{xy})$,

(d) $\vec{F}(x,y,z) = (y/(x^2 + y^2), -x/(x^2 + y^2), z)$, $(x,y) \neq (0,0)$,

(e) $\vec{F}(x,y,z) = (xr^{-3}, yr^{-3}, zr^{-3})$, $r = \sqrt{x^2 + y^2 + z^2}$, $(x,y,z) \neq (0,0,0)$.

22. For each vector field defined in parts c through e of Exercise 21, let

$$(u, v, w) = \vec{F}(x, y, z).$$

Find the Jacobian $\partial(u, v, w)/\partial(x, y, z)$ and express $du\, dv\, dw$ in terms of $dx\, dy\, dz$.

Exercises 23 through 26 demonstrate that if the mixed partial derivatives $\partial^2 f/\partial y\, \partial x$ and $\partial^2 f/\partial x\, \partial y$ exist at all points sufficiently close to (a, b) and at least one of them is continuous at (a, b) then they are equal at (a, b).

23. Define $\Delta(h, k)$ to be

$$\Delta(h, k) = \frac{f(a + h, b + k) - f(a + h, b) - f(a, b + k) + f(a, b)}{hk}.$$

Prove that

$$\frac{\partial^2 f}{\partial y\, \partial x}(a, b) = \lim_{k \to 0}\left(\lim_{h \to 0} \Delta(h, k)\right)$$

and that

$$\frac{\partial^2 f}{\partial x\, \partial y}(a, b) = \lim_{h \to 0}\left(\lim_{k \to 0} \Delta(h, k)\right).$$

It follows from Exercise 6 in Section 7.2 that if both of these mixed partials exist and if

$$\lim_{(h,k)\to(0,0)} \Delta(h, k)$$

exists, then the mixed partials are equal.

24. Define $F(x) = f(x, b + k) - f(x, b)$ so that

$$\Delta(h, k) = \frac{F(a + h) - F(a)}{hk}.$$

Prove that there is a point a_1 between a and $a + h$ such that

$$\frac{F(a + h) - F(a)}{h} = F'(a_1) = \frac{\partial f}{\partial x}(a_1, b + k) - \frac{\partial f}{\partial x}(a_1, b).$$

25. Prove that there is a point b_1 between b and $b + k$ such that

$$\frac{F(a + h) - F(a)}{hk} = \frac{\partial^2 f}{\partial y\, \partial x}(a_1, b_1).$$

26. Using the previous three exercises, prove that if $\partial^2 f/\partial x\, \partial y$ and $\partial^2 f/\partial y\, \partial x$ exist for all points sufficiently close to (a, b) and if $\partial^2 f/\partial y\, \partial x$ is continuous at (a, b) then

$$\frac{\partial^2 f}{\partial y\, \partial x}(a, b) = \frac{\partial^2 f}{\partial x\, \partial y}(a, b).$$

27. Given a scalar field $f : \mathbf{R}^2 \longrightarrow \mathbf{R}$, consider the following five statements.

A: $f(x, y)$ is continuous at $\vec{c}$.

B: $f(x, y)$ is differentiable at $\vec{c}$.

C: $\frac{\partial f}{\partial x}$ and $\frac{\partial f}{\partial y}$ exist at $\vec{c}$.

D: $\frac{\partial f}{\partial x}$ and $\frac{\partial f}{\partial y}$ exist at $\vec{c}$ and are continuous in some circle with center at $\vec{c}$.

E: $f'(\vec{c}; \vec{u})$ exists at $\vec{c}$ for all unit vectors $\vec{u}$.

In each row of the diagram below, place an X under each statement that is implied by the statement for that row. For example, D implies C and so an X has been placed in row D, column C.

	A	B	C	D	E
A	X				
B		X			
C			X		
D			X	X	
E					X

7.6   The Chain Rule

One of the advantages of recognizing the derivative of $\vec{F}$ at $\vec{c}$ to be a linear transformation is that it makes the chain rule conceptually very simple: *the derivative of a composition is the composition of the derivatives.*

Theorem 7.6 (The Chain Rule) *Let* $\vec{F} : \mathbf{R}^n \longrightarrow \mathbf{R}^m, \vec{G} : \mathbf{R}^m \longrightarrow \mathbf{R}^l$, $l, m, n \geq 1$, *be functions for which the range of* $\vec{F}$ *is contained in the domain of* $\vec{G}$. *If* $\vec{F}$ *is differentiable at* $\vec{c}$ *and* $\vec{G}$ *is differentiable at* $\vec{F}(\vec{c})$, *then the derivative of the composition,* $\vec{G} \circ \vec{F}$, *at* $\vec{c}$ *is the composition of the* *derivatives:*

$$(\vec{G} \circ \vec{F})_c'(\Delta \vec{x}) = \vec{G}_{F(c)}' \circ \vec{F}_c'(\Delta \vec{x}). \tag{7.23}$$

Proof: The differentiability of $\vec{F}$ and $\vec{G}$ implies that

$$\vec{F}(\vec{c} + \Delta \vec{x}) - \vec{F}(\vec{c}) = \vec{F}_c'(\Delta \vec{x}) + |\Delta \vec{x}| \vec{E}_1(\vec{c}, \Delta \vec{x}),$$
$$\lim_{\Delta \vec{x} \to \vec{0}} \vec{E}_1(\vec{c}, \Delta \vec{x}) = \vec{0},$$
$$\vec{G}\left(\vec{F}(\vec{c}) + \Delta \vec{y}\right) - \vec{G}\left(\vec{F}(\vec{c})\right) = \vec{G}_{F(c)}'(\Delta \vec{y}) + |\Delta \vec{y}| \vec{E}_2(\vec{F}(\vec{c}), \Delta \vec{y}),$$
$$\lim_{\Delta \vec{y} \to \vec{0}} \vec{E}_2\left(\vec{F}(\vec{c}), \Delta \vec{y}\right) = \vec{0}.$$

We set

$$\Delta \vec{y} = \vec{F}(\vec{c} + \Delta \vec{x}) - \vec{F}(\vec{c})$$
$$= \vec{F}_c'(\Delta \vec{x}) + |\Delta \vec{x}| \vec{E}_1(\vec{c}, \Delta \vec{x}),$$

so that

$$\lim_{\Delta \vec{x} \to \vec{0}} \Delta \vec{y} = \vec{F}_c'(\vec{0}) + \vec{0}$$
$$= \vec{0}.$$

Putting these pieces together gives us

$$\vec{G} \circ \vec{F}(\vec{c} + \Delta \vec{x}) - \vec{G} \circ \vec{F}(\vec{c})$$

$$= \vec{G}\left(\vec{F}(\vec{c} + \Delta \vec{x})\right) - \vec{G}\left(\vec{F}(\vec{c})\right)$$

$$= \vec{G}_{F(c)}'\left(\vec{F}(\vec{c} + \Delta \vec{x}) - \vec{F}(\vec{c})\right) + |\Delta \vec{y}| \vec{E}_2\left(\vec{F}(\vec{c}), \Delta \vec{y}\right)$$

$$= \vec{G}_{F(c)}'\left(\vec{F}_c'(\Delta \vec{x})\right) + |\Delta \vec{x}| \vec{G}_{F(c)}'\left(\vec{E}_1(\vec{c}, \Delta \vec{x})\right)$$
$$+ |\Delta \vec{y}| \vec{E}_2\left(\vec{F}(\vec{c}), \Delta \vec{y}\right)$$

$$= \vec{G}_{F(c)}' \circ \vec{F}_c'(\Delta \vec{x})$$
$$+ |\Delta \vec{x}| \left(\vec{G}_{F(c)}'\left(\vec{E}_1(\vec{c}, \Delta \vec{x})\right) + \frac{|\Delta \vec{y}|}{|\Delta \vec{x}|} \vec{E}_2\left(\vec{F}(\vec{c}), \Delta \vec{y}\right)\right)$$

$$= \vec{G}_{F(c)}' \circ \vec{F}_c'(\Delta \vec{x}) + |\Delta \vec{x}| \vec{E}(\vec{c}, \Delta \vec{x}),$$

where

$$\vec{E}(\vec{c}, \Delta\vec{x}) = \vec{G}'_{F(c)}\left(\vec{E}_1(\vec{c}, \Delta\vec{x})\right) + \frac{|\Delta\vec{y}|}{|\Delta\vec{x}|}\vec{E}_2\left(\vec{F}(\vec{c}), \Delta\vec{y}\right)$$

$$= \vec{G}'_{F(c)}\left(\vec{E}_1(\vec{c}, \Delta\vec{x})\right)$$

$$+ \left|\vec{F}'_c\left(\frac{\Delta\vec{x}}{|\Delta\vec{x}|}\right) + \vec{E}_1(\vec{c}, \Delta\vec{x})\right| \vec{E}_2\left(\vec{F}(\vec{c}), \Delta\vec{y}\right).$$

It only remains to show that $\vec{E}(\vec{c}, \Delta\vec{x})$ approaches $\vec{0}$ as $\Delta\vec{x}$ approaches $\vec{0}$:

$$\lim_{\Delta\vec{x}\to\vec{0}} \vec{E}(\vec{c}, \Delta\vec{x}) = \vec{G}'_{F(c)}(\vec{0})$$

$$+ \lim_{\Delta\vec{x}\to\vec{0}}\left|\vec{F}'_c\left(\frac{\Delta\vec{x}}{|\Delta\vec{x}|}\right) + \vec{E}_1(\vec{c}, \Delta\vec{x})\right|$$

$$\times \lim_{\Delta\vec{y}\to\vec{0}} \vec{E}_2\left(\vec{F}(\vec{c}), \Delta\vec{y}\right)$$

$$= \vec{0}.$$

In taking this last limit, we used the fact that $\left|\vec{F}'_c(\Delta\vec{x}/|\Delta\vec{x}|)\right|$ is bounded (see Exercise 16, Section 7.8).

Q.E.D.

Corollaries

Corollary 7.2 *The Jacobian matrix of the composition $\vec{G}\circ\vec{F}$ is the product of the Jacobian matrices of $\vec{G}$ and $\vec{F}$.*

If we are given

$$\vec{F}(x, y) = (x^2y, xy, xy^2),$$
$$\vec{G}(r, s, t) = (r^2 - s^2, r^2 - t^2),$$

then the Jacobian matrix of $\vec{G} \circ \vec{F}$ is

$$\mathbf{G'F'} = \begin{pmatrix} 2r & -2s & 0 \\ 2r & 0 & -2t \end{pmatrix}\begin{pmatrix} 2xy & x^2 \\ y & x \\ y^2 & 2xy \end{pmatrix}$$

$$= \begin{pmatrix} 4rxy - 2sy & 2rx^2 - 2sx \\ 4rxy - 4ty^2 & 2rx^2 - 4txy \end{pmatrix}$$

$$= \begin{pmatrix} 4x^3y^2 - 2xy^2 & 2x^4y - 2x^2y \\ 4x^3y - 4xy^4 & 2x^4y - 4x^2y^3 \end{pmatrix}.$$

Since the determinant of a product is the product of the determinants, we also have the following result for Jacobians (the determinants of Jacobian matrices).

Corollary 7.3 *The Jacobian of the composition* $\vec{G} \circ \vec{F}$ *is the product of the Jacobians of* $\vec{G}$ *and* $\vec{F}$.

If we are given

$$\begin{aligned} \vec{F}(r,\theta) &= (r\cos\theta, r\sin\theta), \\ \vec{G}(x,y) &= (x^2 - y^2, x^2 + y^2), \end{aligned}$$

then we have that

$$\begin{aligned} \det(\mathbf{G'F'}) &= \det(\mathbf{G'})\det(\mathbf{F'}) \\ &= \begin{vmatrix} 2x & -2y \\ 2x & 2y \end{vmatrix} \begin{vmatrix} \cos\theta & -r\sin\theta \\ \sin\theta & r\cos\theta \end{vmatrix} \\ &= 8xyr \\ &= 8r^3\cos\theta\sin\theta. \end{aligned}$$

Corollary 7.4 *If* $g(\vec{x})$ *is a scalar field and* $\vec{r}(t)$ *is a vector-valued function of a single variable whose range lies in the domain of* g *and if* $\vec{r}$ *is differentiable at* t_0, g *is differentiable at* $\vec{r}(t_0)$, *then*

$$\frac{d}{dt}(g \circ \vec{r})(t_0) = \nabla g(\vec{r}(t_0)) \cdot \frac{d\vec{r}}{dt}(t_0).$$

Let $\vec{r}(t)$ be the spiral path

$$\vec{r}(t) = (\cos t, \sin t, t)$$

and let $g(x,y,z) = x^2 - y^2 + z^2$, then

$$\begin{aligned} \frac{d}{dt}g(\vec{r}(t)) &= (2x, -2y, 2z) \cdot (-\sin t, \cos t, 1) \\ &= -2x\sin t - 2y\cos t + 2z \\ &= -4\cos t\sin t + 2t. \end{aligned}$$

Corollary 7.5 *If* $g(\vec{y})$ *is a scalar field and* $\vec{F}(\vec{x})$ *is a vector field whose range lies in the domain of* g *and if* $\vec{F}$ *is differentiable at* $\vec{c}$, g *is differentiable at* $\vec{F}(\vec{c})$, *then*

$$\frac{\partial g}{\partial x_j}(\vec{c}) = \nabla g\left(\vec{F}(\vec{c})\right) \cdot \frac{\partial \vec{F}}{\partial x_j}(\vec{c}).$$

If we are given

$$\begin{aligned} \vec{F}(\theta,\phi) &= (\cos\theta\cos\phi, \sin\theta\cos\phi, \sin\phi), \\ &\quad 0 \le \theta \le 2\pi, \quad -\frac{\pi}{2} \le \phi \le \frac{\pi}{2}, \\ g(x,y,z) &= x^2 + y^2 - z^2, \end{aligned}$$

then

$$
\begin{aligned}
\frac{\partial g}{\partial \theta} &= (2x, 2y, -2z) \cdot (-\sin\theta\cos\phi, \cos\theta\cos\phi, 0) \\
&= -2x\sin\theta\cos\phi + 2y\cos\theta\cos\phi \\
&= -2\cos\theta\sin\theta\cos^2\phi + 2\cos\theta\sin\theta\cos^2\phi \\
&= 0, \\
\frac{\partial g}{\partial \phi} &= (2x, 2y, -2z) \cdot (-\cos\theta\sin\phi, -\sin\theta\sin\phi, \cos\phi) \\
&= -2x\cos\theta\sin\phi - 2y\sin\theta\sin\phi - 2z\cos\phi \\
&= -2\cos^2\theta\cos\phi\sin\phi - 2\sin^2\theta\cos\phi\sin\phi - 2\cos\phi\sin\phi \\
&= -4\cos\phi\sin\phi.
\end{aligned}
$$

7.7 Using the Gradient

Before leaving this chapter, we shall take a closer look at the gradient and consider its physical significance. As was shown in Corollary 7.1, if f is differentiable at $\vec{c}$, then

$$
f'(\vec{c}; \vec{u}) = \nabla f(\vec{c}) \cdot \vec{u}.
$$

This implies that $\nabla f(\vec{c})$ points in the direction that maximizes the directional derivative. That is, $\nabla f(\vec{c})$ points in the direction of greatest change. This equation also implies that

$$
-|\nabla f(\vec{c})| \le f'(\vec{c}; \vec{u}) \le |\nabla f(\vec{c})|,
$$

and that if $\vec{u}$ is perpendicular to $\nabla f(\vec{c})$, then $f'(\vec{c}; \vec{u}) = 0$.

Tangent Planes

Let us now restrict our attention to scalar fields on $\mathbf{R}^3$ that are differentiable at every point of the domain. If we set such a scalar field equal to a constant, $f(x, y, z) = c$, we get a level surface, for example (see Figure 7.8),

$$
x^2 - y^2 + z^2 = 1.
$$

Consider a path that is constrained to stay on this surface, for example,

$$
\vec{r}(t) = \left(\cos t\sqrt{1 + t^2},\ t,\ \sin t\sqrt{1 + t^2}\right),
$$

$$
\left(\cos t\sqrt{1 + t^2}\right)^2 - t^2 + \left(\sin t\sqrt{1 + t^2}\right)^2 = 1.
$$

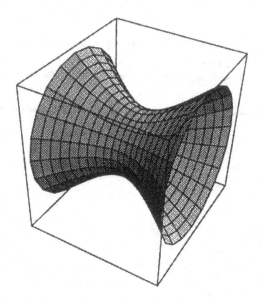

FIGURE 7.8. The level surface, $x^2 - y^2 + z^2 = 1$.

Let $g(t) = f(\vec{r}(t)) = c$ so that

$$\frac{dg}{dt} = 0.$$

By the chain rule, we have

$$0 = \frac{dg}{dt} = \nabla f \cdot \frac{d\vec{r}}{dt} = \nabla f \cdot \vec{v}. \qquad (7.24)$$

We recall that $\vec{v}(t)$ is the tangent to the curve $\vec{r}(t)$. Equation (7.24) implies that if $\vec{r}(t)$ is a differentiable curve lying on the surface $f(x, y, z) = c$, then its tangent is always perpendicular to ∇f. If (x_0, y_0, z_0) is any point on our surface, then every differentiable curve lying on the surface and passing through (x_0, y_0, z_0) has its tangent in the plane perpendicular to $\nabla f(x_0, y_0, z_0)$. For this reason, we make the following definition.

Definition: The plane through the point (x_0, y_0, z_0), perpendicular to the vector $\nabla f(x_0, y_0, z_0)$, is the *tangent plane* to the surface at this point. Its equation is

$$\nabla f(x_0, y_0, z_0) \cdot (x - x_0, y - y_0, z - z_0) = 0. \qquad (7.25)$$

For example, the tangent plane to $x^2 - y^2 + z^2 = 1$ at $(2, 2\sqrt{3}, 3)$ is

$$(4, -4\sqrt{3}, 6) \cdot (x - 2, y - 2\sqrt{3}, z - 3) \;=\; 0,$$
$$4x - (4\sqrt{3})y + 6z \;=\; 2.$$

If a surface is described by the equation

$$z = f(x, y),$$

then $f(x, y) - z = 0$ and the tangent plane is normal to

$$\left(\frac{\partial f}{\partial x}, \frac{\partial f}{\partial y}, -1 \right).$$

As an example, for the paraboloid $z = x^2 + y^2$, the tangent plane at $(1, 2, 5)$ is

$$(2, 4, -1) \cdot (x - 1, y - 2, z - 5) \;=\; 0,$$
$$2x + 4y - z \;=\; 5.$$

7.8   Exercises

1. For each of the following pairs of a vector function, $\vec{r}(t)$, and a scalar field, $f(\vec{x})$, let $F(t) = f(\vec{r}(t))$. Find $F'(t)$:

   (a) $f(x, y) = 1/\sqrt{x^2 + y^2}$, $\vec{r}(t) = (t^2, e^t)$,
   (b) $f(x, y) = x^y$, $\vec{r}(t) = (t^2, \log t)$,
   (c) $f(x, y, z) = xy \tan z$, $\vec{r}(t) = (\cos t, \sin t, t)$.

2. Let x and y be functions of u and v: $x = x(u, v), y = y(u, v)$, and let $f(x, y)$ be a scalar field. Find $\partial f/\partial u$ and $\partial f/\partial v$ in terms of $\partial f/\partial x$, $\partial f/\partial y$, $\partial x/\partial u$, $\partial x/\partial v$, $\partial y/\partial u$, and $\partial y/\partial v$.

3. Find $\partial f/\partial u$ and $\partial f/\partial v$ for each of the following specific functions:

   (a) $f(x, y) = x^2 + xy$, $x = ve^u$, $y = ue^v$,
   (b) $f(x, y) = x^2 + xy$, $x = u^2 - v^2$, $y = uv$,
   (c) $f(x, y) = \sqrt{x + y^2}$, $x = 4uv$, $y = u - v$,
   (d) $f(x, y) = x \log(x^2 + y^2)$, $x = u^2 - v^2$, $y = u^2 + v^2$.

4. For f, x, y as in Exercise 2, express $\partial^2 f/\partial u^2$ in terms of the partial derivatives of f with respect to x and y and the partial derivatives of x and y with respect to u. Assume that $\partial^2 f/\partial x\, \partial y = \partial^2 f/\partial y\, \partial x$.

5. For each set of functions in Exercise 3, find $\partial^2 f/\partial u^2$.

6. Let $f(u)$ be differentiable function of u. Use the chain rule to find

$$\frac{\partial}{\partial x_j} f\left(|\vec{x}|\right).$$

7. An airplane is flying east by northeast (22.5° from due east) at a constant speed of 400 km/h. At the moment when it is 1 km east and 2 km north of a radar station that measures the airplane's position in polar coordinates, (r, θ), find dr/dt and $d\theta/dt$.

8. Let f be a scalar field defined on (x, y) where $x = r\cos\theta$, $y = r\sin\theta$. Show that

$$\nabla f = \frac{\partial f}{\partial r}\vec{u}_r + r^{-1}\frac{\partial f}{\partial \theta}\vec{u}_\theta, \tag{7.26}$$

where $\vec{u}_r = (\cos\theta, \sin\theta)$ and $\vec{u}_\theta = (-\sin\theta, \cos\theta)$. Hint: first find $\partial f/\partial r$ and $\partial f/\partial \theta$ in terms of $\partial f/\partial x$ and $\partial f/\partial y$ and then solve for $\partial f/\partial x$ and $\partial f/\partial y$ in terms of $\partial f/\partial r$ and $\partial f/\partial \theta$.

9. Let f be a scalar field defined on (x, y) where $x = r\cos\theta$, $y = r\sin\theta$ and for which the mixed partials are equal:

$$\frac{\partial^2 f}{\partial x\, \partial y} = \frac{\partial^2 f}{\partial y\, \partial x}.$$

Find $\partial f/\partial r$, $\partial f/\partial \theta$, $\partial^2 f/\partial r^2$, $\partial^2 f/\partial r\, \partial \theta$, and $\partial^2 f/\partial \theta^2$ in terms of the partial derivatives of f with respect to x and y.

10. For f, x, y, and z as defined in Exercise 9, show that

$$\frac{\partial^2 f}{\partial x^2} + \frac{\partial^2 f}{\partial y^2} = \frac{\partial^2 f}{\partial r^2} + \frac{1}{r^2}\frac{\partial^2 f}{\partial \theta^2} + \frac{1}{r}\frac{\partial f}{\partial r}. \tag{7.27}$$

11. Let f be a scalar field defined on (x, y, z) where $x = r\cos\theta\cos\phi$, $y = r\sin\theta\cos\phi$, and $z = r\sin\phi$. Find $\partial f/\partial r$, $\partial f/\partial \theta$, and $\partial f/\partial \phi$ in terms of the partial derivatives of f with respect to x, y, and z.

12. For each of the following scalar fields, find the direction in which the directional derivative is maximized:

   (a) $f(x, y) = x/\sqrt{x^2 + y^2}$ at $(1, -2)$,
   (b) $f(x, y) = \log(x^2 + 2y^2)$ at $(-1, 1)$,
   (c) $f(x, y, z) = x^2 y^2 - xz^3 + y^4 z^2$ at $(1, -1, 2)$,
   (d) $f(x, y, z) = \sin xy - \cos xz$ at $(\pi, 1/2, 1)$.

13. What is the physical significance of those values of $\vec{x}$ for which

$$\nabla f(\vec{x}) = \vec{0}?$$

14. To be done on a computer or programmable calculator: A practical means of finding the approximate location of a local maximum of a scalar field is to take discrete steps in the direction of the gradient. If you are lost in the woods and want to reach a high point, just keep walking uphill. In practice, you select a starting point, (x_0, y_0), and a step size, s, and recursively define

$$(x_{i+1}, y_{i+1}) = (x_i, y_i) + s\frac{\nabla f(x_i, y_i)}{|\nabla f(x_i, y_i)|}.$$

Let

$$f(x, y) = \frac{2}{x^2 - 4x + y^2 + 5} + \frac{1}{x^2 - 4y + y^2 + 5},$$

and investigate what happens if

    (a) $(x_0, y_0) = (0, 0)$, $s = 0.5$,

    (b) $(x_0, y_0) = (0, 1)$, $s = 0.5$,

    (c) $(x_0, y_0) = (2, 3)$, $s = 0.5$,

    (d) $(x_0, y_0) = (3, 2)$, $s = 0.5$,

    (e) $(x_0, y_0) = (1, 1)$, $s = 0.1$,

    (f) $(x_0, y_0) = (2, 0)$, $s = 0.001$,

    (g) $(x_0, y_0) = (0, 2)$, $s = 0.001$.

15. One of the problems of setting up a program to find the approximate value of a local maximum is the determination of the step size s. If s is to small, then it can take a very long time to get to the local maximum. If s is too large, then it is difficult to hit the local maximum with any accuracy. A possible solution is to have a variable step size which gets smaller as you get closer to the local maximum. Write and implement a program that will do this.

16. Let $\vec{F}(\vec{x}) = (f_1(\vec{x}), f_2(\vec{x}), \ldots, f_m(\vec{x}))$ be differentiable at $\vec{c}$. Show that for any unit vector $\vec{u}$

$$|\vec{F}'_c(\vec{u})| \le \sqrt{|\nabla f_1(\vec{c})|^2 + \cdots + |\nabla f_m(\vec{c})|^2},$$

and therefore, $|\vec{F}'_c(\vec{u})|$ is bounded as a function of $\vec{u}$.

17. Find the equation of the tangent plane to each of the given surfaces at the indicated point:

    (a) $x^2 + y^2 - z = 2$ at $(3, -1, 8)$,

    (b) $yz + \sin xy - \cos xz = 5/2$ at $(\pi, 1/2, 1)$,

    (c) $z = x^3 - y^3$ at $(2, 1, 7)$,

    (d) $xe^{yz} + ye^{xz} = 1$ at $(1, 0, -2)$.

18. Show that every tangent plane to the cone $x^2 + y^2 = z^2$ passes through the origin.

19. Find the point or points where $x^2 - 2y^2 + 2z = 1$ is parallel to the plane $3x - 2y + z = 2$.

20. Show that the osculating plane is not necessarily the same as the tangent plane by considering the curve

$$\vec{r}(t) = (\sqrt{2}\cos t, 1, \sqrt{2}\sin t)$$

on the surface $x^2 - y^2 + z^2 = 1$ at $(\sqrt{2}, 1, 0)$, $t = 0$.

8

Integration by Pullback

8.1   Change of Variables

We interrupt our study of differential calculus to apply the results of Section 4 of Chapter 7 to the outstanding problem of evaluating pullbacks. As we saw in Theorem 7.4, if

$$\vec{F}(x_1,\, x_2,\, \ldots,\, x_m) = (y_1,\, y_2,\, \ldots,\, y_m),$$

then

$$dy_1\, dy_2 \cdots dy_m = \frac{\partial(y_1, y_2, \ldots, y_m)}{\partial(x_1, x_2, \ldots, x_m)} dx_1\, dx_2 \cdots dx_m.$$

Furthermore, as we saw in Chapter 6, the Jacobian,

$$\frac{\partial(y_1, y_2, \ldots, y_m)}{\partial(x_1, x_2, \ldots, x_m)} = \begin{vmatrix} \frac{\partial y_1}{\partial x_1} & \frac{\partial y_1}{\partial x_2} & \cdots & \frac{\partial y_1}{\partial x_m} \\ \frac{\partial y_2}{\partial x_1} & \frac{\partial y_2}{\partial x_2} & \cdots & \frac{\partial y_2}{\partial x_m} \\ & & \vdots & \\ \frac{\partial y_m}{\partial x_1} & \frac{\partial y_m}{\partial x_2} & \cdots & \frac{\partial y_m}{\partial x_m} \end{vmatrix},$$

when evaluated at $\vec{c}$, is the signed m-dimensional volume of the parallelepiped spanned by

$$\vec{F}_c'(\vec{i}_1) = \left(\frac{\partial y_1}{\partial x_1}(\vec{c}), \frac{\partial y_1}{\partial x_2}(\vec{c}), \ldots, \frac{\partial y_1}{\partial x_m}(\vec{c}) \right),$$

$$\vec{F}_c'(\vec{i}_2) = \left(\frac{\partial y_2}{\partial x_1}(\vec{c}), \frac{\partial y_2}{\partial x_2}(\vec{c}), \ldots, \frac{\partial y_2}{\partial x_m}(\vec{c}) \right),$$

$$\vdots$$

$$\vec{F}_c'(\vec{i}_m) = \left(\frac{\partial y_m}{\partial x_1}(\vec{c}), \frac{\partial y_m}{\partial x_2}(\vec{c}), \ldots, \frac{\partial y_m}{\partial x_m}(\vec{c}) \right).$$

Therefore, the linear transformation $\vec{F}_c'$ takes a cube of unit volume in $\vec{x}$ space to a parallelepiped whose signed volume is the Jacobian evaluated at $\vec{c}$. We know that

$$\vec{F}(\vec{x}) = \vec{F}(\vec{c}) + \vec{F}_c'(\vec{x} - \vec{c}) + |\vec{x} - \vec{c}|\vec{E},$$

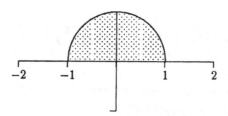

FIGURE 8.1. $D = \{(x, y) \mid x^2 + y^2 \leq 1, \ y \geq 0\}$.

where the error, $\vec{E}$, can be made arbitrarily small by taking $\vec{x}$ sufficiently close to $\vec{c}$. Thus, in the immediate neighborhood of $\vec{c}$, $\vec{F}$ takes a cube of unit volume to an m-dimensional solid of a volume that can be made arbitrarily close to

$$\frac{\partial(y_1, y_2, \ldots, y_m)}{\partial(x_1, x_2, \ldots, x_m)}(\vec{c}).$$

As we shall see in the next chapter, if the Jacobian of $\vec{F}$ at $\vec{c}$ is not zero, then $\vec{F}$ is invertible near $\vec{c}$. The inverse transformation, $\vec{F}^{-1}$, takes a unit cube near $\vec{F}(\vec{c})$ back to an m-dimensional solid of a volume that can be made arbitrarily close to

$$\left(\frac{\partial(y_1, y_2, \ldots, y_m)}{\partial(x_1, x_2, \ldots, x_m)}(\vec{c})\right)^{-1}.$$

It follows that if R is a bounded region in $\vec{y}$ space, then

$$\begin{aligned}
\text{Volume}(R) &= \int_R dy_1 \, dy_2 \cdots dy_m \\
&= \int_{\vec{F}^{-1}(R)} \frac{\partial(y_1, y_2, \ldots, y_m)}{\partial(x_1, x_2, \ldots, x_m)} dx_1 \, dx_2 \cdots dx_m. \quad (8.1)
\end{aligned}$$

Example

As an example, let D be the upper half of the unit disc (Figure 8.1):

$$D = \{(x, y) \mid x^2 + y^2 \leq 1, \ y \geq 0\},$$

and consider the integral

$$\int_D x^2 y \, dx \, dy.$$

We shall integrate this by pullback to ρ, θ space where

$$x = \rho \cos \theta, \quad y = \rho \sin \theta,$$

$$\begin{pmatrix} dx \\ dy \end{pmatrix} = \begin{pmatrix} \cos\theta & -\rho\sin\theta \\ \sin\theta & \rho\cos\theta \end{pmatrix} \begin{pmatrix} d\rho \\ d\theta \end{pmatrix}.$$

The half disc is pulled back to the rectangle

$$R = \{(\rho, \theta) \mid 0 \le \rho \le 1, \ 0 \le \theta \le \pi\}.$$

The Jacobian is

$$\begin{vmatrix} \cos\theta & -\rho\sin\theta \\ \sin\theta & \rho\cos\theta \end{vmatrix} = \rho\cos^2\theta + \rho\sin^2\theta = \rho,$$

and so the pullback of $dx\,dy$ is

$$dx\,dy = \rho\,d\rho\,d\theta.$$

Making our substitutions, the integral becomes

$$\begin{aligned}
\int_D x^2 y\,dx\,dy &= \int_R \rho^2\cos^2\theta\,\rho\sin\theta\,\rho\,d\rho\,d\theta \\
&= \int_0^1 \int_0^\pi \rho^4 \cos^2\theta\,\sin\theta\,d\theta\,d\rho \\
&= \frac{2}{3}\int_0^1 \rho^4\,d\rho \\
&= \frac{2}{15}.
\end{aligned}$$

8.2   Interlude with Lagrange

The first published account of the use of change of variables to evaluate a 3-dimensional integral was Lagrange's 1773 paper "Sur l'Attraction des Sphéroïdes Elliptiques" ("On the Attraction of Elliptical Spheroids") published in *Nouveaux Mémoires de l'Academie royale des Sciences et Belle-Lettres de Berlin*. The first problem addressed was to find the components of the gravitational force generated by an elliptical spheroid acting on a point mass at (a, b, c). Lagrange begins by explaining that if S is the solid generating the force, then the force vector is given by

$$\left(\int_S \frac{x-a}{r^3}\,dx\,dy\,dz, \ \int_S \frac{y-b}{r^3}\,dx\,dy\,dz, \ \int_S \frac{z-c}{r^3}\,dx\,dy\,dz \right),$$

where

$$r = \sqrt{(x-a)^2 + (y-b)^2 + (z-c)^2}.$$

To compute the integrals, he wants to make a change of variables to spherical coordinates with center at (a, b, c). In doing this, he sets out a general method for changing variables in multiple integrals. His language is sufficiently modern, and his approach so ingenious, that I am including his account in hopes that it will motivate the reader to examine the reasoning and appreciate what was happening two and a quarter centuries ago.

Lagrange's explanation is particularly interesting on two counts. First of all, he tackles the problem of what is meant by the product of differentials, probably the first time this was ever done. If his reasoning is slightly obscure, he nevertheless achieves the correct result while describing a method that is valid for any number of dimensions. Second, he accomplishes this at a time when the geometric significance of a 3×3 determinant was not yet understood. In fact, Lagrange's next paper was to be the geometric interpretation of the 3×3 determinant, suggesting that it was his work on change of variables that directly led to the understanding of the determinant as a volume. Or, perhaps it happened in the other order.

I quote Lagrange directly. He uses

$$A, B, C, D, E, F, G, H, I,$$

to denote

$$\frac{\partial x}{\partial p}, \frac{\partial x}{\partial q}, \frac{\partial x}{\partial r}, \frac{\partial y}{\partial p}, \frac{\partial y}{\partial q}, \frac{\partial y}{\partial r}, \frac{\partial z}{\partial p}, \frac{\partial z}{\partial q}, \frac{\partial z}{\partial r},$$

respectively.

PROBLEM II

"*Let us suppose that we have a differential $P\,dx\,dy\,dz$, where P is a given function of x, y, z which should be integrated three times as we successively range over the variables x, y, z, observing the conditions laid out in Problem I; we propose to introduce in place of these variables three other variables p, q, r which are given functions of the others.*

"Since p, q, r are assumed to be functions of x, y, z, we shall also have, reciprocally, that x, y, z can be expressed as functions of p, q, r; we shall therefore first make these substitutions into the quantity P, which will reduce it to a function of p, q, r, and the only difficulty will be with regard to the quantity $dx\,dy\,dz$.

"If we use differentiation to find the values of the differentials dx, dy, dz, then we shall have, in general,

$$\begin{aligned} dx &= A\,dp + B\,dq + C\,dr, \\ dy &= D\,dp + E\,dq + F\,dr, \\ dz &= G\,dp + H\,dq + I\,dr, \end{aligned}$$

$A, B, C, \dots$ being known functions of p, q, r; now it is easy to understand that to find the value of $dx\,dy\,dz$, we should not multiply together the preceding values of dx, dy, dz; because then the differential $P\,dx\,dy\,dz$ would contain terms where the differentials dp, dq, dr are found raised to the square or the cube, and thus triple integration which should be made relative to the three variables p, q, r will no longer be possible; moreover, as $dx\,dy\,dz$ expresses a solid element of the body, it is clear that whatever variables p, q, r we introduce in place of the variables x, y, z, this element can only be represented by the product $dp\,dq\,dr$ of the three differentials dp, dq, dr multiplied by some function of p, q, r. I therefore consider that in the expression of the parallelepiped $dx\,dy\,dz$, the differential dz should be taken while x and y remain constant; that next the differential dy should be taken while considering x and z to be constant, and that finally the differential dx should be taken while supposing dy and dz to be zero; therefore:

1. "To find the value of dz, we shall set $dx = 0$ and $dy = 0$, which will give

$$A\,dp + B\,dq + C\,dr = 0,$$
$$D\,dp + E\,dq + F\,dr = 0,$$

from which we shall pull dp and dq in terms of dr, that is to say,

$$dp = \frac{BF - CE}{AE - BD}\,dr, \quad dq = \frac{CD - AF}{AE - BD}\,dr,$$

and substituting these values into the expression for dz, we shall have

$$dz = \frac{G(BF - CE) + H(CD - AF) + I(AE - BD)}{AE - BD}\,dr.$$

2. "To find the value of dy, we shall set $dx = 0$ and $dz = 0$, which will give

$$dr = 0, \quad A\,dp + B\,dq = 0,$$

from which we pull

$$dp = -\frac{B\,dq}{A},$$

and substituting these values into the expression for dy, we shall have

$$dy = \frac{AE - BD}{A}\,dq.$$

3. "Finally to find the value of dx, we shall set $dy = 0$ and $dz = 0$, which yields

$$dr = 0, \quad dq = 0,$$

from which we shall have

$$dx = A\,dp.$$

"Now, multiplying these values together, we shall have

$$dx\,dy\,dz = [G(BF - CE) + H(CD - AF) + I(AE - BD)]\,dp\,dq\,dr,$$

where we observe that the quantity

$$G(BF - CE) + H(CD - AF) + I(AE - BD)$$

or equivalently

$$AEI + BFG + CDH - AFH - BDI - CEG$$

remains the same, or at most changes sign, if the three systems of quantities A,B,C; D,E,F; G,H,I are respectively exchanged among themselves; from which we shall have the same result if, instead of first looking as we did for the value of dz, then that of dy and of dx, we wish to begin by looking for that of dy or of dx, and then either one of the other two differentials. The only possible change is in the sign, but this will not be of any importance as it does not matter whether the element $\alpha = dx\,dy\,dz$ is taken with a plus or a minus; however, as it is more natural to regard this quantity as positive, we shall always take care to take the value of α positively; this is why we shall assume, in general,

$$\alpha = \pm(AEI + BFG + CDH - AFH - BDI - CEG)\,dp\,dq\,dr."$$

8.3   Exercises

1. Let R be the parallelogram with vertices at $(1, 0)$, $(3, 1)$, $(4, 4)$, $(2, 3)$. Show that

$$\begin{aligned} x &= 1 + 2u + v, \\ y &= u + 3v \end{aligned}$$

maps the unit square with vertices at $(0, 0)$, $(1, 0)$, $(1, 1)$, $(0, 1)$ to R. Use the pullback to evaluate

$$\int_R (2x + y)\,dx\,dy.$$

2. Use a pullback to evaluate

$$\int_R xy\,dx\,dy,$$

where R is the parallelogram with vertices at $(-1, -2)$, $(3, 0)$, $(2, 2)$, $(-2, 0)$.

3. Let R be the region bounded by $y = x^2$, $y = 2x^2$, and $x = 1$. Evaluate

$$\int_R xy^2 \, dx \, dy$$

by setting $x = \sqrt{u}$, $y = v$ and first pulling the integral back to an integral in u, v space.

4. Let R be the region

$$R = \{(x, y) \mid 1 \leq x^2 + y^2 \leq 2, \ y \geq |x|\}.$$

Sketch this region. Let $u = x/y$, $v = x^2 + y^2$ so that R pulls back to the rectangle

$$-1 \leq u \leq 1, \quad 1 \leq v \leq 2.$$

Show that $du \, dv = 2(1 + u^2)dx \, dy$, and thus

$$dx \, dy = \frac{1}{2(1 + u^2)} \, du \, dv.$$

5. Let $C(R)$ be the circle of radius R with center at the origin: $C(R) = \{(x, y) \mid x^2 + y^2 \leq R\}$. Evaluate the integral

$$I(p, R) = \int_{C(R)} \frac{dx \, dy}{(p^2 + x^2 + y^2)^p}.$$

Determine those values of p for which

$$\lim_{R \to +\infty} I(p, R)$$

exists.

6. Define the "error function," $\operatorname{erf}(x)$, by

$$\operatorname{erf}(x) = \int_0^x e^{-u^2} \, du.$$

(a) Show that $2 \operatorname{erf}(x) = \int_{-x}^x e^{-u^2} du$, and thus

$$4 \operatorname{erf}^2(x) = \int_R e^{-u^2 - v^2} \, du \, dv,$$

where R is the rectangle $\{(u, v) \mid -x \leq u, v \leq x\}$.

(b) Show that

$$\int_{C_1} e^{-u^2 - v^2} \, du \, dv \leq \int_R e^{-u^2 - v^2} \, du \, dv \leq \int_{C_2} e^{-u^2 - v^2} \, du \, dv,$$

where C_1 is the largest circle that lies inside R and C_2 is the smallest circle that contains R.

(c) Express the integrals over C_1 and C_2 in terms of polar coordinates and show that

$$\lim_{x \to \infty} \int_R e^{-u^2 - v^2} \, du \, dv = \pi.$$

(d) Use this last result to prove

$$\int_0^\infty e^{-u^2} \, du = \frac{1}{2} \sqrt{\pi}.$$

7. Let R be the region

$$R = \{(x, y) \mid a \le f(x, y) \le b, \ c \le g(x, y) \le d\},$$

where we assume that f and g are differentiable everywhere in R and that the Jacobian

$$\frac{\partial(f, g)}{\partial(x, y)}$$

is never 0 in R. Let $u = f(x, y), v = g(x, y)$. Find the pullback of $dx \, dy$ to u, v space, expressed in terms of the partial derivatives of f and g with respect to x and y.

8. What is the Jacobian of the transformation from cylindrical coordinates:

$$\begin{aligned} x &= \rho \cos \theta, \\ y &= \rho \sin \theta, \\ z &= z? \end{aligned}$$

9. What is the Jacobian of the transformation from spherical coordinates:

$$\begin{aligned} x &= \rho \cos \theta \, \cos \phi, \\ y &= \rho \sin \theta \, \cos \phi, \\ z &= \rho \sin \phi? \end{aligned}$$

In Exercises 10 through 12, we use general spherical coordinates to compute the hypervolume of an n-dimensional sphere (an n sphere) of radius r.

10. Justify the following parametrization of the n sphere:

$$\begin{aligned} x_1 &= \rho \cos \theta \, \cos \phi_1 \cdots \cos \phi_{n-2}, \\ x_2 &= \rho \sin \theta \, \cos \phi_1 \cdots \cos \phi_{n-2}, \end{aligned}$$

$$x_3 = \rho \sin\phi_1 \cos\phi_2 \cdots \cos\phi_{n-2},$$

$$\vdots$$

$$x_{n-1} = \rho \sin\phi_{n-3} \cos\phi_{n-2},$$
$$x_n = \rho \sin\phi_{n-2},$$

where

$$0 \le \rho \le r, \ 0 \le \theta \le 2\pi, \ -\frac{\pi}{2} \le \phi_1 \le \frac{\pi}{2}, \ldots, -\frac{\pi}{2} \le \phi_{n-2} \le \frac{\pi}{2}.$$

11. Prove that the Jacobian of the transformation of Exercise 10 is

$$\frac{\partial(x_1, x_2, \ldots, x_n)}{\partial(\rho, \theta, \phi_1, \ldots, \phi_{n-2})} = \rho^{n-1} \cos\phi_1 \, \cos^2\phi_2 \ldots \cos^{n-2}\phi_{n-2}. \quad (8.2)$$

Hint: add to the first column of the Jacobian matrix an appropriate multiple of the nth column and use induction.

12. Let $S(r)$ be the n sphere of radius r. Integrate the pullback of

$$\int_{S(r)} dx_1 \, dx_2 \ldots dx_n$$

and so prove that the volume of $S(r)$ is

$$\frac{2^{n/2}\pi^{n/2}r^n}{2 \cdot 4 \cdot 6 \cdot \cdots \cdot n} = \frac{\pi^{n/2}}{(n/2)!}r^n, \quad n \text{ even}, \quad (8.3)$$

$$\frac{2^{(n+1)/2}\pi^{(n-1)/2}r^n}{1 \cdot 3 \cdot 5 \cdot \cdots \cdot n} = \frac{[(n-1)/2]!}{n!}2^n\pi^{(n-1)/2}r^n, \quad n \text{ odd}. \, (8.4)$$

13. Prove that Lagrange's method will work for a change of variables in any number of dimensions: Let $\vec{F} : \mathbf{R}^n \longrightarrow \mathbf{R}^n$ be a differentiable transformation,

$$\vec{y} = \vec{F}(\vec{x}).$$

We want to relate $dy_1 \, dy_2 \ldots dy_n$ to $dx_1 \, dx_2 \ldots dx_n$.

(a) Set up the differential equations

$$dy_1 = \frac{\partial y_1}{\partial x_1}dx_1 + \frac{\partial y_1}{\partial x_2}dx_2 + \cdots + \frac{\partial y_1}{\partial x_n}dx_n,$$

$$dy_2 = \frac{\partial y_2}{\partial x_1}dx_1 + \frac{\partial y_2}{\partial x_2}dx_2 + \cdots + \frac{\partial y_2}{\partial x_n}dx_n,$$

$$\vdots$$

$$dy_n = \frac{\partial y_n}{\partial x_1}dx_1 + \frac{\partial y_n}{\partial x_2}dx_2 + \cdots + \frac{\partial y_n}{\partial x_n}dx_n.$$

Prove that if $dy_1 = dy_2 = \cdots = dy_{n-1} = 0$, then

$$dx_1 = (-1)^n \frac{\partial(y_1, y_2, \ldots, y_{n-1})/\partial(x_2, x_3, \ldots, x_n)}{\partial(y_1, y_2, \ldots, y_{n-1})/\partial(x_1, x_2, \ldots, x_{n-1})} \, dx_n,$$

$$dx_2 = (-1)^{n-1} \frac{\partial(y_1, y_2, \ldots, y_{n-1})/\partial(x_1, x_3, \ldots, x_n)}{\partial(y_1, y_2, \ldots, y_{n-1})/\partial(x_1, x_2, \ldots, x_{n-1})} \, dx_n,$$

$$\vdots$$

$$dx_{n-1} = -\frac{\partial(y_1, y_2, \ldots, y_{n-1})/\partial(x_1, x_2, \ldots, x_{n-2}, x_n)}{\partial(y_1, y_2, \ldots, y_{n-1})/\partial(x_1, x_2, \ldots, x_{n-1})} \, dx_n.$$

(b) Continuing from part a, show that

$$dy_n = \frac{\partial(y_1, y_2, \ldots, y_n)/\partial(x_1, x_2, \ldots, x_n)}{\partial(y_1, y_2, \ldots, y_{n-1})/\partial(x_1, x_2, \ldots, x_{n-1})} \, dx_n.$$

(c) Assume that we have proven for some $j > 1$ that if $j < i \leq n$ then

$$dy_i = \frac{\partial(y_1, y_2, \ldots, y_i)/\partial(x_1, x_2, \ldots, x_i)}{\partial(y_1, y_2, \ldots, y_{i-1})/\partial(x_1, x_2, \ldots, x_{i-1})} \, dx_i. \qquad (8.5)$$

Part b proved this result when $j = n - 1$. Prove that if we now set

$$dy_1 = dy_2 = \cdots = dy_{j-1} = dy_{j+1} = \cdots = dy_n = 0,$$

then

$$dy_j = \frac{\partial(y_1, y_2, \ldots, y_j)/\partial(x_1, x_2, \ldots, x_j)}{\partial(y_1, y_2, \ldots, y_{j-1})/\partial(x_1, x_2, \ldots, x_{j-1})} \, dx_j.$$

(d) Assuming that Equation (8.5) holds for all $i > 1$, prove that if we set

$$dy_2 = dy_3 = \cdots = dy_n = 0,$$

then

$$dy_1 = \frac{\partial y_1}{\partial x_1} \, dx_1.$$

(e) Complete the proof that Lagrange's approach will always yield the relationship

$$dy_1 \, dy_2 \ldots dy_n = \frac{\partial(y_1, y_2, \ldots, y_n)}{\partial(x_1, x_2, \ldots, x_n)} \, dx_1 \, dx_2 \ldots dx_n,$$

in any number of dimensions.

14. Why does Lagrange's approach work?

8.4   The Surface Integral

The integral of a 2-form over a surface is defined in terms of a parametrization as the integral of the pullback. As an example, let us integrate

$$\int_S x^2 \, dy \, dz - yz^3 \, dz \, dx + y^3 \, dx \, dy$$

over the surface of the hemisphere

$$S = \{(x, y, z) \mid x^2 + y^2 + z^2 = 1, x \geq 0\},$$

oriented so that the positive direction is away from the origin. We parametrize this surface using latitude (ϕ) and longitude (θ):

$$
\begin{aligned}
x &= \cos\theta \, \cos\phi, \\
y &= \sin\theta \, \cos\phi, \\
z &= \sin\phi, \\
-\frac{\pi}{2} \leq \theta \leq \frac{\pi}{2}, \quad & -\frac{\pi}{2} \leq \phi \leq \frac{\pi}{2},
\end{aligned}
$$

and pull back each of our 2-forms to θ, ϕ space.

Determining Orientation

Before going any further with our example, we need to determine whether it pulls back to θ, ϕ space or to ϕ, θ space. That is, which order on the variables preserves the sense of positive flow? The most consistently applicable method for determining orientation is to choose three noncollinear points on the surface and order them so that if we travel around them, the right-hand rule points us in the direction of positive flow. We now pull those three points back to θ, ϕ space and see if traveling from one to the next takes us around a triangle in the counterclockwise (positive) direction. If yes, then the pullback is to θ, ϕ space, if no, then the pullback should be to ϕ, θ space, where the three points will necessarily be traversed in counterclockwise rotation.

In practice, we choose three points for which the pullback is as simple as possible. Let us take $(1, 0, 0)$, $(0, 1, 0)$, and $(0, 0, 1)$. In that order, the right-hand rule describes the direction of positive flow (see Figure 8.2). These three points correspond in the θ, ϕ plane to $(0, 0)$, $(\pi/2, 0)$, $(0, \pi/2)$, respectively. Since the triangle $[(0,0), (\pi/2, 0), (0, \pi/2)]$ has positive orientation, we have chosen the order of the pullback correctly.

Note that if we pulled back to the ϕ, θ plane, then $[(1,0,0),(0,1,0),(0,0,1)]$ would pull back to $[(0,0), (0, \pi/2), (\pi/2, 0)]$, which has a negative orientation.

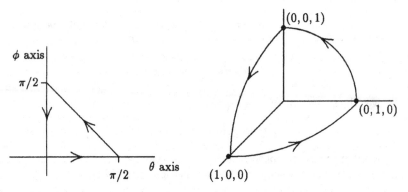

FIGURE 8.2. Direction of positive flow must match.

Our example is now completed by finding the relationship between the differentials of x, y, and z and those of θ and ϕ, substituting and simplifying, and then evaluating the resulting multiple integral. Note that once we have pulled our integral back to the θ, ϕ plane and have rewritten it as a multiple integral, then we are free to change the order of integration.

$$\int_S x^2 \, dy \, dz - yz^3 \, dz \, dx + y^3 \, dx \, dy$$

$$= \int_R \cos^2 \theta \, \cos^2 \phi \, (\cos \theta \, \cos \phi \, d\theta - \sin \theta \, \sin \phi \, d\phi)(\cos \phi \, d\phi)$$

$$- \int_R \sin \theta \, \cos \phi \, \sin^3 \phi \, (\cos \phi \, d\phi)$$

$$\times (-\sin \theta \, \cos \phi \, d\theta - \cos \theta \, \sin \phi \, d\phi)$$

$$+ \int_R \sin^3 \theta \, \cos^3 \phi \, (-\sin \theta \, \cos \phi \, d\theta - \cos \theta \, \sin \phi \, d\phi)$$

$$\times (\cos \theta \, \cos \phi \, d\theta - \sin \theta \, \sin \phi \, d\phi)$$

$$= \int_R \cos^2 \theta \, \cos^2 \phi \, (\cos \theta \, \cos^2 \phi \, d\theta \, d\phi)$$

$$- \int_R \sin \theta \, \cos \phi \, \sin^3 \phi \, (\sin \theta \, \cos^2 \phi \, d\theta \, d\phi)$$

$$+ \int_R \sin^3 \theta \, \cos^3 \phi \, (\cos \phi \, \sin \phi \, d\theta \, d\phi)$$

$$= \int_{-\pi/2}^{\pi/2} \int_{-\pi/2}^{\pi/2} (\cos^3 \theta \, \cos^4 \phi - \sin^2 \theta \, \cos^3 \phi \, \sin^3 \phi$$

$$+ \sin^3 \theta \, \cos^4 \phi \, \sin \phi) \, d\theta \, d\phi$$

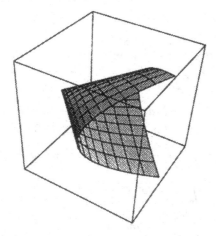

FIGURE 8.3. The surface $x = u^2 - v^2$, $y = u^2 + v^2$, $z = uv$.

$$= \int_{-\pi/2}^{\pi/2} \int_{-\pi/2}^{\pi/2} (\cos^3\theta \, \cos^4\phi - \sin^2\theta \, \cos^3\phi \, \sin^3\phi$$
$$+ \sin^3\theta \, \cos^4\phi \, \sin\phi) \, d\phi \, d\theta$$
$$= \int_{-\pi/2}^{\pi/2} \frac{3\pi}{8} \cos^3\theta \, d\theta = \frac{\pi}{2}.$$

Another Example

Consider the surface S (Figure 8.3) parametrized by

$$x = u^2 - v^2,$$
$$y = u^2 + v^2,$$
$$z = uv,$$
$$0 \le u \le 1, \quad 0 \le v \le 1.$$

We want to compute the rate at which the constant flow $(1,1,1)$ crosses this surface. We take the direction with the positive z coordinate as the positive direction. The triangle $[(0,0),(1,0),(0,1)]$ has positive orientation in the u, v plane. It corresponds to the triangle $[(0,0,0),(1,1,0),(-1,1,0)]$ on our surface. This triangle has a *positive* orientation, and so the pullback should be to u, v space:

$$\int_S dy\,dz + dz\,dx + dx\,dy$$

$$= \int_R (2u\,du + 2v\,dv)(v\,du + u\,dv)$$
$$+ (v\,du + u\,dv)(2u\,du - 2v\,dv)$$
$$+ (2u\,du - 2v\,dv)(2u\,du + 2v\,dv)$$

$$= \int_R (2u^2 - 2v^2 - 2u^2 - 2v^2 + 4uv + 4uv)\,du\,dv$$

$$= \int_0^1 \int_0^1 -4v^2 + 8uv\,du\,dv$$

$$= \int_0^1 \left(-\frac{4}{3} + 4u\right)\,du = \frac{2}{3}.$$

Integration with Respect to Surface Area

Given a surface in $\mathbf{R}^3$ with a differentiable and invertible parametrization,

$$(x, y, z) = \vec{F}(u, v)$$
$$= (f_1(u, v), f_2(u, v), f_3(u, v)),$$

we can use pullbacks to compute the area of a bounded piece of this surface. To do this, we need to calculate the magnification effect of our transformation $\vec{F}$ at a particular spot in the u, v plane, for example at $\vec{c}$.

The differentiability of $\vec{F}$ implies that for small values of Δu and Δv

$$\vec{F}(\vec{c} + (\Delta u, \Delta v)) \simeq \vec{F}(\vec{c}) + \vec{F}'_c(\Delta u, \Delta v), \qquad (8.6)$$

where

$$\vec{F}'_c(\Delta u, \Delta v)$$
$$= \left(\frac{\partial f_1}{\partial u}\Delta u + \frac{\partial f_1}{\partial v}\Delta v, \frac{\partial f_2}{\partial u}\Delta u + \frac{\partial f_2}{\partial v}\Delta v, \frac{\partial f_3}{\partial u}\Delta u + \frac{\partial f_3}{\partial v}\Delta v\right)$$
$$= \frac{\partial \vec{F}}{\partial u}\Delta u + \frac{\partial \vec{F}}{\partial v}\Delta v.$$

The partial derivatives are each evaluated at $\vec{c}$, and we have introduced a convenient shorthand in the last line: if $\vec{F} = (f_1, f_2, \ldots, f_n)$, then

$$\frac{\partial \vec{F}}{\partial u} = \left(\frac{\partial f_1}{\partial u}, \frac{\partial f_2}{\partial u}, \ldots, \frac{\partial f_n}{\partial u}\right). \qquad (8.7)$$

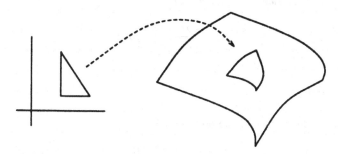

FIGURE 8.4. Mapping a triangle from u, v space to the surface.

The approximate equality of Equation (8.6) can be made arbitrarily close by taking Δu and Δv sufficiently small.

The triangle in the u, v plane with vertices at

$$\vec{c}, \quad \vec{c} + (\Delta u, 0), \quad \vec{c} + (0, \Delta v)$$

is approximately mapped to the triangle in $\mathbf{R}^3$ (Figure 8.4) with vertices at

$$\vec{F}(\vec{c}), \quad \vec{F}(\vec{c}) + \frac{\partial \vec{F}}{\partial u} \Delta u, \quad \vec{F}(\vec{c}) + \frac{\partial \vec{F}}{\partial v} \Delta v.$$

The triangle in the u, v plane has area

$$\frac{1}{2} \Delta u \, \Delta v.$$

The triangle on the surface has area

$$\frac{1}{2} \left| \frac{\partial \vec{F}}{\partial u} \Delta u \times \frac{\partial \vec{F}}{\partial v} \Delta v \right| = \frac{1}{2} \left| \frac{\partial \vec{F}}{\partial u} \times \frac{\partial \vec{F}}{\partial v} \right| \Delta u \, \Delta v;$$

the magnification factor is

$$\left| \frac{\partial \vec{F}}{\partial u}(\vec{c}) \times \frac{\partial \vec{F}}{\partial v}(\vec{c}) \right|.$$

It follows that if S is the bounded piece of surface whose area is to be calculated, then

$$\text{Area}(S) = \int_{\vec{F}^{-1}(S)} \left| \frac{\partial \vec{F}}{\partial u} \times \frac{\partial \vec{F}}{\partial v} \right| du \, dv. \tag{8.8}$$

It is often convenient to let $\int d\sigma$ denote integration with respect to surface area. The area of S is

$$\int_S d\sigma = \int_{\vec{F}^{-1}(S)} \left| \frac{\partial \vec{F}}{\partial u} \times \frac{\partial \vec{F}}{\partial v} \right| du \, dv, \tag{8.9}$$

from which we define the pullback of $d\sigma$:

$$d\sigma = \left| \frac{\partial \vec{F}}{\partial u} \times \frac{\partial \vec{F}}{\partial v} \right| du\, dv. \tag{8.10}$$

Example

Let S be the surface of the unit sphere:

$$
\begin{aligned}
x &= \cos\theta \,\cos\phi, \\
y &= \sin\theta \,\cos\phi, \\
z &= \sin\phi.
\end{aligned}
$$

This surface pulls back to a rectangle in the θ, ϕ plane:

$$R = \left\{ (\theta,\phi) \mid 0 \le \theta \le 2\pi,\ -\frac{\pi}{2} \le \phi \le \frac{\pi}{2} \right\}.$$

The area of S is

$$
\begin{aligned}
\int_S d\sigma &= \int_R |(-\sin\theta \,\cos\phi, \cos\theta \,\cos\phi, 0) \\
&\qquad \times (-\cos\theta \,\sin\phi, -\sin\theta \,\sin\phi, \cos\phi)|\, d\theta\, d\phi \\
&= \int_R |(\cos\theta \,\cos^2\phi, \sin\theta \,\cos^2\phi, \cos\phi \,\sin\phi)|\, d\theta\, d\phi \\
&= \int_R |\cos\phi|\, d\theta\, d\phi \\
&= \int_0^{2\pi} \int_{-\pi/2}^{\pi/2} \cos\phi \,d\phi\, d\theta = 4\pi.
\end{aligned}
$$

Since the pullback of $d\sigma$ is defined as the absolute value of the cross product of $\partial\vec{F}/\partial u$ and $\partial\vec{F}/\partial v$, it does not matter whether we pull back to the θ, ϕ plane or the ϕ, θ plane.

If the unit sphere has a density given by

$$
\begin{aligned}
\rho(x,y,z) &= x^2 + y^2 \\
&= \cos^2\theta \,\cos^2\phi + \sin^2\theta \,\cos^2\phi \\
&= \cos^2\phi,
\end{aligned}
$$

then the total mass is

$$
\begin{aligned}
\int_S \rho(x,y,z)\, d\sigma &= \int_R \cos^2\phi \,|\cos\phi|\, d\theta\, d\phi \\
&= \int_0^{2\pi} \int_{-\pi/2}^{\pi/2} \cos^3\phi \,d\phi\, d\theta = \frac{8\pi}{3}.
\end{aligned}
$$

A Second Look at Flows

Let S be a bounded piece of surface in $\mathbf{R}^3$ and let $\vec{v} : \mathbf{R}^3 \longrightarrow \mathbf{R}^3$ be a vector field describing a flow: $\vec{v}(x, y, z)$ is the magnitude and direction of the flow from (x, y, z). As we saw in Chapter 4, the rate at which our flow crosses S is given by

$$\text{rate of flow across } S = \int_S v_1 \, dy \, dz + v_2 dz \, dx + v_3 dx \, dy. \qquad (8.11)$$

We can use integration with respect to surface area to further justify this formula.

Let S be parametrized by

$$(x, y, z) = \vec{F}(u, v), \quad S = \vec{F}(R).$$

Let $\vec{n} = \vec{n}(x, y, z)$ be the unit normal to S at (x, y, z):

$$\vec{n} = \frac{(\partial \vec{F}/\partial u) \times (\partial \vec{F}/\partial v)}{\left|(\partial \vec{F}/\partial u) \times (\partial \vec{F}/\partial v)\right|}.$$

We take the integral of our 2-form and pull it back to u, v space:

$$
\begin{aligned}
\int_S v_1 \, dy \, dz + v_2 \, dz \, dx + v_3 \, dx \, dy &= \int_R v_1 \left(\frac{\partial f_2}{\partial u} \frac{\partial f_3}{\partial v} - \frac{\partial f_2}{\partial v} \frac{\partial f_3}{\partial u} \right) \\
&\quad + v_2 \left(\frac{\partial f_3}{\partial u} \frac{\partial f_1}{\partial v} - \frac{\partial f_3}{\partial v} \frac{\partial f_1}{\partial u} \right) \\
&\quad + v_3 \left(\frac{\partial f_1}{\partial u} \frac{\partial f_2}{\partial v} - \frac{\partial f_1}{\partial v} \frac{\partial f_2}{\partial u} \right) du \, dv \\
&= \int_R \vec{v} \cdot \left(\frac{\partial \vec{F}}{\partial u} \times \frac{\partial \vec{F}}{\partial v} \right) du \, dv \\
&= \int_R \vec{v} \cdot \vec{n} \left| \frac{\partial \vec{F}}{\partial u} \times \frac{\partial \vec{F}}{\partial v} \right| du \, dv \\
&= \int_S \vec{v} \cdot \vec{n} \, d\sigma. \qquad (8.12)
\end{aligned}
$$

The final surface integral has an obvious physical interpretation: the integrand $\vec{v} \cdot \vec{n}$ is the magnitude of the component of $\vec{v}$ in the direction of the normal to S, signed positive if it is in the same direction as $\vec{n}$, negative if in the opposite direction. Multiplying this by $d\sigma$, the element of surface area, and integrating generalizes the definition of constant flow across a triangular region: the signed magnitude of the component of flow in the normal direction times the area of the triangle.

This last integral

$$\int_S \vec{v} \cdot \vec{n} \, d\sigma$$

is often seen in textbooks and is very useful for remembering the physical meaning of the integral

$$\int_S v_1 \, dy \, dz + v_2 \, dz \, dx + v_3 \, dx \, dy.$$

Nevertheless, we shall work with the integral of the explicit 2-form in dx, dy, and dz. The latter has two overwhelming advantages. The first is computability. Given the parametrization of S, we only need to remember how to multiply differentials in order to evaluate it. The second is extendability. In order to tackle Maxwell's equations and Einstein's special relativity, we shall need to consider flows in 4-dimensional space–time. The proper analog of the first integral is not at all apparent. Viewed as a 2-form, the proper generalization is evident: we shall be integrating differential forms in x, y, z, t space.

8.5   Heat Flow

We can pull together a great deal of what has been done in this chapter and the previous one if we investigate the properties of the flow of heat from a point source at the origin. To speak of the "flow" of heat illuminates the fact that we are viewing heat as a fictitious fluid. Following this analogy, temperature is the measurement of heat density. Let $T(x, y, z)$ denote the temperature at (x, y, z). This is a scalar field. A reasonable model for a homogeneous medium is that the temperature is inversely proportional to the distance from the heat source:

$$T(x, y, z) = \frac{c}{\sqrt{x^2 + y^2 + z^2}}.$$

Under normal circumstances, heat flows in the direction of the greatest temperature difference, from higher to lower temperatures, with a velocity that is proportional to the difference in temperature. That is, the vector field describing the flow of heat is given by

$$-k \, \nabla T = ck(xr^{-3}, yr^{-3}, zr^{-3}),$$

where $r = \sqrt{x^2 + y^2 + z^2}$.

We now compute the rate at which heat crosses the sphere of radius a, $x^2 + y^2 + z^2 = a^2$, per unit time. Let S be the surface of our sphere, oriented so that an outward flow is positive. The rate at which our fictitious fluid crosses this surface is given by the integral of the 2-form:

$$\int_S \frac{ckx}{r^3} \, dy \, dz + \frac{cky}{r^3} \, dz \, dx + \frac{ckz}{r^3} \, dx \, dy.$$

To evaluate this integral, we find a parametrization for S:

$$
\begin{aligned}
x &= a\cos\theta\cos\phi, \\
y &= a\sin\theta\cos\phi, \\
z &= a\sin\phi,
\end{aligned}
$$

where $0 \le \theta \le 2\pi$, $-\pi/2 \le \phi \le \pi/2$. With this relationship, we can pull our integral back to the integral over the rectangle:

$$
R = \{(\theta,\phi)\,|\,0 \le \theta \le 2\pi, \ -\pi/2 \le \phi \le \pi/2\}.
$$

The rate at which heat crosses the unit sphere is

$$
\begin{aligned}
\int_S \frac{ckx}{r^3}\,dy\,dz &+ \frac{cky}{r^3}\,dz\,dx + \frac{ckz}{r^3}\,dx\,dy \\
&= \frac{ck}{a^3}\int_R (a^3\cos^2\theta\cos^3\phi + a^3\sin^2\theta\cos^3\phi + a^3\cos\phi\sin^2\phi)\,d\theta\,d\phi \\
&= \frac{ck}{a^3}a^3\int_R \cos\phi\,d\theta\,d\phi \\
&= ck\int_0^{2\pi}\int_{-\pi/2}^{\pi/2} \cos\phi\,d\phi\,d\theta = 4\pi ck.
\end{aligned}
$$

The result is somewhat surprising. We have shown that the rate at which fluid crosses the surface is $4\pi ck$ regardless of which radius we select. This means that if we consider concentric spherical shells, each with its center at the origin, then the rate at which fluid flows into the region between any two shells is equal to the rate at which the fluid flows out of that region. We have reached a steady state.

But, the beauty of what we have done lies in the great generality of our methods. We could have started with any differentiable scalar field describing temperature and any surface or piece of surface that could be parametrized by differentiable functions. For a nonhomogeneous medium, we merely replace the constant k with a function $k(x,y,z)$ describing heat conductivity at that point. The rate at which heat crosses the surface S is then given by

$$
\int_S k\frac{\partial T}{\partial x}\,dy\,dz + k\frac{\partial T}{\partial y}\,dz\,dx + k\frac{\partial T}{\partial z}\,dx\,dy. \tag{8.13}
$$

As we shall see in Chapter 10, we can actually use this integral to characterize the equilibrium state of a heated body.

8.6   Exercises

1. For each of the following parametrized surfaces, oriented so that the normal vector points away from the origin, determine the order of the two parametrizing variables so that positive orientation in the pullback corresponds to the positive orientation of the surface:

   (a) upper hemisphere:
   $x = \cos\theta \cos\phi$, $y = \sin\theta \cos\phi$, $z = \sin\phi$,
   $0 \le \theta \le 2\pi$, $0 \le \phi \le \pi/2$;

   (b) paraboloid opening downward with vertex at $(0, 0, 2)$:
   $x = r\cos\theta$, $y = r\sin\theta$, $z = 2 - 2r^2$,
   $0 \le r \le 1$, $0 \le \theta \le 2\pi$;

   (c) cone opening downward with vertex at (0,0,3):
   $x = r\cos\theta$, $y = r\sin\theta$, $z = 3 - 3r$,
   $0 \le r \le 1$, $0 \le \theta \le 2\pi$;

   (d) lower hemisphere:
   $x = \cos\theta \cos\phi$, $y = \sin\theta \cos\phi$, $z = -\sin\phi$,
   $0 \le \theta \le 2\pi$, $0 \le \phi \le \pi/2$;

   (e) paraboloid opening upward with vertex at (0,0,-2):
   $x = r\cos\theta$, $y = r\sin\theta$, $z = -2 + 2r^2$,
   $0 \le r \le 1$, $0 \le \theta \le 2\pi$,

   (f) cone opening upward with vertex at (0,0,-3):
   $x = r\cos\theta$, $y = r\sin\theta$, $z = -3 + 3r$,
   $0 \le r \le 1$, $0 \le \theta \le 2\pi$.

2. Evaluate
$$\int_S x \, dy \, dz - z \, dz \, dx + z^2 \, dx \, dy$$
over each of the oriented surfaces of Exercise 1.

3. Repeat Exercise 2 for the following integral:
$$\int_S y^2 \, dy \, dz + 2z \, dz \, dx - dx \, dy.$$

4. All six surfaces in Exercise 1 have the same boundary. Ponder the difference between the results of Exercises 2 and 3.

5. Find the rate at which the flow
$$\vec{v} = \left(\frac{x}{x^2 + y^2}, \frac{y}{x^2 + y^2}, \frac{z}{x^2 + y^2} \right)$$
crosses the cylindrical surface
$$S = \{(x, y, z) \mid x^2 + y^2 = 4, \, 0 \le z \le 2\},$$
oriented so that the positive direction is away from the z axis.

6. Given a charged surface, S, of charge density ρ, the electrostatic potential at a point (a, b, c) not on S is

$$\int_S \frac{\rho}{\sqrt{(x-a)^2 + (y-b)^2 + (z-c)^2}} \, d\sigma.$$

If S is the cylindrical surface $x^2 + y^2 = 4$, $0 \le z \le 1$, and ρ is constant, find the electrostatic potential at the origin.

7. Find the electrostatic potential at the origin generated by the charged hemisphere $x^2 + y^2 + z^2 = 1$, $z \ge 0$, assuming that ρ is constant.

8. Show that if a surface is described by $S = \{(x, y, z) | z = f(x, y), (x, y) \in R\}$ then

$$\int_S d\sigma = \int_R \sqrt{1 + \left(\frac{\partial f}{\partial x}\right)^2 + \left(\frac{\partial f}{\partial y}\right)^2} \, dx \, dy.$$

9. Find the area of the surface

$$z = x^{3/2} + y^{3/2},$$

located above the square $0 \le x, y \le 1$.

10. Find the area of that portion of the cylinder $x^2 + y^2 = 4$ that lies above the x, y plane and below the plane $z = y$.

11. Show that if a surface is described by

$$\begin{aligned} x &= r\cos\theta, \\ y &= r\sin\theta, \\ z &= f(r, \theta) \end{aligned}$$

then

$$d\sigma = \sqrt{1 + \left(\frac{\partial f}{\partial r}\right)^2 + r^{-2}\left(\frac{\partial f}{\partial \theta}\right)^2} \; r \, dr \, d\theta.$$

12. Find the area of the portion of the paraboloid $z = x^2 + y^2$ that lies inside the cylinder $x^2 + y^2 = 16$.

13. Find the surface area of the torus $(a > b)$:

$$\begin{aligned} x &= (a + b\cos\theta)\sin\phi, \\ y &= (a + b\cos\theta)\cos\phi, \\ z &= b\sin\theta. \end{aligned}$$

14. Let S denote the parallelogram in $\mathbf{R}^3$ with vertices at $(2,1,0)$, $(3,4,2)$, $(0,3,3)$, $(-1,0,1)$, oriented so that the triangle

$$[(2,1,0),(3,4,2),(-1,0,1)]$$

has positive orientation. Find a parametrization for S and use it to evaluate the flow

$$\int_S x\,dy\,dz - y^2\,dz\,dx + (y+z)\,dx\,dy.$$

15. Assume that temperature is proportional to some power of the distance from the origin:

$$T = cr^\alpha, \quad r = \sqrt{x^2 + y^2 + z^2},$$

where α is some real constant. Show that if we have a homogeneous medium and if the rate at which heat flows through the surface $x^2 + y^2 + z^2 = a^2$ is independent of a then either the temperature is constant or it is inversely proportional to the distance from the origin.

9

Techniques of Differential Calculus

9.1 Implicit Differentiation

Consider a level surface in three dimensions,

$$f(x, y, z) = c.$$

If we are constrained to stay on this surface starting at a specified point and changing two of our variables in a carefully controlled fashion, then we can expect this to determine the way in which the third variable changes. Specifically, if we hold one variable constant, $z = z_0$, then we are constrained to stay on a particular contour line. It makes sense to ask for the rate at which y changes as we vary x, a relationship denoted by the partial derivative: $\partial y / \partial x$.

As an example (Figure 9.1), consider

$$f(x, y, z) = x^2 - y^2 + z^2.$$

For our level surface, we shall choose

$$x^2 - y^2 + z^2 = 12$$

and take $(3, -1, 2)$ as our starting point on this surface. If we are to hold z constant, for example $z = 2$, then we are constrained to the curve

$$x^2 - y^2 = 8.$$

In this case, we can actually solve for y as a function of x near $(3, -1, 2)$:

$$y = -\sqrt{x^2 - 8},$$

so that

$$\frac{\partial y}{\partial x} = \frac{x}{-\sqrt{x^2 - 8}} = \frac{x}{y},$$

which is -3 at $(3, -1, 2)$. Implicit differentiation will provide a means for finding $\partial y / \partial x$ without explicitly expressing y as a function of x.

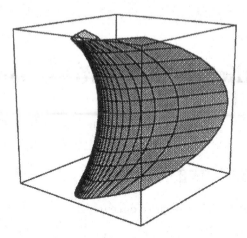

FIGURE 9.1. Contour lines on the surface $x^2 - y^2 + z^2 = 12$.

Using Differentials

If f is a function in (x, y, z), then from Equation (7.11) we know that its differential is

$$df = \frac{\partial f}{\partial x}\, dx + \frac{\partial f}{\partial y}\, dy + \frac{\partial f}{\partial z}\, dz.$$

On the other hand, since f is constant, its gradient is always $\vec{0}$, and so

$$df = 0.$$

Combining these two equations, we can solve for dy in terms of dx and dz, provided $\partial f / \partial y$ is not zero:

$$\frac{\partial f}{\partial y}\, dy = -\frac{\partial f}{\partial x}\, dx - \frac{\partial f}{\partial z}\, dz,$$

$$dy = -\frac{\partial f / \partial x}{\partial f / \partial y}\, dx - \frac{\partial f / \partial z}{\partial f / \partial y}\, dz. \tag{9.1}$$

Now, if the value of y does depend on x and z, then

$$dy = \frac{\partial y}{\partial x}\, dx + \frac{\partial y}{\partial z}\, dz. \tag{9.2}$$

Comparing Equations (9.1) and (9.2), we see that if $\partial y / \partial x$ is well defined then it must be

$$\frac{\partial y}{\partial x} = -\frac{\partial f / \partial x}{\partial f / \partial y}. \tag{9.3}$$

In our example,
$$2x\,dx - 2y\,dy + 2z\,dz = 0,$$

$$dy = \frac{x}{y}dx + \frac{z}{y}dz, \quad y \neq 0,$$

$$\frac{\partial y}{\partial x} = \frac{x}{y}, \quad \frac{\partial y}{\partial z} = \frac{z}{y}.$$

At $y = 0$, $\partial y/\partial x$ is not well defined. For example, at $(2\sqrt{2}, 0, 2)$, a positive change in x could accompany either a positive or a negative change in y.

Implicit Differentiation on Level Curves

If f is a function of two variables, x and y, then restricting (x, y) to a level curve of f gives us a direct relationship between the rates at which x and y change. If
$$f(x, y) = c,$$

then
$$\frac{\partial f}{\partial x}\,dx + \frac{\partial f}{\partial y}\,dy = 0,$$

and so
$$\frac{dy}{dx} = -\frac{\partial f/\partial x}{\partial f/\partial y},$$

provided $\partial f/\partial y \neq 0$.

As an example, if
$$(x + y)e^{xy} = 1,$$

then
$$[e^{xy} + y(x + y)e^{xy}]\,dx + [e^{xy} + x(x + y)e^{xy}]\,dy = 0,$$

$$\frac{dy}{dx} = -\frac{1 + xy + y^2}{1 + xy + x^2},$$

provided that $1 + xy + x^2 \neq 0$. At $(1, 0)$, we have
$$\frac{dy}{dx} = -\frac{1}{2}.$$

More Than One Constraint

What if we have a relationship defined by more than one equation? As an example, consider the following system of equations:

$$\begin{aligned} x^2 + xu - v^2 - yv &= 0, \\ xuv + xyv &= 2. \end{aligned}$$

These define two linear equations in the four differentials dx, dy, du, and dv:

$$(2x + u)\,dx - v\,dy + x\,du - (2v + y)\,dv \;=\; 0,$$
$$(uv + yv)\,dx + xv\,dy + xv\,du + (xu + xy)\,dv \;=\; 0.$$

At $(1, 1, 1, 1)$, these equations become

$$3\,dx - dy + du - 3\,dv \;=\; 0,$$
$$2\,dx + dy + du + 2\,dv \;=\; 0.$$

We can solve for du and dv in terms of dx and dy by using Cramer's rule:

$$du - 3\,dv \;=\; -3\,dx + dy,$$
$$du + 2\,dv \;=\; -2\,dx - dy,$$

$$du \;=\; \frac{\begin{vmatrix} -3dx + dy & -3 \\ -2dx - dy & 2 \end{vmatrix}}{\begin{vmatrix} 1 & -3 \\ 1 & 2 \end{vmatrix}} \;=\; \frac{-12dx - dy}{5},$$

$$dv \;=\; \frac{\begin{vmatrix} 1 & -3dx + dy \\ 1 & -2dx - dy \end{vmatrix}}{\begin{vmatrix} 1 & -3 \\ 1 & 2 \end{vmatrix}} \;=\; \frac{dx - 2dy}{5}.$$

Since

$$du \;=\; \frac{\partial u}{\partial x}dx + \frac{\partial u}{\partial y}dy,$$
$$dv \;=\; \frac{\partial v}{\partial x}dx + \frac{\partial v}{\partial y}dy,$$

it follows that, at $x = y = u = v = 1$,

$$\frac{\partial u}{\partial x} = -\frac{12}{5}, \quad \frac{\partial u}{\partial y} = -\frac{1}{5}, \quad \frac{\partial v}{\partial x} = \frac{1}{5}, \quad \frac{\partial v}{\partial y} = -\frac{2}{5}.$$

Warning

There is a trap lying for us here. Instead of expressing u and v as implicit functions of x and y, let us consider u and y as implicit functions of v and x:

$$du - dy \;=\; 3\,dv - 3\,dx,$$
$$du + dy \;=\; -2\,dv - 2\,dx,$$

$$du = \frac{1}{2} dv - \frac{5}{2} dx,$$

$$dy = \frac{-5}{2} dv + \frac{1}{2} dx,$$

so that at $x = y = u = v = 1$ we have

$$\frac{\partial u}{\partial v} = \frac{1}{2}, \quad \frac{\partial u}{\partial x} = -\frac{5}{2}, \quad \frac{\partial y}{\partial v} = -\frac{5}{2}, \quad \frac{\partial y}{\partial x} = \frac{1}{2}.$$

Something is wrong. We have shown that at the same point we have

$$\frac{\partial u}{\partial x} = -\frac{12}{5} \quad \text{and} \quad \frac{\partial u}{\partial x} = -\frac{5}{2}.$$

The problem is not with our mathematics. The problem lies in our notation. *The first $\partial u / \partial x$ and the second $\partial u / \partial x$ do not mean the same thing.* In the first, we have held y constant and determined the effect on u of varying x. In the second, we have held v constant and determined the effect on u of varying x.

Unfortunately, there is no generally accepted notation for describing which variables are being held constant. Fortunately, this is seldom a problem except in textbook examples. It is usually clear from the context of the computation which variables are being held constant. But, you need to be aware that the notation of the partial derivative implies that some variables are being held constant and that the partial derivative is not well defined until you know which ones these are.

A Celestial Example

We shall take as an example of the type of problem described above the description of a satellite orbiting the earth. We let r be the distance from the center of the earth at perigee, v the speed at perigee, a the mean distance of the orbit, and ε the eccentricity of the orbit. Using Equations (3.46) and (3.54) from pages 69 and 72, these four variables are connected by two equations (where we have assumed that GM is the constant 4×10^{14} m^3/s^2):

$$\varepsilon = 2.5 \times 10^{-15} r v^2 - 1,$$
$$a(1 - \varepsilon^2) = 2.5 \times 10^{-15} r^2 v^2.$$

Our differentials are related by the equations:

$$d\varepsilon = 2.5 \times 10^{-15} v^2 \, dr + 5 \times 10^{-15} r v \, dv,$$
$$(1 - \varepsilon^2) \, da - 2a\varepsilon \, d\varepsilon = 5 \times 10^{-15} r v^2 \, dr + 5 \times 10^{-15} r^2 v \, dv.$$

Let us take as our initial point $r = 10^7$, $v = 8 \times 10^3$, $\varepsilon = 0.6$, $a = 2.5 \times 10^7$. At this point, our differential equations become

$$d\varepsilon = 1.6 \times 10^{-7} \, dr + 4 \times 10^{-4} \, dv,$$
$$.64 \, da - 3 \times 10^7 \, d\varepsilon = 3.2 \, dr + 4 \times 10^3 \, dv.$$

We are interested in finding the effect on mean distance of a change in the minimal distance: $\frac{\partial a}{\partial r}$. If we are to hold the speed at perigee constant, then we need to solve for da in terms of dr and dv:

$$.64\, da = 8\, dr + 1.6 \times 10^4\, dv,$$

and so

$$\frac{\partial a}{\partial r} = \frac{8}{.64} = 12.5.$$

If, on the other hand, we want to keep the eccentricity constant, then we need to express da in terms of dr and $d\varepsilon$:

$$.64\, da - 4 \times 10^7\, d\varepsilon = 1.6\, dr,$$

and so

$$\frac{\partial a}{\partial r} = \frac{1.6}{.64} = 2.5.$$

In the first case, a 1 m change in minimal distance results in approximately a 12.5 m change in mean distance. In the second case, a 1 m change in minimal distance results in approximately a 2.5 m change in mean distance. It makes a considerable difference whether it is the speed at perigee or the eccentricity that is held constant.

9.2   Invertibility

Consider a vector field $\vec{F} : \mathbf{R}^n \longrightarrow \mathbf{R}^n$ that we represent as a system of equations:

$$
\begin{aligned}
y_1 &= f_1(x_1, x_2, \ldots, x_n), \\
y_2 &= f_2(x_1, x_2, \ldots, x_n), \\
&\vdots \\
y_n &= f_n(x_1, x_2, \ldots, x_n).
\end{aligned}
\tag{9.4}
$$

We know that if $n = 1$ and $F'(c) \neq 0$, then F is invertible over some interval containing c and the inverse is differentiable. The following theorem gives us the multidimensional extension of this result. A scalar field is said to be *continuously differentiable* at $\vec{c}$ if its partial derivatives exist and are continuous in some ball with its center at $\vec{c}$. A vector field is *continuously differentiable* at $\vec{c}$ if each of its component scalar fields is continuously differentiable at $\vec{c}$.

Theorem 9.1 *Let $\vec{F} : \mathbf{R}^n \longrightarrow \mathbf{R}^n$ be a vector field continuously differentiable at $\vec{c}$, described by the system of equations given in 9.4. If the Jacobian at $\vec{c}$ is not zero,*

$$\frac{\partial(f_1, f_2, \ldots, f_n)}{\partial(x_1, x_2, \ldots, x_n)}(\vec{c}) \neq 0,$$

then $\vec{F}$ is invertible in some ball with its center at $\vec{c}$. Specifically, there exist scalar fields $g_1, g_2, \ldots, g_n$ that are differentiable at $\vec{F}(\vec{c})$ and such that for all $\vec{y}$ sufficiently close to $\vec{F}(\vec{c})$

$$
\begin{aligned}
x_1 &= g_1(y_1, y_2, \ldots, y_n), \\
x_2 &= g_2(y_1, y_2, \ldots, y_n), \\
&\vdots \\
x_n &= g_n(y_1, y_2, \ldots, y_n).
\end{aligned}
$$

Before proving this theorem, we need a lemma which may appear obvious but is actually a deep result. The proof of this lemma may be found in Edwards' *Advanced Calculus*.

Lemma 9.1 *Let R be a bounded region in $\mathbf{R}^n$ that includes the points on its boundary and let $f : \mathbf{R}^n \longrightarrow \mathbf{R}$ be a scalar field that is continuous on R, then there exist points $\vec{a}$ and $\vec{b}$ in R such that for all $\vec{x}$ in R*

$$
f(\vec{a}) \leq f(\vec{x}) \leq f(\vec{b}). \tag{9.5}
$$

The value of $f(\vec{a})$ is called the absolute minimum *of f on R, $f(\vec{b})$ is the* absolute maximum *of f on R.*

Proof of Invertibility

Let $\vec{L}_c$ be the derivative of $\vec{F}$ at $\vec{c}$ so that

$$
\vec{F}(\vec{c} + \Delta\vec{x}) - \vec{F}(\vec{c}) = \vec{L}_c(\Delta\vec{x}) + |\Delta\vec{x}|\vec{E}(\vec{c}, \Delta\vec{x}),
$$

$$
\lim_{\Delta\vec{x}\to\vec{0}} \vec{E}(\vec{c}, \Delta\vec{x}) = \vec{0}.
$$

Let $\vec{a}$ and $\vec{b}$ be two distinct points that are close to $\vec{c}$. (It is permitted that one of them be $\vec{c}$.) How close they must be will be determined later. We want to show that $\vec{F}(\vec{a})$ is not equal to $\vec{F}(\vec{b})$. We consider

$$
\begin{aligned}
\vec{F}(\vec{a}) - \vec{F}(\vec{b}) &= \vec{F}(\vec{a}) - \vec{F}(\vec{c}) - \left(\vec{F}(\vec{b}) - \vec{F}(\vec{c})\right) \\
&= \vec{L}_c(\vec{a} - \vec{c}) + |\vec{a} - \vec{c}|\,\vec{E}(\vec{c}, \vec{a} - \vec{c}) \\
&\quad - \vec{L}_c(\vec{b} - \vec{c}) - |\vec{b} - \vec{c}|\,\vec{E}(\vec{c}, \vec{b} - \vec{c}) \\
&= \vec{L}_c(\vec{a} - \vec{b}) + |\vec{a} - \vec{c}|\,\vec{E}(\vec{c}, \vec{a} - \vec{c}) - |\vec{b} - \vec{c}|\,\vec{E}(\vec{c}, \vec{b} - \vec{c}).
\end{aligned}
\tag{9.6}
$$

To prove that $\vec{F}(\vec{a}) - \vec{F}(\vec{b})$ is not zero, we shall show that $\left|\vec{L}_c(\vec{a} - \vec{b})\right|$ is strictly larger than

$$
\left| |\vec{a} - \vec{c}|\vec{E}(\vec{c}, \vec{a} - \vec{c}) - |\vec{b} - \vec{c}|\,\vec{E}(\vec{c}, \vec{b} - \vec{c}) \right|.
$$

Let $\vec{u}$ be the unit vector in the direction of $\vec{a} - \vec{b}$:

$$\vec{u} = \frac{\vec{a} - \vec{b}}{|\vec{a} - \vec{b}|}.$$

It follows that

$$\vec{L}_c(\vec{a} - \vec{b}) = |\vec{a} - \vec{b}| \, \vec{L}_c\left(\frac{\vec{a} - \vec{b}}{|\vec{a} - \vec{b}|}\right) = |\vec{a} - \vec{b}| \, \vec{L}_c(\vec{u}). \qquad (9.7)$$

Now, the set of all unit vectors in $\mathbf{R}^n$ forms a bounded region that includes all points on its boundary. By Lemma 9.1, there is some unit vector, $\vec{u}_0$, such that for any unit vector, $\vec{u}$,

$$\left|\vec{L}_c(\vec{u})\right| \geq \left|\vec{L}_c(\vec{u}_0)\right|.$$

The Jacobian of $\vec{F}$ at $\vec{c}$ is the determinant of $\mathbf{L}_c$. Since the determinant is not zero, Theorem 6.6 tells us that $\vec{L}_c$ is invertible. Since $\vec{u}_0 \neq \vec{0}$, this implies that $\vec{L}_c(\vec{u}_0) \neq \vec{0}$. Define μ to be

$$\mu = \left|\vec{L}_c(\vec{u}_0)\right|,$$

so that

$$\left|\vec{L}_c(\vec{a} - \vec{b})\right| \geq |\vec{a} - \vec{b}| \, \mu. \qquad (9.8)$$

We have our lower bound on $|\vec{L}_c(\vec{a} - \vec{b})|$. The fact that μ is not zero is critically important.

We define the scalar field

$$f(\Delta\vec{x}) = |\Delta\vec{x}| \left|\vec{E}(\vec{c}, \Delta\vec{x})\right| = \left|\vec{F}(\vec{c} + \Delta\vec{x}) - \vec{F}(\vec{c}) - \vec{L}_c(\Delta\vec{x})\right|. \qquad (9.9)$$

Since $\vec{F}$ is continuously differentiable at $\vec{c}$, $f(\Delta\vec{x})$ is continuously differentiable at $\vec{0}$. The quantity for which we want to find an upper bound is

$$\left| |\vec{a} - \vec{c}| \, \vec{E}(\vec{c}, \vec{a} - \vec{c}) - |\vec{b} - \vec{c}| \, \vec{E}(\vec{c}, \vec{b} - \vec{c}) \right| = \left| f(\vec{a} - \vec{c}) - f(\vec{b} - \vec{c}) \right|. \qquad (9.10)$$

The mean value theorem (Theorem 7.2) tells us that

$$\frac{f(\vec{a} - \vec{c}) - f(\vec{b} - \vec{c})}{|\vec{a} - \vec{b}|} = f'\left(\vec{d}; \frac{\vec{a} - \vec{b}}{|\vec{a} - \vec{b}|}\right) = f'(\vec{d}; \vec{u})$$

for some $\vec{d}$ on the line segment between $\vec{a} - \vec{c}$ and $\vec{b} - \vec{c}$. We note that $f'(\vec{0}; \vec{u})$ is zero:

$$f'(\vec{0}; \vec{u}) = \lim_{h \to 0} \frac{f(h\vec{u}) - f(\vec{0})}{h}$$

$$= \lim_{h \to 0} \frac{|h| \, |\vec{E}(\vec{c}, h\vec{u})| - 0}{h} = 0.$$

Since $f'(\vec{x}; \vec{u})$ is continuous in some ball with its center at $\vec{0}$, we can find a ball with its center at $\vec{0}$ in which

$$|f'(\vec{x}; \vec{u})| \leq \frac{\mu}{2}. \tag{9.11}$$

Since $\vec{d}$ lies between $\vec{a} - \vec{c}$ and $\vec{b} - \vec{c}$, if we restrict $\vec{a}$ and $\vec{b}$ so that $\vec{a} - \vec{c}$ and $\vec{b} - \vec{c}$ lie inside this ball, then so will $\vec{d}$:

$$\left| \frac{f(\vec{a} - \vec{c}) - f(\vec{b} - \vec{c})}{|\vec{a} - \vec{b}|} \right| = \left| f'(\vec{d}; \vec{u}) \right| \leq \frac{\mu}{2},$$

$$\left| f(\vec{a} - \vec{c}) - f(\vec{b} - \vec{c}) \right| \leq \frac{\mu}{2} |\vec{a} - \vec{b}|.$$

By Equation (9.10), we have

$$\left| |\vec{a} - \vec{c}| \, \vec{E}(\vec{c}, \vec{a} - \vec{c}) - |\vec{b} - \vec{c}| \, \vec{E}(\vec{c}, \vec{b} - \vec{c}) \right| \leq \frac{\mu}{2} |\vec{a} - \vec{b}|. \tag{9.12}$$

Combining Equations (9.6), (9.8), and (9.12), we see that

$$\begin{aligned}
\left| \vec{F}(\vec{a}) - \vec{F}(\vec{b}) \right| &= \left| \vec{L}_c(\vec{a} - \vec{b}) + |\vec{a} - \vec{c}| \vec{E}(\vec{c}, \vec{a} - \vec{c}) - |\vec{b} - \vec{c}| \vec{E}(\vec{c}, \vec{b} - \vec{c}) \right| \\
&\geq \left| \vec{L}_c(\vec{a} - \vec{b}) \right| - \left| |\vec{a} - \vec{c}| \vec{E}(\vec{c}, \vec{a} - \vec{c}) - |\vec{b} - \vec{c}| \vec{E}(\vec{c}, \vec{b} - \vec{c}) \right| \\
&\geq \mu |\vec{a} - \vec{b}| - \frac{\mu}{2} |\vec{a} - \vec{b}| = \frac{\mu}{2} |\vec{a} - \vec{b}| > 0.
\end{aligned}$$

Proof of Differentiability of the Inverse

The differentiability of the inverse comes without too much more work. Let

$$\begin{aligned}
\vec{y} &= \vec{F}(\vec{x}), \\
\vec{k} &= \vec{F}(\vec{c}), \\
\Delta \vec{y} &= \vec{F}(\vec{c} + \Delta \vec{x}) - \vec{F}(\vec{c}), \\
\Delta \vec{x} &= \vec{F}^{-1}(\vec{k} + \Delta \vec{y}) - \vec{F}^{-1}(\vec{k}).
\end{aligned}$$

The differentiability of $\vec{F}$ at $\vec{c}$ means that

$$\Delta \vec{y} = \vec{L}_c(\Delta \vec{x}) + |\Delta \vec{x}| \, \vec{E}(\vec{c}, \Delta \vec{x}), \tag{9.13}$$

$$\lim_{\Delta \vec{x} \to \vec{0}} \vec{E}(\vec{c}, \Delta \vec{x}) = \vec{0}.$$

Since $\vec{L}_c$ is invertible, we can apply $\vec{L}_c^{-1}$ to both sides of Equation (9.13):

$$\begin{aligned}
\vec{L}_c^{-1}(\Delta \vec{y}) &= \Delta \vec{x} + \vec{L}_c^{-1}\left(|\Delta \vec{x}| \, \vec{E}(\vec{c}, \Delta \vec{x}) \right), \\
\Delta \vec{x} &= \vec{L}_c^{-1}(\Delta \vec{y}) - \vec{L}_c^{-1}\left(|\Delta \vec{x}| \, \vec{E}(\vec{c}, \Delta \vec{x}) \right) \\
&= \vec{L}_c^{-1}(\Delta \vec{y}) - |\Delta \vec{y}| \frac{|\Delta \vec{x}|}{|\Delta \vec{y}|} \vec{L}_c^{-1}\left(\vec{E}(\vec{c}, \Delta \vec{x}) \right). \tag{9.14}
\end{aligned}$$

If we can show that

$$\lim_{\Delta \vec{y} \to \vec{0}} \frac{|\Delta \vec{x}|}{|\Delta \vec{y}|} \, \vec{L}_c^{-1} \left(\vec{E}(\vec{c}, \Delta \vec{x}) \right) = \vec{0},$$

then it follows that $\vec{F}^{-1}$ is differentiable at $\vec{k}$ with its derivative equal to $\vec{L}_c^{-1}$. Note that since $\vec{F}$ is 1 to 1, $\Delta \vec{x} \longrightarrow \vec{0}$ if and only if $\Delta \vec{y} \longrightarrow \vec{0}$.

Since $\vec{L}_c^{-1}$ is a linear transformation, we have

$$\lim_{\Delta \vec{x} \to \vec{0}} \vec{L}_c^{-1} \left(\vec{E}(\vec{c}, \Delta \vec{x}) \right) = \vec{L}_c^{-1} \left(\lim_{\Delta \vec{x} \to \vec{0}} \vec{E}(\vec{c}, \Delta \vec{x}) \right)$$
$$= \vec{L}_c^{-1}(\vec{0}) = \vec{0}. \tag{9.15}$$

All we need to do is verify that $|\Delta \vec{x}|/|\Delta \vec{y}|$ stays bounded as $|\Delta \vec{x}|$ approaches 0. Equivalently, we need to show that $|\Delta \vec{y}|/|\Delta \vec{x}|$ stays away from 0 as $|\Delta \vec{x}|$ approaches 0. We use the differentiability of $\vec{F}$:

$$\frac{|\Delta \vec{y}|}{|\Delta \vec{x}|} = \frac{|\vec{L}_c(\Delta \vec{x}) + |\Delta \vec{x}| \, \vec{E}(\vec{c}, \Delta \vec{x})|}{|\Delta \vec{x}|}$$
$$= \left| \vec{L}_c \left(\frac{\Delta \vec{x}}{|\Delta \vec{x}|} \right) + \vec{E}(\vec{c}, \Delta \vec{x}) \right|$$
$$\geq \left| \vec{L}_c \left(\frac{\Delta \vec{x}}{|\Delta \vec{x}|} \right) \right| - \left| \vec{E}(\vec{c}, \Delta \vec{x}) \right|.$$

We know that $\left| \vec{L}_c(\Delta \vec{x}/|\Delta \vec{x}|) \right| \geq \mu$. If we restrict our region so that

$$\left| \vec{E}(\vec{c}, \Delta \vec{x}) \right| \leq \frac{\mu}{2},$$

then

$$\frac{|\Delta \vec{y}|}{|\Delta \vec{x}|} \geq \frac{\mu}{2},$$

and so

$$\frac{|\Delta \vec{x}|}{|\Delta \vec{y}|} \leq \frac{2}{\mu}.$$

Q.E.D.

Example

Let $(x, y, z) = \vec{F}(u, v, w)$ be defined by

$$x = u^2 + v^2 + w^2,$$
$$y = uv + uw + vw,$$
$$z = uvw,$$

so that

$$
\begin{aligned}
dx &= 2u\,du + 2v\,dv + 2w\,dw, \\
dy &= (v+w)\,du + (u+w)\,dv + (u+v)\,dw, \\
dz &= vw\,du + uw\,dv + uv\,dw.
\end{aligned}
$$

The Jacobian is

$$
\begin{aligned}
\frac{\partial(x,y,z)}{\partial(u,v,w)} &=
\begin{vmatrix}
2u & 2v & 2w \\
v+w & u+w & u+v \\
vw & uw & uv
\end{vmatrix} \\
&= 2u^2v(u+w) - 2u^2w(u+v) \\
&\quad + 2v^2w(u+v) - 2uv^2(v+w) \\
&\quad + 2uw^2(v+w) - 2vw^2(u+v) \\
&= 2(u-v)(v-w)(u-w)(u+v+w).
\end{aligned}
$$

Our vector field, $\vec{F}$, is invertible as long as we stay away from the four planes:

$$
u = v, \quad v = w, \quad u = w, \quad \text{and} \quad u + v + w = 0.
$$

Away from these planes we can solve for the partial derivatives of u, v, and w with respect to x, y, and z. For example, we can use Cramer's rule to solve for du:

$$
du = \frac{\begin{vmatrix} dx & 2v & 2w \\ dy & u+w & u+v \\ dz & uw & uv \end{vmatrix}}{\partial(x,y,z)/\partial(u,v,w)}
$$

$$
= \frac{\partial(y,z)/\partial(v,w)}{\partial(x,y,z)/\partial(u,v,w)}dx - \frac{\partial(x,z)/\partial(v,w)}{\partial(x,y,z)/\partial(u,v,w)}dy
$$

$$
+ \frac{\partial(x,y)/\partial(v,w)}{\partial(x,y,z)/\partial(u,v,w)}dz.
$$

This implies that

$$
\begin{aligned}
\frac{\partial u}{\partial x} &= \frac{\partial(y,z)/\partial(v,w)}{\partial(x,y,z)/\partial(u,v,w)} \\
&= \frac{u^2v - u^2w}{2(u-v)(v-w)(u-w)(u+v+w)} \\
&= \frac{u^2}{2(u-v)(u-w)(u+v+w)}, \\
\frac{\partial u}{\partial y} &= \frac{-\partial(x,z)/\partial(v,w)}{\partial(x,y,z)/\partial(u,v,w)}
\end{aligned}
$$

$$= \frac{2uw^2 - 2uv^2}{2(u-v)(v-w)(u-w)(u+v+w)}$$

$$= \frac{-u(v+w)}{(u-v)(u-w)(u+v+w)},$$

$$\frac{\partial u}{\partial z} = \frac{\partial(x,y)/\partial(v,w)}{\partial(x,y,z)/\partial(u,v,w)}$$

$$= \frac{2uv + 2v^2 - 2uw - 2w^2}{2(u-v)(v-w)(u-w)(u+v+w)}$$

$$= \frac{1}{(u-v)(u-w)}.$$

9.3 Exercises

1. Use implicit differentiation to express dy/dx as a function of x and y:

   (a) $x^2 y + x^3 y^4 = 1$,

   (b) $(x+y)/\sqrt{x^2 + y^2} = 2$,

   (c) $xe^y + ye^x = 2e$,

   (d) $\log(x^2 + xy) = 1$.

2. Find $\partial y/\partial x$ at the specified point, if it exists, all other variables being held constant:

   (a) $x^3 + y^3 + z^3 = 10$ at $(1,2,1)$,

   (b) $(x^2 + y^2)/(y^2 + z^2) = 1$ at $(-1,3,1)$,

   (c) $\log(xyz) = 3$ at $(e, e^2, 1)$,

   (d) $\sin x \cos y - \cos x \sin z = 0$ at $(\pi, 0, \pi/2)$.

3. Let $f(x,y,z)$ be differentiable at (x_0, y_0, z_0). If f and z are held constant, then $\Delta f = \Delta z = 0$. If we set $\Delta x = x - x_0, \Delta y = y - y_0$, then

   $$0 = \frac{\partial f}{\partial x}(x_0, y_0, z_0)\, \Delta x + \frac{\partial f}{\partial y}(x_0, y_0, z_0)\, \Delta y + \sqrt{\Delta x^2 + \Delta y^2}\, E,$$

   where

   $$\lim_{(\Delta x, \Delta y) \to (0,0)} E = 0.$$

   Show that if

   $$\frac{\partial f}{\partial y}(x_0, y_0, z_0) \neq 0$$

and $\Delta y/\Delta x$ stays bounded for $(\Delta x, \Delta y)$ sufficiently close to $(0,0)$ and subject to the constraint

$$f(x,y,z_0) = f(x_0,y_0,z_0),$$

then

$$\lim_{\Delta x \to 0} \frac{\Delta y}{\Delta x} = -\frac{\frac{\partial f}{\partial x}(x_0,y_0,z_0)}{\frac{\partial f}{\partial y}(x_0,y_0,z_0)}.$$

4. Let x, y, u, v be related by the pair of equations

$$xy = 2e^{uv},$$
$$x + y = e^{u+v}.$$

If v is held constant, find $\partial x/\partial u$ at $(x,y,u,v) = (1,2,0,\log 3)$. If y is held constant, find $\partial x/\partial u$ at the same point.

5. Given the system of equations

$$x^2 + y^2 = u + v,$$
$$xy = u - v,$$

find $\partial u/\partial x$ at $x = 3$, $y = 1$, $u = 6.5$, $v = 3.5$, when

(a) y is held constant.

(b) v is held constant.

6. If $(x-z)^2 + (y-z)^2 = 2$, find $\partial^2 y/\partial x \, \partial z$ in terms of x, y, and z.

7. Show that if $f(x,y,z) = 0$, $\partial f/\partial z \neq 0$, and all second partial derivatives are continuous, then

$$\frac{\partial^2 z}{\partial x^2} = \frac{-\left(\frac{\partial^2 f}{\partial z^2}\right)\left(\frac{\partial f}{\partial x}\right)^2 + 2\left(\frac{\partial^2 f}{\partial x \partial z}\right)\left(\frac{\partial f}{\partial x}\right)\left(\frac{\partial f}{\partial z}\right) - \left(\frac{\partial f}{\partial z}\right)^2\left(\frac{\partial^2 f}{\partial x^2}\right)}{\left(\frac{\partial f}{\partial z}\right)^3}.$$

$$(9.16)$$

8. Show that if $f(x,y,z) = 0$, $\partial f/\partial z \neq 0$, and all second partial derivatives are continuous, then

$$\frac{\partial^2 z}{\partial x \, \partial y} = \frac{-\frac{\partial^2 f}{\partial z^2}\frac{\partial f}{\partial x}\frac{\partial f}{\partial y} + \frac{\partial^2 f}{\partial y \partial z}\frac{\partial f}{\partial x}\frac{\partial f}{\partial z} + \frac{\partial^2 f}{\partial x \partial z}\frac{\partial f}{\partial y}\frac{\partial f}{\partial z} - \frac{\partial^2 f}{\partial x \partial y}\left(\frac{\partial f}{\partial z}\right)^2}{\left(\frac{\partial f}{\partial z}\right)^3}.$$

$$(9.17)$$

9. Let x, y, r, θ be related by

$$x = r \cos \theta,$$
$$y = r \sin \theta.$$

Prove that

$$\frac{\partial r}{\partial x} = \cos \theta, \qquad \frac{\partial \theta}{\partial x} = -r^{-1} \sin \theta,$$

$$\frac{\partial r}{\partial y} = \sin \theta, \qquad \frac{\partial \theta}{\partial y} = r^{-1} \cos \theta.$$

10. Using the chain rule and Exercise 9, show that if $f(x, y) = g(r, \theta)$ and the mixed partial derivatives of g are equal $(\partial^2 g / \partial r \, \partial \theta = \partial^2 g / \partial \theta \, \partial r)$ then

$$\frac{\partial^2 f}{\partial x^2} = \cos^2 \theta \frac{\partial^2 g}{\partial r^2} - 2r^{-1} \sin \theta \cos \theta \frac{\partial^2 g}{\partial r \partial \theta} + r^{-2} \sin^2 \theta \frac{\partial^2 g}{\partial \theta^2}$$
$$+ r^{-1} \sin^2 \theta \frac{\partial g}{\partial r} + 2r^{-2} \sin \theta \cos \theta \frac{\partial g}{\partial \theta}.$$

11. Using the chain rule and Exercise 9, show that if $f(x, y) = g(r, \theta)$ and the mixed partial derivatives of g are equal then

$$\frac{\partial^2 f}{\partial x^2} + \frac{\partial^2 f}{\partial y^2} = \frac{\partial^2 g}{\partial r^2} + r^{-2} \frac{\partial^2 g}{\partial \theta^2} + r^{-1} \frac{\partial g}{\partial r}.$$

12. Let x, y, z, r, θ, ϕ be related by

$$x = r \cos \theta \cos \phi,$$
$$y = r \sin \theta \cos \phi,$$
$$z = r \sin \phi.$$

Find $\partial r / \partial x$, $\partial r / \partial y$, $\partial r / \partial z$, $\partial \theta / \partial x$, $\partial \theta / \partial y$, $\partial \theta / \partial z$, $\partial \phi / \partial x$, $\partial \phi / \partial y$, and $\partial \phi / \partial z$ in terms of r, θ, and ϕ,

13. Prove that if $f(x, y, z) = g(r, \theta, \phi)$ and the mixed partial derivatives of g are equal then

$$\frac{\partial^2 f}{\partial x^2} + \frac{\partial^2 f}{\partial y^2} + \frac{\partial^2 f}{\partial z^2} = \frac{\partial^2 g}{\partial r^2} + r^{-2} \sec^2 \phi \frac{\partial^2 g}{\partial \theta^2} + r^{-2} \frac{\partial^2 g}{\partial \phi^2}$$
$$+ 2r^{-1} \frac{\partial g}{\partial r} - r^{-2} \tan \phi \frac{\partial g}{\partial \phi}.$$

14. Let f and g be differentiable functions of the four variables x, y, u, v such that u and v are implicitly defined as functions of x and y by the equations

$$f(x, y, u, v) = c_1,$$
$$g(x, y, u, v) = c_2.$$

Show that

$$\frac{\partial u}{\partial x} = -\frac{\begin{vmatrix} \frac{\partial f}{\partial x} & \frac{\partial f}{\partial v} \\ \\ \frac{\partial g}{\partial x} & \frac{\partial g}{\partial v} \end{vmatrix}}{\begin{vmatrix} \frac{\partial f}{\partial u} & \frac{\partial f}{\partial v} \\ \\ \frac{\partial g}{\partial u} & \frac{\partial g}{\partial v} \end{vmatrix}}$$

$$= -\frac{\partial(f,g)/\partial(x,v)}{\partial(f,g)/\partial(u,v)}. \tag{9.18}$$

This representation of $\partial u/\partial x$ as a ratio of Jacobians is actually the preferred notation. It simultaneously shows how to compute $\partial u/\partial x$ and specifies which variables are being held constant (all those that are missing, in this case only y).

15. Let $f_1, f_2, \ldots, f_n$ be differentiable functions of the $2n$ variables

$$x_1, x_2, \ldots, x_n, u_1, u_2, \ldots, u_n.$$

If we set n constraints:

$$f_1 = c_1, f_2 = c_2, \ldots, f_n = c_n,$$

then $(u_1, u_2, \ldots, u_n)$ is implicitly defined as a function of $(x_1, x_2, \ldots, x_n)$. Show that

$$\frac{\partial u_i}{\partial x_j} = \frac{-\begin{vmatrix} \frac{\partial f_1}{\partial u_1} & \cdots & \frac{\partial f_1}{\partial u_{i-1}} & \frac{\partial f_1}{\partial x_j} & \frac{\partial f_1}{\partial u_{i+1}} & \cdots & \frac{\partial f_1}{\partial u_n} \\ \vdots & & \vdots & \vdots & \vdots & & \vdots \\ \frac{\partial f_n}{\partial u_1} & \cdots & \frac{\partial f_n}{\partial u_{i-1}} & \frac{\partial f_n}{\partial x_j} & \frac{\partial f_n}{\partial u_{i+1}} & \cdots & \frac{\partial f_n}{\partial u_n} \end{vmatrix}}{\begin{vmatrix} \frac{\partial f_1}{\partial u_1} & \frac{\partial f_1}{\partial u_2} & \cdots & \frac{\partial f_1}{\partial u_n} \\ \vdots & & & \vdots \\ \frac{\partial f_n}{\partial u_1} & \frac{\partial f_n}{\partial u_2} & \cdots & \frac{\partial f_n}{\partial u_n} \end{vmatrix}}$$

$$= \frac{-\partial(f_1, f_2, \ldots, f_n)/\partial(u_1, \ldots, u_{i-1}, x_j, u_{i+1}, \ldots, u_n)}{\partial(f_1, f_2, \ldots, f_n)/\partial(u_1, u_2, \ldots, u_n)}. \tag{9.19}$$

16. Consider the transformation

$$u = -x + \sqrt{x^2 + y^2},$$
$$v = -x - \sqrt{x^2 + y^2}.$$

Show that it is not invertible everywhere. Find a region of the x, y plane on which it is invertible.

17. Consider the transformation

$$u = x^3 - y,$$
$$v = 3x^3 + 2y.$$

Show that the Jacobian is zero everywhere on the y axis. Show that this transformation *is* invertible over the entire x, y plane.

18. Consider the transformation

$$u = e^x \cos y,$$
$$v = e^x \sin y.$$

Prove that the Jacobian is never zero and therefore this transformation is invertible in some neighborhood of each point. Show that it is *not* 1 to 1 (and thus not invertible) when considered over all of $\mathbf{R}^2$.

19. Let f be a differentiable function defined on $\mathbf{R}$ with $|f'(x)| < 1$ for all x in $\mathbf{R}$. Prove that $\vec{g}(u, v) = (u + f(v), v + f(u))$ is invertible in some neighborhood of each point in $\mathbf{R}^2$.

9.4 Locating Extrema

Once again, let us wander over a differentiable surface in $\mathbf{R}^3$. The problem we shall address in this section is that of finding the local extrema, the maxima and minima, the hill tops and valley bottoms. In precise terms, the scalar field f has a *local maximum* at $\vec{x}_0$ if, for all $\vec{x}$ sufficiently close to $\vec{x}_0$, $f(\vec{x})$ is less than or equal to $f(\vec{x}_0)$. Even more precisely, $f(\vec{x})$ has a local maximum at $\vec{x}_0$ if there is some positive δ such that if $|\vec{x} - \vec{x}_0| < \delta$, then $f(\vec{x}) \leq f(\vec{x}_0)$. The scalar field f has a *local minimum* at $\vec{x}_0$ if $\vec{x}$ sufficiently close to $\vec{x}_0$ implies that $f(\vec{x})$ is greater than or equal to $f(\vec{x}_0)$.

How do we recognize a local maximum? How do we know when we have reached the top of a mountain? We have reached the summit when we cannot climb any higher. If we have found the position of a local maximum, then no directional derivative is strictly positive. Proposition 7.1 on page 182 implies that if any directional derivative is strictly negative, then the directional derivative in the opposite direction is strictly positive. This means that where a differentiable scalar field achieves a local maximum all directional derivatives must be zero. This can only happen if the gradient is the zero vector: if f has a local maximum at $\vec{x}_0$, then $\nabla f(\vec{x}_0) = \vec{0}$. The same is true if f has a local minimum at $\vec{x}_0$. The values of $\vec{x}$ for which

$$\nabla f(\vec{x}) = \vec{0}$$

are called the *stationary points*. This is where we shall find the extrema of a differentiable scalar field.

Theorem 9.2 *If $f(\vec{x})$ is differentiable at $\vec{x}_0$ and $\nabla f(\vec{x}_0)$ is not equal to $\vec{0}$, then given any positive δ, we can find a point $\vec{x}_1$ within δ of $\vec{x}_0$, $|\vec{x}_1 - \vec{x}_0| < \delta$, for which*

$$f(\vec{x}_1) > f(\vec{x}_0).$$

Similarly, there is an $\vec{x}_2$ within δ of $\vec{x}_0$ for which

$$f(\vec{x}_2) < f(\vec{x}_0).$$

Proof: By the definition of differentiability,

$$f(\vec{x}) - f(\vec{x}_0) = \nabla f(\vec{x}_0) \cdot (\vec{x} - \vec{x}_0) + |\vec{x} - \vec{x}_0| E(\vec{x}_0, \vec{x} - \vec{x}_0).$$

Since $\lim_{\vec{x} \to \vec{x}_0} E = 0$ and $\nabla f(\vec{x}_0) \neq \vec{0}$, we can force $|E(\vec{x}_0, \vec{x} - \vec{x}_0)|$ to be strictly less than $|\nabla f(\vec{x}_0)|/2$ by taking $\vec{x}$ sufficiently close to $\vec{x}_0$. In particular, we can find an $\vec{x}_1 \neq \vec{x}_0$ such that

1. $|\vec{x}_1 - \vec{x}_0| < \delta$,

2. $\vec{x}_1 - \vec{x}_0$ points in the same direction as $\nabla f(\vec{x}_0)$, and

3. $|E(\vec{x}_0, \vec{x}_1 - \vec{x}_0)| < |\nabla f(\vec{x}_0)|/2$.

It follows that

$$
\begin{aligned}
f(\vec{x}_1) &= f(\vec{x}_0) + \nabla f(\vec{x}_0) \cdot (\vec{x}_1 - \vec{x}_0) + |\vec{x}_1 - \vec{x}_0| E(\vec{x}_0, \vec{x}_1 - \vec{x}_0) \\
&= f(\vec{x}_0) + |\nabla f(\vec{x}_0)| \, |\vec{x}_1 - \vec{x}_0| + |\vec{x}_1 - \vec{x}_0| E(\vec{x}_0, \vec{x}_1 - \vec{x}_0) \\
&> f(\vec{x}_0) + |\nabla f(\vec{x}_0)| \, |\vec{x}_1 - \vec{x}_0| - |\vec{x}_1 - \vec{x}_0| \frac{|\nabla f(\vec{x}_0)|}{2} \\
&= f(\vec{x}_0) + \frac{1}{2} |\nabla f(\vec{x}_0)| \, |\vec{x}_1 - \vec{x}_0| > f(\vec{x}_0).
\end{aligned}
$$

The point $\vec{x}_2$ is taken so that $\vec{x}_2 - \vec{x}_0$ points in the opposite direction from $\nabla f(\vec{x}_0)$.

Q.E.D.

Corollary 9.1 *If f is differentiable at $\vec{x}_0$ and has a local extremum (maximum or minimum) at $\vec{x}_0$, then*

$$\nabla f(\vec{x}_0) = \vec{0}.$$

Examples

Let

$$f(x, y) = 3y^2 - 4xy + 4x^2 + 2y + 4x$$

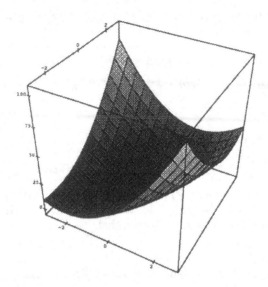

FIGURE 9.2. $f(x,y) = 3y^2 - 4xy + 4x^2 + 2y + 4x$.

(see Figure 9.2). Any local extrema must occur where both $\partial f/\partial x$ and $\partial f/\partial y$ are 0. Our partial derivatives are

$$\frac{\partial f}{\partial x} = -4y + 8x + 4,$$

$$\frac{\partial f}{\partial y} = 6y - 4x + 2.$$

The first partial derivative is zero when $y = 2x + 1$, the second is zero when $3y = 2x - 1$. They are both zero when $(x, y) = (-1, -1)$. By comparing values of f near $(-1, -1)$ with the value $f(-1, -1) = -3$, we see that our scalar field has a local minimum at $(-1, -1)$. A more satisfying proof that we have a local minimum at $(-1, -1)$ will be given in Section 9.5.

Let

$$f(x, y) = x^3 - 3xy^2 + 2y^3$$

(see Figure 9.3). Any local extrema must occur where both $\partial f/\partial x$ and $\partial f/\partial y$ are 0. Our partial derivatives are

$$\frac{\partial f}{\partial x} = 3x^2 - 3y^2,$$

$$\frac{\partial f}{\partial y} = -6xy + 6y^2.$$

The first partial derivative is zero when

$$y = x \quad \text{or} \quad y = -x;$$

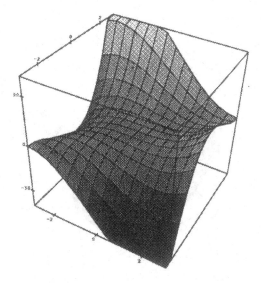

FIGURE 9.3. $f(x, y) = x^3 - 3xy^2 + 2y^3$.

the second is zero when

$$y = x \ \text{ or } \ y = 0.$$

It follows that all candidates for local extrema lie on the line $y = x$, where

$$f(x, x) = x^3 - 3x^3 + 2x^3 = 0.$$

If we factor $f(x, y)$, we can see what is happening:

$$f(x, y) = (x - y)^2 (x + 2y).$$

This is at least 0 when x and y are positive, at most 0 when x and y are negative. It follows that if

$$y = x > 0,$$

then we have a relative minimum; if

$$y = x < 0,$$

then we have a relative maximum; and if

$$y = x = 0,$$

then we have neither a relative maximum nor a relative minimum because there are points arbitrarily close to $(0, 0)$ for which f is positive and points arbitrarily close to $(0, 0)$ for which f is negative.

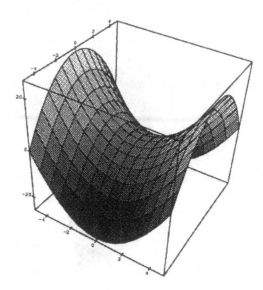

FIGURE 9.4. $f(x, y) = x^2 - y^2$.

Saddle Points

As we saw in the last example, the fact that all partial derivatives are zero does not guarantee a local extremum. A stationary point such as (0,0) for

$$f(x, y) = x^3 - 3xy^2 + 2y^3,$$

where the gradient is $\vec{0}$ but we do not have a local extremum, is called a *saddle point*. The name derives from the classic example:

$$f(x, y) = x^2 - y^2,$$

which has a saddle point at the origin and looks like a saddle (Figure 9.4). We note that if we restrict this function to the x axis (set $y = 0$), then we have local minimum at the origin. If restricted to the y axis ($x = 0$), we get a local maximum at the origin.

Examples

Another example of a surface with a saddle point at the origin is the "monkey saddle" (Figure 9.5):

$$f(x, y) = x^3 - 3xy^2,$$

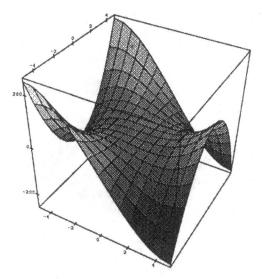

FIGURE 9.5. $f(x, y) = x^3 - 3xy^2$.

so called because three valleys descend from the origin, two for the legs and one for the tail. A third example is the scalar field

$$f(x, y) = y^2 - 3yx^2 + 2x^4.$$

If we restrict this function to any straight line through the origin: $x = 0$, $y = 0$, or $y = mx$ $(m \neq 0)$, then it has a local minimum at $(0,0)$. However, the origin is not a local minimum for this function because there is a parabolic valley following the curve

$$y = \frac{3}{2}x^2$$

along which

$$f\left(x, \frac{3}{2}x^2\right) = -\frac{1}{4}x^4,$$

reaching a local maximum at the origin (Figure 9.6).

While we have restricted our examples to scalar fields defined on $\mathbf{R}^2$, the same technique for locating possible extrema works for any number of variables: in any region where the function is differentiable, look for those points where all of the partial derivatives are simultaneously zero.

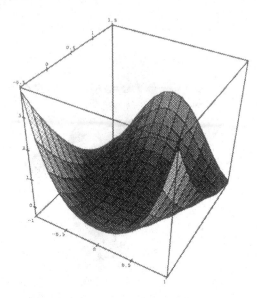

FIGURE 9.6. $f(x,y) = y^2 - 3yx^2 + 2x^4$.

9.5 Taylor's Formula in Several Variables

How do we know if we have an extremum or what kind we have found? In this section, we shall develop an analog of the second derivative test for real-valued functions of a single variable. We recall that if $f'(x_0) = 0$ and $f''(x_0) > 0$, then f has local minimum at x_0, and if $f'(x_0) = 0$ and $f''(x_0) < 0$, then f has a local maximum at x_0. For scalar fields, the analog of the second derivative is the *Hessian* matrix of second partial derivatives, which is the Jacobian matrix of the gradient of f:

$$
\mathbf{H} =
\begin{pmatrix}
\frac{\partial^2 f}{\partial x_1^2} & \frac{\partial^2 f}{\partial x_2 \partial x_1} & \cdots & \frac{\partial^2 f}{\partial x_n \partial x_1} \\[2ex]
\frac{\partial^2 f}{\partial x_1 \partial x_2} & \frac{\partial^2 f}{\partial x_2^2} & \cdots & \frac{\partial^2 f}{\partial x_n \partial x_2} \\
\vdots & \vdots & & \vdots \\[1ex]
\frac{\partial^2 f}{\partial x_1 \partial x_n} & \frac{\partial^2 f}{\partial x_2 \partial x_n} & \cdots & \frac{\partial^2 f}{\partial x_n^2}
\end{pmatrix}.
$$

The Hessian derives its name from Ludwig Otto Hesse, who in 1844 applied it to find solutions of systems of algebraic equations. It had been used over 40 years earlier, and in the context in which we need it, by Joseph-Louis Lagrange, whom we have already met as the discoverer of the geometric significance of the determinant and who, before the end of the eighteenth century, solved the problem of finding the extrema of differentiable functions of several variables. In this section, we shall investigate his determination of the nature of the extrema. In Section 8.7, we shall consider his now classical solution to the problem of finding extrema when constraints are placed on the domain of the function. Lagrange wrote all of this in two textbooks: *Théorie des Fonctions Analytiques*, first published in 1797, a revised version appearing posthumously in 1813, and *Leçons sur le Calcul des Fonctions*, based on lectures given in 1799, published in 1806.

Historical Digression

The key idea is to find the several variable analog of the Taylor series:

$$f(a + h) - f(a) = f'(a)h + f''(a)\frac{h^2}{2} + f'''(a)\frac{h^3}{2 \cdot 3} + \cdots . \qquad (9.20)$$

This series takes its name from the English mathematician Brook Taylor (1685–1731), who published it in 1715 in his treatise *Methodus Incrementorum Directa et Inversa*. But, it appears in the correspondence of the Scottish mathematician James Gregory (1638–1675) in 1670 and again in the correspondence of Leibniz. Newton gives a very similar formula (in which the derivatives are replaced by finite differences) as Lemma 5 of Book III in *Principia*, and in fact Taylor refers to this formula as the starting point for the derivation of his own. But, the chief problem with referring to Equation (9.20) as Taylor's Formula is that Jean Bernoulli (1667–1748) published it in the *Acta Eruditorum* in 1694. The attribution of this formula to Taylor only exacerbated the existing rancor between the English mathematicians who credited Newton with the invention of calculus and the German mathematicians who claimed Leibniz as discoverer.

As a further digression, Jean Bernoulli is also known as Johann, Johannes, Johannis, and John. The confusion arises from the fact that he was Swiss German, born in Basel, where he eventually taught, but of a French-speaking Huguenot family. When he wrote in French, he signed himself Jean, when in German, Johann, and when in Latin, Johannes or Johannis. Because of this, many English writers have simply anglicized his name and call him John, the one name that he never used. There is a similar difficulty with his equally eminent brother Jacques, Jakob, Jacob, Jacobus, or James (1654–1705). However, when the two brothers wrote to each other, they wrote in French, and so I shall adopt Jean and Jacques as their preferred choices.

The special case of Equation (9.20) in which a has been set equal to 0 is generally known as "MacLaurin's series," after the same Colin MacLaurin who should have received the credit for Cramer's rule. MacLaurin described this series in his *Treatise of Fluxions* of 1742, well after Bernoulli and Taylor had published the more general form. This misattribution was not his fault, as he cites both Taylor's publication and the proof of this particular case by James Stirling (1692–1770) in 1730.

From a modern point of view, the problem with Taylor's formula, especially as it was used by Taylor and his contemporaries, is that there is no recognition of the question of convergence: when is such an infinite summation valid? Lagrange was one of the first to recognize and address this difficulty. He proved that

$$f(a+h) - f(a) = f'(a)h + f''(a)\frac{h^2}{2!} + \cdots + f^{(n)}(a)\frac{h^n}{n!} + E_n(a,h), \quad (9.21)$$

where $n! = 1 \cdot 2 \cdot 3 \cdot \cdots \cdot n$ and

$$E_n(a,h) = f^{(n+1)}(b)\frac{h^{n+1}}{(n+1)!} \quad (9.22)$$

for some b between a and $a+h$. The infinite summation makes sense if and only if

$$\lim_{n \to \infty} E_n(a,h) = 0.$$

Equation (9.22) is known as the Lagrange form of the remainder. Other mathematicians were to find other ways of expressing $E_n(a,h)$. For our purposes, Lagrange's form is sufficient.

Second-Order Taylor Formula

In the following derivation of Taylor's formula, it is convenient to use one of our alternate notations for the partial derivative:

$$D_i F \quad \text{in place of} \quad \frac{\partial F}{\partial x_i},$$

$$D_{ij} F \quad \text{in place of} \quad \frac{\partial^2 F}{\partial x_i \, \partial x_j}.$$

Theorem 9.3 Let $F : \mathbf{R}^n \longrightarrow \mathbf{R}$ be a scalar field, differentiable at $\vec{a}$, with continuous second partial derivatives at all points sufficiently close to $\vec{a}$. Let $\mathbf{H}_a$ denote the Hessian of F evaluated at $\vec{a}$ and $\vec{H}_a$ the corresponding linear transformation. We then have

$$F(\vec{a} + \Delta\vec{x}) - F(\vec{a}) = \nabla F(\vec{a}) \cdot \Delta\vec{x} + \frac{1}{2}\Delta\vec{x} \cdot \vec{H}_a(\Delta\vec{x}) + |\Delta\vec{x}|^2 E(\vec{a}, \Delta\vec{x}), \quad (9.23)$$

where, by definition,

$$\Delta\vec{x} \cdot \vec{H}_a(\Delta\vec{x}) = \sum_{i=1}^{n}\sum_{j=1}^{n} D_{ij}F(\vec{a})\, \Delta x_i \, \Delta x_j,$$

and

$$\lim_{\Delta\vec{x} \to \vec{0}} E(\vec{a}, \Delta\vec{x}) = 0.$$

Proof: We begin the proof by setting

$$f(h) = F(\vec{a} + h\,\Delta\vec{x}).$$

Using the chain rule, we see that

$$
\begin{aligned}
f'(h) &= \nabla F(\vec{a} + h\,\Delta\vec{x}) \cdot \Delta\vec{x} \\
&= [D_1 F(\vec{a} + h\,\Delta\vec{x}), D_2 F(\vec{a} + h\,\Delta\vec{x}), \ldots, D_n F(\vec{a} + h\,\Delta\vec{x})] \cdot \Delta\vec{x}
\end{aligned}
$$

and

$$
\begin{aligned}
f''(h) &= \frac{d}{dh}\,[D_1 F(\vec{a} + h\,\Delta\vec{x}), D_2 F(\vec{a} + h\,\Delta\vec{x}), \ldots, \\
&\qquad\qquad D_n F(\vec{a} + h\,\Delta\vec{x})] \cdot \Delta\vec{x} \\
&= [D_{11} F(\vec{a} + h\,\Delta\vec{x})\,\Delta x_1 + \cdots + D_{n1} F(\vec{a} + h\,\Delta\vec{x})\,\Delta x_n, \\
&\qquad D_{12} F(\vec{a} + h\,\Delta\vec{x})\,\Delta x_1 + \cdots + D_{n2} F(\vec{a} + h\,\Delta\vec{x})\,\Delta x_n, \\
&\qquad\qquad\vdots \\
&\qquad D_{1n} F(\vec{a} + h\,\Delta\vec{x})\,\Delta x_1 + \cdots + D_{nn} F(\vec{a} + h\,\Delta\vec{x})\,\Delta x_n] \cdot \Delta\vec{x} \\
&= \sum_{i=1}^{n}\sum_{j=1}^{n} D_{ij} F(\vec{a} + h\,\Delta\vec{x})\,\Delta x_i\,\Delta x_j \\
&= \Delta\vec{x} \cdot \vec{H}_{a+h\Delta x}(\Delta\vec{x}).
\end{aligned}
$$

Using Lagrange's form of the Taylor formula for f with $a = 0$ and $h = 1$ yields

$$
\begin{aligned}
F(\vec{a} + \Delta\vec{x}) - F(\vec{a}) &= f(1) - f(0) \\
&= f'(0) + \frac{1}{2}\,f''(b) \\
&= \nabla F(\vec{a}) \cdot \Delta\vec{x} + \frac{1}{2}\Delta\vec{x} \cdot \vec{H}_{a+b\Delta x}(\Delta\vec{x}) \\
&= \nabla F(\vec{a}) \cdot \Delta\vec{x} + \frac{1}{2}\Delta\vec{x} \cdot \vec{H}_{a}(\Delta\vec{x}) \\
&\qquad + |\Delta\vec{x}|^2 E(\vec{a}, \Delta\vec{x}),
\end{aligned}
$$

where $0 < b < 1$ and

$$
\begin{aligned}
|\Delta\vec{x}|^2\,|E(\vec{a}, \Delta\vec{x})| &= \frac{1}{2}\left|\Delta\vec{x} \cdot \vec{H}_{a+b\Delta x}(\Delta\vec{x}) - \Delta\vec{x} \cdot \vec{H}_{a}(\Delta\vec{x})\right| \\
&= \frac{1}{2}\left|\sum_{i=1}^{n}\sum_{j=1}^{n} \{D_{ij} F(\vec{a} + b\,\Delta\vec{x}) - D_{ij} F(\vec{a})\}\,\Delta x_i\,\Delta x_j\right| \\
&\leq \frac{1}{2}\sum_{i=1}^{n}\sum_{j=1}^{n} |D_{ij} F(\vec{a} + b\,\Delta\vec{x}) - D_{ij} F(\vec{a})|\,|\Delta\vec{x}|\,|\Delta\vec{x}|.
\end{aligned}
$$

It follows from the continuity of $D_{ij}F$ that

$$\lim_{\Delta\vec{x}\to\vec{0}}|E(\vec{a},\Delta\vec{x})| \le \frac{1}{2}\lim_{\Delta\vec{x}\to\vec{0}}\sum_{i=1}^{n}\sum_{j=1}^{n}|D_{ij}F(\vec{a}+b\Delta\vec{x})-D_{ij}F(\vec{a})| = 0.$$

Q.E.D.

The Nature of Extrema

If $\nabla F(\vec{a}) = \vec{0}$ and we set

$$\vec{u} = \frac{\Delta\vec{x}}{|\Delta\vec{x}|},$$

then

$$
\begin{aligned}
F(\vec{a}+\Delta\vec{x}) &= F(\vec{a}) + \frac{1}{2}\Delta\vec{x}\cdot\vec{H}_a(\Delta\vec{x}) + |\Delta\vec{x}|^2 E(\vec{a},\Delta\vec{x}) \\
&= F(\vec{a}) + \frac{1}{2}|\Delta\vec{x}|^2\left(\vec{u}\cdot\vec{H}_a(\vec{u}) + 2E(\vec{a},\Delta\vec{x})\right).
\end{aligned}
$$

If $\vec{u}\cdot\vec{H}_a(\vec{u})$ is positive for every unit vector $\vec{u}$, then it has a minimal positive value. If we choose $\Delta\vec{x}$ sufficiently small, then $2|E(\vec{a},\Delta\vec{x})|$ will be strictly less than this minimal value. This implies that if $\Delta\vec{x}$ is sufficiently small, but not $\vec{0}$, then $F(\vec{a}+\Delta\vec{x})$ will be strictly larger than $F(\vec{a})$. We have a local minimum at $\vec{a}$. Similarly, if $\vec{u}\cdot\vec{H}_a(\vec{u})$ is negative for every unit vector $\vec{u}$, then F has a local maximum at $\vec{a}$. If $\vec{u}\cdot\vec{H}_a(\vec{u})$ is positive for some $\vec{u}$ and negative for others, then we must have a saddle point at $\vec{a}$.

Theorem 9.4 *Let $F:\mathbf{R}^2 \longrightarrow \mathbf{R}$ be a scalar field, differentiable at $\vec{a}$, with continuous second partial derivatives at all points sufficiently close to $\vec{a}$, and assume that*

$$\nabla F(\vec{a}) = \vec{0}.$$

If $\det(\mathbf{H}_a)$ is strictly positive, then

1. *$\partial^2 F/\partial x^2 > 0$ at $\vec{a}$ implies that F has a local minimum at $\vec{a}$,*

2. *$\partial^2 F/\partial x^2 < 0$ at $\vec{a}$ implies that F has a local maximum at $\vec{a}$.*

If $\det(\mathbf{H}_a)$ is strictly negative, then F has a saddle point at $\vec{a}$.

Note that if $\det(\mathbf{H}_a)$ is strictly positive, then $\partial^2 F/\partial x^2$ cannot be zero at $\vec{a}$. If $\det(\mathbf{H}_a)$ equals zero, then this approach tells us nothing about the nature of F at $\vec{a}$.

Proof: If $\vec{u} = (1,0)$, then

$$\vec{u}\cdot\vec{H}_a(\vec{u}) = \frac{\partial^2 F}{\partial x^2}(\vec{a}).$$

For this value of $\vec{u}$, the sign of $\partial^2 F/\partial x^2$ at $\vec{a}$ determines the sign of $\vec{u} \cdot \vec{H}_a(\vec{u})$. If $\vec{u} = (u_1, u_2)$, $u_2 \neq 0$, then we set $A = \partial^2 F/\partial x^2$, $B = \partial^2 F/\partial x \, \partial y$, and $C = \partial^2 F/\partial y^2$, each evaluated at $\vec{a}$, so that

$$\mathbf{H}_a = \begin{pmatrix} A & B \\ B & C \end{pmatrix},$$

and we let $v = u_1/u_2$. We then have that

$$\begin{aligned} \vec{u} \cdot \vec{H}_a(\vec{u}) &= (u_1, u_2) \cdot (Au_1 + Bu_2, Bu_1 + Cu_2) \\ &= Au_1^2 - 2Bu_1u_2 + Cu_2^2 \\ &= u_2^2(Av^2 - 2Bv + C). \end{aligned}$$

The discriminant of this quadratic polynomial in v is

$$4B^2 - 4AC = -4\det(\mathbf{H}_a).$$

If the discriminant is positive (the determinant is negative), then our polynomial has two real roots, and so it takes on both positive and negative values. This implies that $\vec{u} \cdot \vec{H}_a(\vec{u})$ also takes on both positive and negative values, and so we have a saddle point at $\vec{a}$.

If the discriminant is negative (the determinant is positive), then our polynomial has no real roots, and so it is either always positive or always negative. At $v = 0$, the value of the polynomial is C, so the sign of C is the sign of the polynomial. Since

$$AC - B^2 > 0,$$

AC must be positive, so A and C have the same sign. If A is positive, then our polynomial is positive for all values of v, and if A is negative, then our polynomial is negative for all values of v.

Q.E.D.

Examples

If

$$f(x, y) = 3y^2 - 4xy + 4x^2 + 2y + 4x,$$

then, as we have seen, the gradient is zero at $(-1, -1)$. The Hessian is a constant matrix

$$\mathbf{H}_{(x,y)} = \begin{pmatrix} 8 & -4 \\ -4 & 6 \end{pmatrix}$$

of determinant 32, which is positive. Since $D_{11}f = 8 > 0$, we have a local minimum at $(-1, -1)$.

If
$$f(x,y) = x^3 - 3xy^2 + 2y^3,$$
then the gradient is zero whenever $y = x$. The Hessian is
$$\mathbf{H}_{(x,y)} = \begin{pmatrix} 6x & -6y \\ -6y & -6x + 12y \end{pmatrix},$$
which has the determinant
$$\det(\mathbf{H}_{(x,y)}) = -36x^2 + 72xy - 36y^2 = -36(x - y)^2.$$
If $y = x$, the determinant is zero, and our theorem tells us nothing about the nature of f at these points.

If
$$f(x,y) = \frac{3}{5}x^5 - 3xy^2 + 3y,$$
then the first partial derivative is zero when
$$y = x^2 \quad \text{or} \quad y = -x^2,$$
and the second is zero when
$$y = \frac{1}{2x}.$$
The gradient is $\vec{0}$ when
$$(x,y) = (2^{-1/3}, 2^{-2/3}) \quad \text{or} \quad (-2^{-1/3}, -2^{-2/3}).$$
The Hessian is
$$\mathbf{H}_{(x,y)} = \begin{pmatrix} 12x^3 & -6y \\ -6y & -6x \end{pmatrix},$$
which has the determinant $-72x^4 - 36y^2$, which is negative at both stationary points. It follows that both of these are saddle points (Figure 9.7).

More than 2 Dimensions

In more than two dimensions, we shall need results from the theory of linear transformations that go beyond what we have done in Chapter 6. I shall merely state the results without proof. The reader who is truly familiar with these concepts should be able to modify the proof of Theorem 9.4 to prove the general case given as Theorem 9.5. An outline of the proof can be found in Exercises 12 through 15 in Section 9.6.

If $\mathbf{H}$ is any symmetric $n \times n$ matrix with a nonzero determinant, then there exist n linear independent unit vectors, $\vec{u}_1, \vec{u}_2, \ldots, \vec{u}_n$, and n nonzero real numbers, $e_1, e_2, \ldots, e_n$, such that for $i = 1, 2, \ldots, n$,
$$\vec{H}(\vec{u}_i) = e_i\vec{u}_i.$$

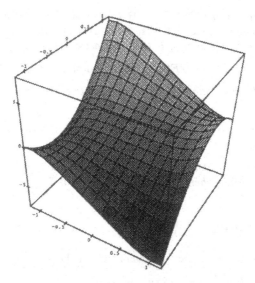

FIGURE 9.7. $f(x,y) = \frac{3}{5}x^5 - 3xy^2 + 3y$.

The $\vec{u}_i$ are called *characteristic vectors* or *eigenvectors* of $\mathbf{H}$ ("eigen" is the German prefix meaning "particular" or "characteristic"); the e_i are the *characteristic values* or *eigenvalues* of $\mathbf{H}$.

If

$$\mathbf{H} = \left(\begin{array}{cc} A & B \\ B & C \end{array} \right)$$

and $AC - B^2 \neq 0$, then the eigenvalues of $\mathbf{H}$ are

$$e_1 = \frac{A + C + \sqrt{(A-C)^2 + 4B^2}}{2}, \quad e_2 = \frac{A + C - \sqrt{(A-C)^2 + 4B^2}}{2}.$$

(9.24)

It is left as an exercise to verify that

$$e_1 + e_2 = A + C, \tag{9.25}$$

$$e_1 e_2 = AC - B^2; \tag{9.26}$$

and therefore, the determinant of $\mathbf{H}$ is negative if and only if e_1 and e_2 have opposite signs, and it is positive if and only if e_1 and e_2 have the same sign. If e_1 and e_2 share the same sign then that is the same as the sign of A. In general, the sum of the eigenvalues is always equal to the sum of the terms on the main diagonal, and the product of the eigenvalues is always equal to the determinant. This second property implies that if the determinant is not zero, then none of the eigenvalues is zero.

Theorem 9.5 *Let $F : \mathbf{R}^n \longrightarrow \mathbf{R}$ be a scalar field, differentiable at $\vec{a}$, with continuous second partial derivatives at all points sufficiently close to $\vec{a}$. Let $\mathbf{H}_a$ be the Hessian of F at $\vec{a}$. We assume that*

$$\nabla F(\vec{a}) = \vec{0} \quad and \quad \det(\mathbf{H}_a) \neq 0.$$

1. *If the eigenvalues of $\mathbf{H}_a$ are all positive, then F has a local minimum at $\vec{a}$.*

2. *If the eigenvalues of $\mathbf{H}_a$ are all negative, then F has a local maximum at $\vec{a}$.*

3. *If $\mathbf{H}_a$ has at least one positive eigenvalue and at least one negative eigenvalue, then F has a saddle point at $\vec{a}$.*

9.6   Exercises

1. Find the stationary points of each of the following scalar fields:

   (a) $f(x, y) = x^2 - 2xy - y^2 + 4x - 2y$,
   (b) $f(x, y) = x^2 + 4xy + y^2 - 3x + y$,
   (c) $f(x, y) = -3x^2 + 4xy - 3y^2 - 3x + 3y$,
   (d) $f(x, y) = 3x^2 - 6xy + 3y^2 + 2x - 2y$,
   (e) $f(x, y) = 5x^2 + 2xy + y^2 + 2x - 2y$,
   (f) $f(x, y) = \frac{3}{2}x^4 - 3x^2y^2 + y^3$,
   (g) $f(x, y) = \sin x \cos y$,
   (h) $f(x, y) = y^2 - \sin^2 x$,
   (i) $f(x, y) = (x^2 - y^2)e^{-(x^2-y^2)}$,
   (j) $f(x, y, z) = x^2 - 4y^2 + 2z^2 - 3xy + 4yz + 2y - 3z$,
   (k) $f(x, y, z) = x^3y + xyz^2$,
   (l) $f(x, y, z) = y^2 - x^2y - yz^2 + x^2z^2$.

2. Describe the nature of the stationary points in Exercise 1.

3. Recall that f has an *absolute maximum* at $\vec{x}_0$ if $f(\vec{x}) \leq f(\vec{x}_0)$ for all $\vec{x}$ in the domain of f. Similarly, it has an *absolute minimum* at $\vec{x}_0$ if $f(\vec{x}) \geq f(\vec{x}_0)$ for all $\vec{x}$ in the domain of f. For each scalar field in Exercise 1, find the points at which f has an absolute maximum or an absolute minimum, or state that such points do not exist.

4. Verify that if we restrict the scalar field

$$f(x, y) = y^2 - 3yx^2 + 2x^4$$

to any line through the origin, then it has a local minimum at the origin.

5. Show that the second-order Taylor formula for a scalar field in two variables with continuous second partial deviatives at (a_1, a_2) is

$$F(x, y) = F(a_1, a_2) + \frac{\partial F}{\partial x}(x - a_1) + \frac{\partial F}{\partial y}(y - a_2)$$

$$+ \frac{1}{2}\frac{\partial^2 F}{\partial x^2}(x - a_1)^2 + \frac{\partial^2 F}{\partial x\, \partial y}(x - a_1)(y - a_2)$$

$$+ \frac{1}{2}\frac{\partial^2 F}{\partial y^2}(y - a_2)^2 + \left[(x - a_1)^2 + (y - a_2)^2\right] E(x, y), \tag{9.27}$$

where each partial derivative is evaluated at (a_1, a_2) and

$$\lim_{(x, y) \to (a_1, a_2)} E(x, y) = 0.$$

6. Use the expansion of Exercise 5 with $(a_1, a_2) = (0, 0)$ to find the quadratic polynomial in x and y that best approximates e^{x+y}.

7. Use the expansion of Exercise 5 with $(a_1, a_2) = (0, 0)$ to find the quadratic polynomial in x and y that best approximates $\cos x \cos y$.

8. Prove that if $F : \mathbf{R}^n \longrightarrow \mathbf{R}$ is a scalar field for which all partial derivatives up to and including the mth partial derivatives exist and are continuous at all points sufficiently close to $\vec{a}$, then

$$F(\vec{a} + \Delta \vec{x}) - F(\vec{a})$$

$$= \nabla F(\vec{a}) \cdot \Delta \vec{x} + \frac{1}{2} \sum_{i_1=1}^{n} \sum_{i_2=1}^{n} D_{i_1 i_2} F(\vec{a})\, \Delta x_{i_1}\, \Delta x_{i_2}$$

$$+ \frac{1}{3!} \sum_{i_1=1}^{n} \sum_{i_2=1}^{n} \sum_{i_3=1}^{n} D_{i_1 i_2 i_3} F(\vec{a})\, \Delta x_{i_1}\, \Delta x_{i_2}\, \Delta x_{i_3} + \cdots$$

$$+ \frac{1}{m!} \sum_{i_1=1}^{n} \sum_{i_2=1}^{n} \cdots \sum_{i_m=1}^{n} D_{i_1 i_2 \cdots i_m} F(\vec{a})\, \Delta x_{i_1}\, \Delta x_{i_2} \cdots \Delta x_{i_m}$$

$$+ |\Delta \vec{x}|^m E_m(\vec{a}, \Delta \vec{x}), \tag{9.28}$$

where

$$\lim_{\Delta \vec{x} \to \vec{0}} E_m(\vec{a}, \Delta \vec{x}) = 0.$$

9. Prove that e_1 and e_2 as defined in Equation (9.24) satisfy

$$e_1 + e_2 = A + C, \tag{9.29}$$

$$e_1 e_2 = AC - B^2. \tag{9.30}$$

10. Continuing Exercise 9, prove that if $AC - B^2$ is positive, then e_1 is positive if and only if A is positive.

11. Continuing Exercise 9, show that if $B = 0$, then $(1,0)$ and $(0,1)$ are eigenvectors.

12. Let $\mathbf{A}$ and $\mathbf{B}$ be $n \times n$ matrices, $\vec{A}$ and $\vec{B}$ the corresponding transformations. Prove that if $\det(\mathbf{A} - \mathbf{B}) = 0$, then there exists a unit vector, $\vec{u}$, such that $\vec{A}(\vec{u}) = \vec{B}(\vec{u})$. Also show that if we can find a unit vector, $\vec{u}$, for which $\vec{A}(\vec{u}) = \vec{B}(\vec{u})$, then $\det(\mathbf{A} - \mathbf{B}) = 0$.

13. Let $\mathbf{B}$ be the matrix with each term on the main diagonal equal to e and all other entries 0. Show that

$$\vec{B}(\vec{u}) = e\vec{u}. \tag{9.31}$$

Let $\mathbf{X}$ be the matrix with each term on the main diagonal equal to x and all other entries 0. Use Exercise 12 and Equation (9.31) to show that e is an eigenvalue of $\mathbf{A}$ if and only if it is a root of the polynomial

$$\det(\mathbf{A} - \mathbf{X}).$$

14. Prove that e_1, e_2, as defined in Equation (9.24) are the roots of

$$\begin{vmatrix} A - x & B \\ B & C - x \end{vmatrix} = (A - x)(C - x) - B^2.$$

15. Let $e_1, \ldots, e_n$ be the eigenvalues of $\mathbf{A}$ (with repetitions permitted according to the multiplicity of the roots of $\det(\mathbf{A} - \mathbf{X})$). We can find n linearly independent unit vectors, $\vec{u}_1, \ldots, \vec{u}_n$, such that each $\vec{u}_j$ is an eigenvector for e_j. Each vector in $\mathbf{R}^n$ can be uniquely written as a linear combination of $\vec{u}_1, \ldots, \vec{u}_n$. Use this information to prove Theorem 9.5.

16. Find the eigenvalues of

$$\begin{pmatrix} 1 & 2 & 1 \\ 2 & 2 & 3 \\ 1 & 0 & 2 \end{pmatrix}.$$

17. Find the stationary points of

$$F(x, y, z) = x^2 + xy + y^2 - 2z^2 + 3x - 2y + z.$$

If possible, use Theorem 9.5 to characterize them.

18. Find the stationary points of

$$F(x, y, z) = x^3 + xy^2 - y^2 + z^3 - 3z.$$

If possible, use Theorem 9.5 to characterize them.

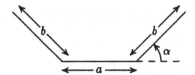

FIGURE 9.8. Exercise 19.

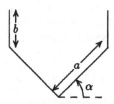

FIGURE 9.9. Exercise 20.

19. A long flat piece of sheet metal, 12 in. wide, is to be bent to form a long trough of trapezoidal cross section (see Figure 9.8). Find the values of a, b, and α that maximize the area of the cross section (and thus maximize the volume of the trough).

20. A long flat piece of sheet metal, 12 in. wide, is to be bent to form a long trough of pentagonal cross section (see Figure 9.9). Find the values of a, b, and α that maximize the area of the cross section (and thus maximize the volume of the trough).

21. A common real world problem is trying to describe an unknown function, f, that we have sampled or evaluated at n points:

$$x_1, x_2, \ldots, x_n.$$

If we want to find a linear function, $y = ax + b$, that approximates f, we can look for the linear function that minimizes the sum of the squares of the differences between $f(x_i)$ and $ax_i + b$:

$$E(a,b) = \sum_{i=1}^{n} [f(x_i) - (ax_i + b)]^2.$$

Find the values of a and b [in terms of $\sum x_i$, $\sum x_i^2$, $\sum f(x_i)$, and $\sum x_i f(x_i)$] that minimize $E(a,b)$.

22. Extend the method of the previous exercise to find the best quadratic

approximation. Specifically, find values of a, b, and c that minimize

$$E(a, b, c) = \sum_{i=1}^{n} [f(x_i) - (ax_i^2 + bx_i + c)]^2.$$

9.7   Lagrange Multipliers

Many problems of maximization or minimization do not permit us to wander over the entire domain of our scalar field but restrict us to some finite subset. In one dimension, this was never a problem. To find the extrema of $f(x) = x^3 - x$ over the interval $[0,2]$, we evaluate f at those points in the interval where the derivative is zero, $x = \sqrt{3}/3$, and at the endpoints, $x = 0, 2$:

$$f(0) = 0, \quad f\left(\sqrt{3}/3\right) = \frac{-2\sqrt{3}}{9}, \quad f(2) = 6.$$

The absolute maximum is 6, occurring at $x = 2$; the absolute minimum is $-2\sqrt{3}/9$, occurring at $x = \sqrt{3}/3$.

In more than one dimension, the situation is more complicated because the boundary of our restricted domain will contain infinitely many points. Consider the problem of finding the extrema of

$$f(x, y) = x^2 y - y^3,$$

restricted to the unit disc

$$x^2 + y^2 \leq 1$$

(Figure 9.10). We can think of the cylinder $x^2 + y^2 = 1$ as a cookie cutter that comes down through the surface $z = x^2 y - y^3$. We restrict our attention to that portion of the surface inside the cookie cutter and ask where it reaches its maximum and mimimum.

There is one stationary point inside, at $(0, 0)$. This is a saddle point. As we saw in Lemma 9.1, a bounded region that includes all of its boundary points must have points at which it achieves both an absolute maximum and an absolute minimum. Since this does not happen inside the disc, it must happen on the boundary:

$$x^2 + y^2 = 1.$$

Obviously, we cannot test each point of the unit circle. In this case, we can solve for x^2 in terms of y^2:

$$x^2 = 1 - y^2,$$

and then substitute into the equation for f:

$$f(x, y) = y - 2y^3, \quad -1 \leq y \leq 1.$$

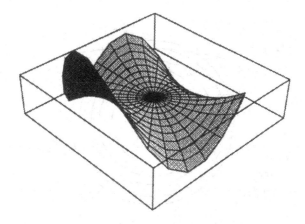

FIGURE 9.10. $f(x, y) = x^2 y - y^3$ over $x^2 + y^2 \leq 1$.

We have reduced the problem to the 1-dimensional case in which we can use standard techniques:

$$\frac{d}{dy}(y - 2y^3) = 1 - 6y^2,$$

which is zero when

$$y = \pm\frac{1}{\sqrt{6}}.$$

As a function of y, the candidates for extrema are

$$y = -1, \; \frac{-1}{\sqrt{6}}, \; \frac{1}{\sqrt{6}}, \; 1.$$

This yields six possible places on the boundary where the extrema might lie:

$$\begin{aligned}
f(0, -1) &= 1, \\
f\left(\sqrt{5/6}, -\sqrt{1/6}\right) &= \frac{-2}{3}\sqrt{1/6} \simeq -0.27, \\
f\left(-\sqrt{5/6}, -\sqrt{1/6}\right) &= \frac{-2}{3}\sqrt{1/6}, \\
f\left(\sqrt{5/6}, \sqrt{1/6}\right) &= \frac{2}{3}\sqrt{1/6}, \\
f\left(-\sqrt{5/6}, \sqrt{1/6}\right) &= \frac{2}{3}\sqrt{1/6}, \\
f(0, 1) &= -1.
\end{aligned}$$

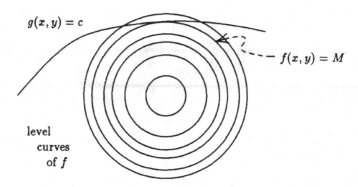

$g(x,y) = c$

$f(x,y) = M$

level
curves
of f

FIGURE 9.11. Level curves of $f(x,y)$ against $g(x,y) = c$.

The absolute maximum is 1, achieved at $(0,-1)$; the absolute minimum is -1, achieved at $(0,1)$.

A General Approach

We were lucky with this example because we could eliminate one of our variables. The problem before us is what to do if we cannot easily eliminate a variable. We want a general technique for maximizing or minimizing $f(x,y)$ when (x,y) is restricted to a level curve:

$$g(x,y) = c.$$

Let us assume that both f and g are differentiable functions and let M be the unknown maximum value of $f(x,y)$ when restricted to the level curve. We consider two curves:

$$f(x,y) = M, \quad g(x,y) = c.$$

These curves intersect. Let us call the point of intersection (x_0, y_0). If we increase M, no matter how little, we get a level curve of f that does not intersect $g(x,y) = c$. This implies that at (x_0, y_0) either M is a local maximum for the unrestricted scalar field $f(x,y)$ or the tangents to $f(x,y) = M$ and $g(x,y) = c$ are parallel. If they were not parallel, then moving in one direction along $g(x,y) = c$, we would cross lower level curves of f; moving in the opposite direction, we would cross higher level curves (Figure 9.11).

If M is a local maximum of the unrestricted scalar field, then

$$\nabla f(x_0, y_0) = \vec{0}.$$

If the tangents are parallel, then $\nabla f(x_0, y_0)$ is parallel to $\nabla g(x_0, y_0)$. (Recall that ∇f is perpendicular to the tangent.) In terms of the differentials of f and g, this says that either $df = 0$ or $df = \lambda\, dg$ for some λ. Since the first equation is merely the case $\lambda = 0$ of the second, it is enough to seek values of x, y, and λ for which $g(x, y) = c$ and $df = \lambda\, dg$. This yields three equations in our three unknowns:

$$\frac{\partial f}{\partial x} = \lambda \frac{\partial g}{\partial x},$$
$$\frac{\partial f}{\partial y} = \lambda \frac{\partial g}{\partial y},$$
$$g(x, y) = c.$$

Second Look at Our Example

Going back to

$$f(x, y) = x^2 y - y^3$$

on the level curve

$$g(x, y) = x^2 + y^2 = 1, \tag{9.32}$$

the equation $df = \lambda\, dg$ becomes

$$2xy\, dx + (x^2 - 3y^2)\, dy = \lambda(2x\, dx + 2y\, dy).$$

This yields two equations

$$2xy = 2x\lambda, \tag{9.33}$$
$$x^2 - 3y^2 = 2y\lambda. \tag{9.34}$$

Equation (9.33) implies that either $x = 0$ (in which case $y = \pm 1$) or $y = \lambda$. In the latter case, Equation (9.34) becomes

$$x^2 - 5y^2 = 0,$$

which, combined with Equation (9.32), implies that

$$x = \pm\sqrt{\frac{5}{6}}, \quad y = \pm\sqrt{\frac{1}{6}},$$

yielding exactly the same six critical points we saw earlier.

Higher Dimensions

This technique carries over into higher dimensions. In general, we find our extrema by trying to express the differential of the scalar field we are to maximize or minimize as a linear combination of the differentials of the scalar fields used in the restrictions. To maximize or minimize $f(\vec{x})$ subject to

$$
\begin{aligned}
g_1(\vec{x}) &= c_1, \\
g_2(\vec{x}) &= c_2, \\
&\vdots \\
g_k(\vec{x}) &= c_k,
\end{aligned}
$$

we introduce k new unknowns, $\lambda_1, \lambda_2, \ldots, \lambda_k$, called *Lagrange multipliers*, and create n more equations by setting

$$df = \lambda_1 \, dg_1 + \lambda_2 \, dg_2 + \ldots + \lambda_k \, dg_k.$$

To justify this approach, we need to return to Theorem 9.1, which implies the following result. Our theorem actually states where we do not need to look for extrema.

Theorem 9.6 *Let $f : \mathbf{R}^n \longrightarrow \mathbf{R}$ be a scalar field and let $\vec{a}$ be a point of the domain of f that satisfies the system of equations*

$$
\begin{aligned}
g_1(x_1, x_2, \ldots, x_n) &= c_1, \\
g_2(x_1, x_2, \ldots, x_n) &= c_2, \\
&\vdots \\
g_k(x_1, x_2, \ldots, x_n) &= c_k.
\end{aligned}
$$

If $f, g_1, g_2, \ldots, g_k$ are each continuously differentiable at $\vec{a}$ and if

$$df \, dg_1 \, dg_2 \ldots dg_k$$

is not zero at $\vec{a}$, then there exist points $\vec{a}_1$ and $\vec{a}_2$ in the domain of f and arbitrarily close to $\vec{a}$ that also satisfy our system of equations and for which

$$f(\vec{a}_1) < f(\vec{a}) < f(\vec{a}_2).$$

In particular, if the product of the differentials is not zero at $\vec{a}$, then f on the restricted domain does not have a local or an absolute extremum at $\vec{a}$.

Of course, if the product of the differentials is zero, this does not guarantee an extremum. But, this condition usually restricts our candidates to a finite set of points that can be individually checked.

Proof: Let us first assume that $k = n-1$ and define a vector field $\vec{y} = \vec{F}(\vec{x})$ by

$$\begin{aligned}
y_1 &= f(\vec{x}), \\
y_2 &= g_1(\vec{x}), \\
y_3 &= g_2(\vec{x}), \\
&\vdots \\
y_n &= g_{n-1}(\vec{x}).
\end{aligned}$$

Recalling Equation (7.17) on page 191, we have

$$\begin{aligned}
df\, dg_1\, dg_2 \cdots dg_k &= dy_1\, dy_2 \cdots dy_n \\
&= \frac{\partial(y_1, y_2, \ldots, y_n)}{\partial(x_1, x_2, \ldots, x_n)} dx_1\, dx_2 \cdots dx_n.
\end{aligned}$$

The product of the differentials at $\vec{a}$ is zero if and only if the Jacobian of $\vec{F}$ at $\vec{a}$ is zero. If this Jacobian is not zero, then we know by Theorem 9.1 that $\vec{F}$ is invertible with $\vec{F}^{-1}$ defined for all $\vec{y}$ sufficiently close to $\vec{F}(\vec{a})$. We find a small positive number, $\epsilon > 0$, so that both

$$\vec{b}_1 = [f(\vec{a}) - \epsilon, g_1(\vec{a}), \ldots, g_{n-1}(\vec{a})]$$

and

$$\vec{b}_2 = [f(\vec{a}) + \epsilon, g_1(\vec{a}), \ldots, g_{n-1}(\vec{a})]$$

lie in the domain of $\vec{F}^{-1}$. For $i = 1, 2, \ldots, n - 1$,

$$\begin{aligned}
g_i\left(\vec{F}^{-1}(\vec{b}_1)\right) &= g_i(\vec{a}) = c_i, \\
g_i\left(\vec{F}^{-1}(\vec{b}_2)\right) &= g_i(\vec{a}) = c_i,
\end{aligned}$$

and so both $\vec{F}^{-1}(\vec{b}_1)$ and $\vec{F}^{-1}(\vec{b}_2)$ lie in the restricted domain of points satisfying our system of equations. We also have

$$\begin{aligned}
f\left(\vec{F}^{-1}(\vec{b}_1)\right) &= f(\vec{a}) - \epsilon < f(\vec{a}), \\
f\left(\vec{F}^{-1}(\vec{b}_2)\right) &= f(\vec{a}) + \epsilon > f(\vec{a}).
\end{aligned}$$

The continuity of $\vec{F}$ and $\vec{F}^{-1}$ guarantee that we can choose $\vec{b}_1$ and $\vec{b}_2$ so that $\vec{F}^{-1}(\vec{b}_1)$ and $\vec{F}^{-1}(\vec{b}_2)$ are within any specified positive distance of $\vec{a}$.

If $k \geq n$, then the product of our differentials must be zero and the hypothesis of the theorem does not apply.

If $k < n - 1$, then $df\, dg_1 \cdots dg_k$ is a $k+1$-form. Given any $k+1$ variables: $x_{i_1}, x_{i_2}, \ldots, x_{i_{k+1}}$, the coefficient of $dx_{i_1} dx_{i_2} \cdots dx_{i_{k+1}}$ in this differential form is the Jacobian

$$\frac{\partial(f, g_1, \ldots, g_k)}{\partial(x_{i_1}, x_{i_2}, \ldots, x_{i_{k+1}})}.$$

Since our $k+1$-form at $\vec{a}$ is not zero, at least one of coefficients is not zero. If necessary, relabel the variables so that

$$\frac{\partial(f, g_1, \ldots, g_k)}{\partial(x_1, x_2, \ldots, x_{k+1})}(\vec{a}) \neq 0.$$

We define the vector field $\vec{F} : \mathbf{R}^{k+1} \longrightarrow \mathbf{R}^{k+1}$ by

$$
\begin{aligned}
y_1 &= f(x_1, x_2, \ldots, x_{k+1}), \\
y_2 &= g_1(x_1, x_2, \ldots, x_{k+1}), \\
&\;\;\vdots \\
y_{k+1} &= g_k(x_1, x_2, \ldots, x_{k+1}),
\end{aligned}
$$

and we proceed exactly as we did when k was $n - 1$.

Q.E.D.

Multipliers

The product $df\, dg_1 \cdots dg_k$ is zero if and only if these $k + 1$ differential 1-forms are linearly dependent, that is, if and only if there exist constants $b_1, b_2, \ldots, b_{k+1}$, not all zero, for which

$$b_1\, df + b_2\, dg_1 + \cdots + b_{k+1}\, dg_k = 0.$$

If b_1 is not zero, then we can solve for df is terms of the other 1-forms:

$$df = \lambda_1\, dg_1 + \lambda_2\, dg_2 + \cdots + \lambda_k\, dg_k, \qquad (9.35)$$

where $\lambda_i = -b_{i+1}/b_1$. This finishes the justification of the use of Lagrange multipliers. If we equate coefficients of $dx_1, dx_2, \ldots, dx_n$ in Equation (9.35), we obtain n equations in $n + k$ unknowns: $x_1, x_2, \ldots, x_n, \lambda_1, \lambda_2, \ldots, \lambda_k$. Combining this with our k restrictions:

$$g_1(\vec{x}) = c_1,\; g_2(\vec{x}) = c_2,\; \ldots,\; g_k(\vec{x}) = c_k,$$

we have the same number of equations as unknowns, and hopefully, we can solve for $x_1, x_2, \ldots, x_n$.

Examples

Find the extrema of
$$f(x, y, z) = x^2 + y^2 + z^2$$
subject to the constraint
$$(x - 1)^2 + (y - 2)^2 + (z - 3)^2 = 4. \tag{9.36}$$

We introduce the Lagrange multiplier, λ, and set up our differential equation:

$$2x\, dx + 2y\, dy + 2z\, dz = \lambda[2(x - 1)\, dx + 2(y - 2)\, dy + 2(z - 3)\, dz].$$

Comparing coefficients of dx, dy, and dz yields three more equations:

$$
\begin{aligned}
x &= \lambda(x - 1), & (9.37) \\
y &= \lambda(y - 2), & (9.38) \\
z &= \lambda(z - 3). & (9.39)
\end{aligned}
$$

Solving Equations (9.37) through (9.39) for x, y, and z in terms of λ and then substituting these values into Equation (9.36) produces

$$\left(\frac{-\lambda}{1 - \lambda} - 1\right)^2 + \left(\frac{-2\lambda}{1 - \lambda} - 2\right)^2 + \left(\frac{-3\lambda}{1 - \lambda} - 3\right)^2 = 4,$$

$$14 = 4(1 - \lambda)^2,$$

$$\lambda = 1 \pm \frac{\sqrt{14}}{2},$$

yielding two solutions:

$$\left(1 + \frac{2}{\sqrt{14}}, 2 + \frac{4}{\sqrt{14}}, 3 + \frac{6}{\sqrt{14}}\right) \quad \text{and} \quad \left(1 - \frac{2}{\sqrt{14}}, 2 - \frac{4}{\sqrt{14}}, 3 - \frac{6}{\sqrt{14}}\right).$$

We have an absolute maximum of $18 + 12\sqrt{14}/7$ at the first point, an absolute minimum of $18 - 12\sqrt{14}/7$ at the second.

Find the extrema of

$$f(x, y, z) = x^2 + y^2 + z^2$$

subject to the constraints

$$
\begin{aligned}
(x - 1)^2 + (y - 2)^2 + (z - 3)^2 &= 9, & (9.40) \\
x - 2z &= 0. & (9.41)
\end{aligned}
$$

While this looks very similar to the previous example, the extra constraint makes it more complicated. We introduce two Lagrange multipliers and set up our differential equation:

$$2x\,dx + 2y\,dy + 2z\,dz = \lambda_1[2(x-1)\,dx + 2(y-2)\,dy + 2(z-3)\,dz] + \lambda_2(dx - 2dz).$$

This yields three additional equations:

$$2x = 2(x-1)\lambda_1 + \lambda_2, \qquad (9.42)$$
$$y = (y-2)\lambda_1, \qquad (9.43)$$
$$z = (z-3)\lambda_1 - \lambda_2. \qquad (9.44)$$

Adding Equations (9.42) and (9.44) and then using Equation (9.41) to substitute $2z$ for x , we get

$$5z = (5z - 5)\lambda_1,$$

so that

$$z = \frac{-\lambda_1}{1 - \lambda_1},$$
$$y = \frac{-2\lambda_1}{1 - \lambda_1},$$
$$x = \frac{-2\lambda_1}{1 - \lambda_1}.$$

Substituting into Equation (9.40), we can solve for λ_1:

$$\lambda_1 = -\frac{1}{2} \text{ or } \frac{5}{2},$$

yielding two solutions:

$$\left(\frac{2}{3}, \frac{2}{3}, \frac{1}{3}\right) \text{ and } \left(\frac{10}{3}, \frac{10}{3}, \frac{5}{3}\right).$$

We have an absolute minimum of 1 at the first point, an absolute maximum of 25 at the second.

We have two attracting particles in the x, y plane. The first is constrained to lie on the ellipse

$$x^2 + 9y^2 = 9,$$

and the second is constrained to lie on the line

$$x + 5y = 10,$$

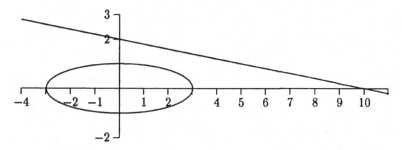

FIGURE 9.12. $x^2 + 9y^2 = 9$; $x + 5y = 10$.

(Figure 9.12). Let (r, s) be the position of the first particle, (u, v) the position of the second. We want to minimize the distance between them subject to our given constraints. Equivalently, we want to minimize the square of the distance between them:

$$f(r, s, u, v) = (r - u)^2 + (s - v)^2,$$

subject to the constraints

$$\begin{aligned} g(r, s, u, v) &= r^2 + 9s^2 = 9, & (9.45) \\ h(r, s, u, v) &= u + 5v = 10. & (9.46) \end{aligned}$$

The relationship

$$df = \lambda_1\, dg + \lambda_2\, dh$$

yields four equations when we compare coefficients of dr, ds, du, and dv:

$$\begin{aligned} 2(r - u) &= 2r\lambda_1, & (9.47) \\ 2(s - v) &= 18s\lambda_1, & (9.48) \\ -2(r - u) &= \lambda_2, & (9.49) \\ -2(s - v) &= 5\lambda_2. & (9.50) \end{aligned}$$

Combining Equation (9.47) with (9.49) and (9.48) with (9.50) yields

$$\begin{aligned} 2r\lambda_1 &= -\lambda_2, & (9.51) \\ 18s\lambda_1 &= -5\lambda_2, & (9.52) \end{aligned}$$

and therefore,

$$\lambda_1(5r - 9s) = 0. \qquad (9.53)$$

If $\lambda_1 = 0$, then $r = u$ and $s = v$. From Equation (9.46),

$$u = 10 - 5v. \qquad (9.54)$$

Making these substituions into Equation (9.45) yields

$$(10 - 5v)^2 + 9v^2 = 9,$$

or equivalently,

$$34v^2 - 100v + 91 = 0,$$

which has no solutions. Therefore $\lambda_1 \neq 0$, and so Equation (9.53) implies that

$$9s = 5r. \tag{9.55}$$

Making this substitution into Equation (9.45) produces

$$r^2 + \tfrac{25}{9}r^2 = 9,$$

$$r = \pm\frac{9}{\sqrt{34}}. \tag{9.56}$$

Combining Equations (9.49) and (9.50) gives us

$$5r - 5u = s - v$$
$$= \frac{5}{9}r - v,$$
$$v = 5u \mp \frac{40}{\sqrt{34}}. \tag{9.57}$$

Putting this together with Equation (9.46), we can solve for u and v, obtaining two possible locations:

$$(r, s, u, v) = \left(\frac{\pm 9}{\sqrt{34}}, \frac{\pm 5}{\sqrt{34}}, \frac{5}{13} \pm \frac{100}{13\sqrt{34}}, \frac{25}{13} \mp \frac{20}{13\sqrt{34}} \right).$$

It is the first choice that minimizes the distance between our two particles:

$$\sqrt{f\left(\frac{9}{\sqrt{34}}, \frac{5}{\sqrt{34}}, \frac{100 + 5\sqrt{34}}{13\sqrt{34}}, \frac{-20 + 25\sqrt{34}}{13\sqrt{34}} \right)} = \frac{10 - \sqrt{34}}{\sqrt{26}} \simeq 0.818.$$

A Bad Example

We do need some caution in using Lagrange multipliers because it is possible that the linear relationship among $df, dg_1, \ldots, dg_k$ does not, in fact, involve df. As an example of this, consider the problem of minimizing the scalar field

$$f(x, y, z) = x^2 + y^2$$

subject to the constraints

$$g(x, y, z) = z = 0, \tag{9.58}$$
$$h(x, y, z) = z^2 - (y - 1)^3 = 0. \tag{9.59}$$

The three additional equations involving our Lagrange multipliers are

$$2x = 0, \tag{9.60}$$

$$2y = -3\lambda_2(y-1)^2, \tag{9.61}$$

$$0 = \lambda_1 + 2z\lambda_2. \tag{9.62}$$

Equations (9.58) and (9.59) imply that $y = 1$. Substituting this value of y into Equation (9.61) leaves us in an untenable position:

$$2 = -3\lambda_2(1-1)^2.$$

There is no value of λ_2 satisfying this equation. We can avoid this difficulty by looking for the points at which the product of the differentials is zero:

$$\begin{aligned}
0 &= df \, dg \, dh \\
&= (2x \, dx + 2y \, dy)(dz)(-3(y-1)^2 \, dy + 2z \, dz) \\
&= 6x(y-1)^2 \, dx \, dy \, dz,
\end{aligned}$$

$$x = 0 \ \text{ or } \ y = 1.$$

This has not advanced us very far since Equations (9.58) and (9.59) already imply that $y = 1$, but at least it is correct.

9.8   Exercises

1. Find the extrema of f subject to the constraints listed:

  (a) $f(x,y) = x^3 - xy^2$, $\quad x^2 + y^2 = 1$,

  (b) $f(x,y) = x^2 + y^2$, $\quad x^3 - xy^2 = 1$,

  (c) $f(x,y) = 2x + 3y$, $\quad x^2 - 2xy + 2y^2 = 1$,

  (d) $f(x,y,z) = x^2 + y^2 + z^2$, $\quad 3x - y + 2z = 14$,

  (e) $f(x,y,z) = x^2 + y^2 + z^2$, $\quad (x-y)^2 = 1$, $xyz = 1$.

2. Find the point on $x^2 - xy + y^2 = 1$ that is closest to the origin.

3. Find the point on $z^2 = (x-2)^2 + (y+3)^2$ that is closest to $(1,1,1)$.

4. Find the point on the intersection of $(x+1)^2 + (y-3)^2 + (z-12)^2 = 4$ and $x - 2y + z = 5$ that is closest to the origin. Find the point that is farthest from the origin.

5. Find the points on $y^2 = 1 + x^2$ and $2y = x$ for which the distance between the points is minimized.

6. Find the minimal distance between the point $(2,0,3)$ and the line $x = y = z$.

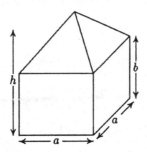

FIGURE 9.13. The tent in Exercise 10.

7. Find the dimensions of the rectangular parallelepiped of maximum volume that has edges parallel to the axes and that can be inscribed in the ellipsoid

$$x^2 + \frac{y^2}{4} + \frac{z^2}{9} = 1.$$

8. Find the minimal distance from the point $(2, 1, 3)$ to the hyperboloid $x^2 + y^2 - z^2 = 4$.

9. Find the minimal distance from the origin to the line defined by the intersection of the planes $x + y + 2z = 4$ and $2x - y + z = 2$.

10. A tent is to be made in the shape of a rectangular block with a square base, surmounted by a pyramid (Figure 9.13). If the total surface area (neglecting the bottom) is 400 m^2, find the dimensions that maximize the volume.

11. Use Lagrange multipliers to find the relative dimensions that maximize the volume of an open-topped box of fixed surface area ($lw + 2lh + 2wh = $ constant).

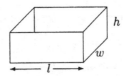

12. What is the minimal value of $f(x, y, z) = x^2 + y^2$ subject to the constraints $z = 0$ and $z^2 - (y - 1)^3 = 0$?

10

The Fundamental Theorem of Calculus

10.1   Overview

For scalar functions of one variable, the fundamental theorem of calculus is a powerful tool for integration. It says that there exists an associated function, often called an *antiderivative*, such that the integral of the scalar function can be evaluated by looking at the antiderivative at the endpoints of the interval. Specifically, it is the following theorem.

Theorem 10.1 The Fundamental Theorem of Calculus for Scalar Functions of One Variable *Let $f(x)$ be an integrable function over the interval $[a, b]$, then*

1. *there exists a function $F(x)$ defined on $[a, b]$ such that*

$$\frac{d}{dx}F(x) = f(x);$$

2. *if $\frac{d}{dx}F(x) = f(x)$, then*

$$\int_a^b f(x)\, dx = F(b) - F(a).$$

As mentioned in Chapter 1, this theorem was known to Isaac Barrow before Newton, and perhaps even to Gregory and Pierre Fermat (1601–1665). But, Newton was the first to exploit it to compute an area by first finding an antiderivative. A natural direction in which to generalize this result is to 1-forms in several variables. If

$$\omega(\vec{x}) = f_1(\vec{x})\, dx_1 + f_2(\vec{x})\, dx_2 + \cdots + f_n(\vec{x})\, dx_n$$

is a 1-form and C is a curve from $\vec{a}$ to $\vec{b}$, can we conclude that there is a scalar field $F(\vec{x})$ such that

$$\int_C \omega(\vec{x}) = F(\vec{b}) - F(\vec{a})\ ?$$

The answer is no and yes. Such a scalar field cannot always exist, because if it did, then line integrals would always be independent of the path. Exercises 1 and 6, Section 5.3, should have demonstrated that the integrals of

$$xy^2\,dx + y\,dy$$

and

$$\frac{-y\,dx + x\,dy}{x^2 + y^2}$$

do depend on the choice of path. However, Exercise 3 of the same section gives a 1-form,

$$xy^2\,dx + x^2y\,dy,$$

whose integral does appear to be independent of the path. If you were clever enough to realize that

$$xy^2\,dx + x^2y\,dy = d\left(\frac{1}{2}x^2y^2\right),$$

you may have noticed that the integral over each of the six paths from (0,0) to (1,1) yielded the value

$$\left.\frac{1}{2}x^2y^2\right|_{(0,0)}^{(1,1)} = \frac{1}{2}.$$

The truth of the matter is that an antiderivative is not promised; the first part of the fundamental theorem of calculus does not hold in higher dimensions. But, if the 1-form is the differential of a scalar field, then we can evaluate the line integral by computing the scalar field at the endpoints. This next theorem will be proven in Section 10.2. To say that a curve is *differentiable* means that it has a differentiable parametrization.

Theorem 10.2 Fundamental Theorem of Calculus for 1-Forms *If C is a differentiable curve from $\vec{a}$ to $\vec{b}$ and $\omega(\vec{x})$ is a 1-form for which the line integral of ω over C exists and if there is a scalar field $F(\vec{x})$ such that*

$$\omega(\vec{x}) = dF(\vec{x}),$$

then

$$\int_C \omega(\vec{x}) = F(\vec{b}) - F(\vec{a}).$$

The Fundamental Theorem of Calculus for Multiple Integrals

By the early 1800s scientists were fairly conversant with multiple and surface integrals, and these were playing an increasingly important role. Phenomena as diverse as gravity, heat, light, mechanical vibrations, electricity, and magnetism all came to be viewed as flows for which one wanted to measure the rate of flow through a given surface, what is called the *flux*. The gradual discovery that the fundamental theorem of calculus also exists in these situations opened the way to physical insights that will be explored in greater depth in the next chapter.

Here is where the language of differential forms really comes into its own for it serves to unify a collection of integral identities discovered in the first half of the nineteenth century by C. F. Gauss, George Green (1793–1841), Michel Ostrogradski (1801–1861), and William Thomson (1824–1907), who is better known as Lord Kelvin, a title bestowed upon him in 1892 for his scientific achievements. In performing this unification, the language of differential forms also points the way to the fundamental theorem of calculus in arbitrarily many dimensions. We shall need the fundamental theorem of calculus in 4-dimensional space–time if we are to develop special relativity.

We first define the differential of a differential form. This is done by specifying that if ω is a constant form (such as dx or $dx\,dy$) then its differential is 0:

$$d\omega = 0, \tag{10.1}$$

the differential operator is additive on differential forms:

$$d(\omega_1 + \omega_2) = d\omega_1 + d\omega_2, \tag{10.2}$$

and if f is a scalar field and ω is any differential form, then the usual product rule holds:

$$d(f\omega) = df\,\omega + f\,d\omega. \tag{10.3}$$

In particular, if ω is a constant differential form, then $d(f\omega) = df\,\omega$. As an example, consider the differential of the following 1-form in $\mathbf{R}^3$:

$$
\begin{aligned}
d(x^2 y\,&dx + xyz\,dy + xy^2\,dz) \\
&= d(x^2 y)\,dx + d(xyz)\,dy + d(x^2 y)\,dz \\
&= (2xy\,dx + x^2\,dy)\,dx + (yz\,dx + xz\,dy + xy\,dz)\,dy \\
&\quad + (y^2\,dx + 2xy\,dy)\,dz \\
&= -x^2\,dx\,dy + yz\,dx\,dy - xy\,dy\,dz - y^2\,dz\,dx + 2xy\,dy\,dz \\
&= xy\,dy\,dz - y^2\,dz\,dx - (x^2 - yz)\,dx\,dy.
\end{aligned}
$$

The differential of a 1-form is a 2-form. In general, the differential of a k-form is a $k+1$-form.

We also need to introduce the notion of an oriented boundary. Let S be a bounded oriented surface. This simply means that it lies inside some ball of finite radius and has well-defined positive and negative sides (the Möbius strip, Figure 10.1, does not qualify as it has only one side). The boundary of S, denoted by ∂S, is a curve that comes back to its starting point. Such a curve is called a *closed* curve. Recall that the right-hand rule relates the

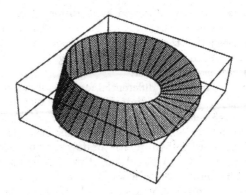

FIGURE 10.1. The Möbius strip.

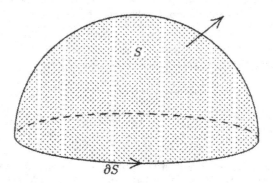

FIGURE 10.2. Boundary of oriented hemisphere.

direction in which we traverse the boundary to the orientation of S. The boundary is followed in a counterclockwise direction as we look down on the surface from the positive side. Thus, if S is the hemisphere

$$S = \{(x,\, y,\, z) \mid x^2 + y^2 + z^2 = 1,\, z \geq 0\},$$

oriented so that the normal with positive z coordinate is in the positive direction, then

$$\partial S = \{(x,\, y,\, z) \mid x^2 + y^2 = 1,\, z = 0\},$$

traced counterclockwise as we look down on the x, y plane (Figure 10.2).

For a solid region, R, its boundary, ∂R, is its surface. Note that the boundary of a solid region is again closed: it has no holes or openings. If R has a positive orientation in x, y, z space, then the positive normal for

∂R is the normal pointing out of the interior of R. This idea of boundary can be extended to higher dimensions where we use the word *manifold* to denote the appropriate extension of the notion of a surface or a solid region. We shall call a manifold *differentiable* if it has a differentiable parametrization whose domain is a rectangular region in the appropriate number of dimensions. It is *continuously differentiable* if the derivative of the parametrization is continuous. In other words, it is continuously differentiable if the partial derivatives are continuous. It is *twice continuously differentiable* if the second derivative exists and is continuous. A k-form in $\mathbf{R}^n$ is *continuously differentiable* if, viewed as a vector field from n space to $\binom{n}{k}$ space, it has a continuous derivative. While the next theorem is true in any number of dimensions, it is enough to understand what it says in two or three.

Theorem 10.3 The Fundamental Theorem of Calculus *Let M be a bounded, twice continuously differentiable, oriented, $k+1$-dimensional manifold in $\mathbf{R}^n$, $n \geq k+1$, with a k-dimensional boundary ∂M, and let $\omega(\vec{x})$ be a continuously differentiable k-form in $\mathbf{R}^n$, then $d\omega$ is a $k+1$-form and*

$$\int_{\partial M} \omega = \int_M d\omega. \tag{10.4}$$

Special Case: Line Integrals

If $\omega(\vec{x})$ is a scalar field (a 0-form) and M is a curve from $\vec{a}$ to $\vec{b}$ (a 1-dimensional manifold), then the boundary of M is the pair of points $\vec{a}, \vec{b}$, where $\vec{a}$ has negative orientation and $\vec{b}$ positive. It follows that ω evaluated on ∂M is $\omega(\vec{b}) - \omega(\vec{a})$, and thus

$$\omega(\vec{b}) - \omega(\vec{a}) = \int_M d\omega, \tag{10.5}$$

which is Theorem 10.2.

Special Case: Green's Theorem

If $\omega(\vec{x})$ is a 1-form in $\mathbf{R}^2$,

$$\omega = f\,dx + g\,dy,$$

and M is a region in the x, y plane with a positive orientation, then ∂M is the closed curve running around M in a counterclockwise direction and

$$d\omega = \left(\frac{\partial g}{\partial x} - \frac{\partial f}{\partial y}\right) dx\,dy.$$

The fundamental theorem of calculus becomes

$$\int_{\partial M} f\,dx + g\,dy = \int_M \left(\frac{\partial g}{\partial x} - \frac{\partial f}{\partial y}\right) dx\,dy. \tag{10.6}$$

This theorem was discovered by George Green in 1828 and independently by Michel Ostrogradski three years later. Green had published his result privately, and it was not widely known until 1846 when William Thomson (Lord Kelvin) discovered the manuscript and promulgated it.

Special Case: Divergence or Gauss's Theorem

If $\omega(\vec{x})$ is a 2-form in $\mathbf{R}^3$,

$$\omega = f\,dy\,dz + g\,dz\,dx + h\,dx\,dy$$

and M is a solid region with positive orientation, then ∂M is the closed surface of M for which the outer normal denotes the positive direction and

$$d\omega = \left(\frac{\partial f}{\partial x} + \frac{\partial g}{\partial y} + \frac{\partial h}{\partial z}\right) dx\,dy\,dz.$$

The fundamental theorem of calculus becomes

$$\int_{\partial M} f\,dy\,dz + g\,dz\,dx + h\,dx\,dy = \int_M \left(\frac{\partial f}{\partial x} + \frac{\partial g}{\partial y} + \frac{\partial h}{\partial z}\right) dx\,dy\,dz. \tag{10.7}$$

This theorem was independently discovered by Gauss and Ostrogradski. Note that if we have an incompressible flow described by the 2-form

$$\omega = f\,dy\,dz + g\,dz\,dx + h\,dx\,dy$$

then the total flow across any closed surface, ∂M, must equal 0. If the fluid cannot compress, then flow in equals flow out. It follows from Equation (10.7) that the fluid is incompressible if and only if

$$\int_M \left(\frac{\partial f}{\partial x} + \frac{\partial g}{\partial y} + \frac{\partial h}{\partial z}\right) dx\,dy\,dz = 0$$

for *any* solid M in the region of incompressibility. This in turn implies that the flow is incompressible if and only if

$$\frac{\partial f}{\partial x} + \frac{\partial g}{\partial y} + \frac{\partial h}{\partial z} = 0. \tag{10.8}$$

Recall from Section 8.5 that the 2-form describing heat flow in a homogeneous medium is a constant times

$$\frac{\partial T}{\partial x}\,dy\,dz + \frac{\partial T}{\partial y}\,dz\,dx + \frac{\partial T}{\partial x}\,dx\,dy.$$

If the heat is neither accumulating nor dissipating, in other words, if we have reached a steady state, then our 2-form is incompressible. Temperature has reached a steady state if and only if

$$\frac{\partial^2 T}{\partial x^2} + \frac{\partial^2 T}{\partial y^2} + \frac{\partial^2 T}{\partial z^2} = 0. \tag{10.9}$$

This is known as Laplace's equation, named for Pierre-Simon Laplace (1749–1827). It was Laplace who brought it to general attention in 1782 in his work on celestial mechanics. Euler had discovered it in 1752 in connection with hydrodynamics. As we shall see, it is ubiquitous.

Special Case: Stokes' Theorem

If $\omega(\vec{x})$ is a continuously differentiable 1-form in $\mathbf{R}^3$

$$\omega = f\, dx + g\, dy + h\, dz$$

and M is an oriented surface, then ∂M is the correspondingly oriented closed curve bounding M and

$$d\omega = \left(\frac{\partial h}{\partial y} - \frac{\partial g}{\partial z}\right) dy\, dz + \left(\frac{\partial f}{\partial z} - \frac{\partial h}{\partial x}\right) dz\, dx + \left(\frac{\partial g}{\partial x} - \frac{\partial f}{\partial y}\right) dx\, dy.$$

The fundamental theorem of calculus becomes

$$\int_{\partial M} f\, dx + g\, dy + h\, dz$$
$$= \int_M \left(\frac{\partial h}{\partial y} - \frac{\partial g}{\partial z}\right) dy\, dz + \left(\frac{\partial f}{\partial z} - \frac{\partial h}{\partial x}\right) dz\, dx + \left(\frac{\partial g}{\partial x} - \frac{\partial f}{\partial y}\right) dx\, dy.$$
$$\tag{10.10}$$

This identity has a curious history. It seems it was discovered by William Thomson (Lord Kelvin). He described it in a letter of 1850 to George Gabriel Stokes (1819–1903). Stokes then posed it as an examination problem in 1854 and his name has stuck. Since it is the most complicated of the special cases, the fundamental theorem of calculus in its full generality is also often referred to as Stokes' theorem. Equation (10.10) will have an important role to play in the development of Maxwell's equations.

Proofs

A proof of the fundamental theorem of calculus in its full generality is beyond the scope of this book. It can be found in Edwards' *Advanced Calculus*. The rest of this chapter will consist of proofs of the individual special cases together with examples of how they are used.

10.2   Independence of Path

Proof of Theorem 10.2

Let
$$\omega = f_1(\vec{x})\,dx_1 + f_2(\vec{x})\,dx_2 + \cdots + f_n(\vec{x})\,dx_n$$
be a 1-form to be evaluated on the curve
$$C = \{\vec{x}(t) \mid a \le t \le b\},$$
where each $f_i(\vec{x}(t))$ is a differentiable function of t. We assume that there exists a scalar field, $F(\vec{x})$, such that
$$\omega = dF,$$
equivalently, for $i = 1, 2, \ldots, n$:
$$f_i(\vec{x}) = \frac{\partial F}{\partial x_i}(\vec{x}).$$

The theorem follows from the definition of the line integral as the 1-dimensional integral of the pullback, the chain rule, and the 1-dimensional fundamental theorem of calculus :

$$
\begin{aligned}
\int_C \omega &= \int_a^b \left(f_1 \frac{dx_1}{dt} + f_2 \frac{dx_2}{dt} + \cdots + f_n \frac{dx_n}{dt} \right) dt \\
&= \int_a^b \left(\frac{\partial F}{\partial x_1} \frac{dx_1}{dt} + \frac{\partial F}{\partial x_2} \frac{dx_2}{dt} + \cdots + \frac{\partial F}{\partial x_n} \frac{dx_n}{dt} \right) dt \\
&= \int_a^b \frac{dF}{dt}\, dt = F(\vec{x}(b)) - F(\vec{x}(a)).
\end{aligned}
$$

Q.E.D.

Exact and Closed Forms

We need to determine when a 1-form is the differential of a scalar field and how to find this scalar field if it exists. We begin with some terminology: if ω is a k-form for which there exists a $k - 1$-form, call it ν, such that
$$\omega = d\nu,$$
then we say that ω is *exact*. The question that arises from the fundamental theorem of calculus in all its guises is *when is a differential form exact?* The following proposition gives a good test for when a form is likely to be exact.

Proposition 10.1 *Let ω be any differential form whose coefficients have continuous second partial derivatives, then*

$$d(d\omega) = 0. \tag{10.11}$$

Proof: Since

$$
\begin{aligned}
d(d(\omega_1 + \omega_2)) &= d(d\omega_1 + d\omega_2) \\
&= d(d\omega_1) + d(d\omega_2),
\end{aligned}
$$

it is sufficient to look at one term of ω,

$$f(\vec{x}) \, dx_1 \, dx_2 \, \cdots \, dx_k,$$

where $\vec{x} = (x_1, x_2, \ldots, x_n)$, $n \geq k$. We see that

$$
\begin{aligned}
d(f(\vec{x}) \, dx_1 \, dx_2 \, \cdots \, dx_k) &= \frac{\partial f}{\partial x_{k+1}} \, dx_{k+1} \, dx_1 \, \cdots \, dx_k \\
&\quad + \frac{\partial f}{\partial x_{k+2}} \, dx_{k+2} \, dx_1 \, \cdots \, dx_k \\
&\quad + \cdots \\
&\quad + \frac{\partial f}{\partial x_n} \, dx_n \, dx_1 \, \cdots \, dx_k \\
&= \sum_{j=k+1}^{n} \frac{\partial f}{\partial x_j} \, dx_j \, dx_1 \, \cdots \, dx_k,
\end{aligned}
$$

$$
\begin{aligned}
d[d(f(\vec{x}) \, dx_1 \, dx_2 \, \cdots \, dx_k)] &= \sum_{i=k+1}^{n} \sum_{j=k+1}^{n} \frac{\partial^2 f}{\partial x_i \, \partial x_j} \, dx_i \, dx_j \, dx_1 \, \cdots \, dx_k \\
&= \sum_{k+1 \leq i < j \leq n} \frac{\partial^2 f}{\partial x_i \, \partial x_j} \, dx_i \, dx_j \, dx_1 \, \cdots \, dx_k \\
&\quad + \sum_{k+1 \leq i = j \leq n} \frac{\partial^2 f}{\partial x_i \, \partial x_j} \, dx_i \, dx_j \, dx_1 \, \cdots \, dx_k \\
&\quad + \sum_{k+1 \leq j < i \leq n} \frac{\partial^2 f}{\partial x_i \, \partial x_j} \, dx_i \, dx_j \, dx_1 \, \cdots \, dx_k \\
&= \sum_{k+1 \leq i < j \leq n} \frac{\partial^2 f}{\partial x_i \, \partial x_j} \, dx_i \, dx_j \, dx_1 \, \cdots \, dx_k \\
&\quad + \sum_{k+1 \leq i < j \leq n} \frac{\partial^2 f}{\partial x_j \, \partial x_i} \, dx_j \, dx_i \, dx_1 \, \cdots \, dx_k
\end{aligned}
$$

$$d[d(f(\bar{x})\,dx_1\,dx_2\cdots dx_k)] \;=\; \sum_{k+1\le i<j\le n}\frac{\partial^2 f}{\partial x_i\,\partial x_j}\,dx_i\,dx_j\,dx_1\cdots dx_k$$

$$-\sum_{k+1\le i<j\le n}\frac{\partial^2 f}{\partial x_j\,\partial x_i}\,dx_i\,dx_j\,dx_1\cdots dx_k$$

$$= 0.$$

Q.E.D.

We say that a differential form, ω, is *closed* if

$$d\omega = 0.$$

If ω is exact ($\omega = d\nu$) and has continuous partial derivatives, then ω must be closed ($d\omega = d(d\nu) = 0$). The terminology is borrowed from the boundary operator, ∂. If C is a *closed* curve, then it has no boundary,

$$\partial C = \emptyset.$$

Since the boundary of a simply connected surface is always a closed curve, we see that

$$\partial(\partial S) = \emptyset. \tag{10.12}$$

This is true in any number of dimensions. The boundary of the boundary of a simply connected manifold is always the empty set.

If a differential form is closed, it does not have to be exact. But, an exact form must be closed. Let us take our three examples from the exercises of Section 5.3, beginning with Exercise 1:

$$d(xy^2\,dx + y\,dy) \;=\; 2xy\,dy\,dx + 0\,dx\,dy \;=\; -2xy\,dx\,dy.$$

This is not identically 0. The form $xy^2\,dx + y\,dy$ is not closed and so cannot be exact. From Exercise 3:

$$d(xy^2\,dx + x^2y\,dy) \;=\; 2xy\,dy\,dx + 2xy\,dx\,dy \;=\; (-2xy + 2xy)\,dx\,dy \;=\; 0.$$

This 1-form is closed and, in fact, is exact, as shown in the previous section:

$$xy^2\,dx + x^2y\,dy = d\left(\frac{1}{2}x^2y^2\right).$$

Nonexact Closed Forms

Exercise 6 of Section 5.3 is one of the most interesting cases:

$$d\left(\frac{-y\,dx + x\,dy}{x^2 + y^2}\right) \;=\; \frac{y^2 - x^2}{(x^2 + y^2)^2}\,dy\,dx + \frac{y^2 - x^2}{(x^2 + y^2)^2}\,dx\,dy \;=\; 0.$$

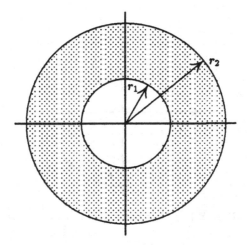

FIGURE 10.3. An annulus or ring.

This differential form is closed, but the line integrals do depend on the path, and so it cannot be exact. The problem arises from the fact that this 1-form is not defined at the origin. We say that a region is *simply connected* if it has no holes in it. Thus, a disc is simply connected but the annulus (Figure 10.3)

$$\{(x,y) \mid r_1 \leq x^2 + y^2 \leq r_2\}$$

is not. The region in which $(-y\,dx + x\,dy)/(x^2 + y^2)$ is defined is not simply connected.

If we take paths whose union does not enclose the origin, then the value of the integral depends only on the endpoints. Of the six distinct paths in this exercise, three pass above the origin and yield the same value, three pass below the origin and give a different common value. We compute four of them here. Note that they all start at $(-1,0)$ and end at $(1,0)$. Curves C_1 and C_3 both pass above the origin, while C_2 and C_4 pass below the origin. Our curves are defined by (Figure 10.4)

$$
\begin{aligned}
C_1 &= \{(t, 1 - |t|) \mid -1 \leq t \leq 1\}, \\
C_2 &= \{(t, |t| - 1) \mid -1 \leq t \leq 1\}, \\
C_3 &= \left\{(\sin t, \cos t) \mid -\frac{\pi}{2} \leq t \leq \frac{\pi}{2}\right\}, \\
C_4 &= \left\{(\sin t, -\cos t) \mid -\frac{\pi}{2} \leq t \leq \frac{\pi}{2}\right\}.
\end{aligned}
$$

The respective integrals are

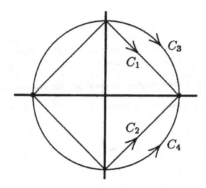

FIGURE 10.4. Four paths from $(-1,0)$ to $(1,0)$.

$$\int_{C_1} \frac{-y\,dx + x\,dy}{x^2 + y^2} = \int_{-1}^{0} \frac{-dt}{2t^2 + 2t + 1} + \int_{0}^{1} \frac{-dt}{2t^2 - 2t + 1}$$

$$= \int_{0}^{1} \frac{-2\,dt}{2t^2 - 2t + 1} = -2\arctan(2t-1)|_0^1 = -\pi,$$

$$\int_{C_2} \frac{-y\,dx + x\,dy}{x^2 + y^2} = \int_{-1}^{0} \frac{dt}{2t^2 + 2t + 1} + \int_{0}^{1} \frac{dt}{2t^2 - 2t + 1} = \pi,$$

$$\int_{C_3} \frac{-y\,dx + x\,dy}{x^2 + y^2} = \int_{-\pi/2}^{\pi/2} -dt = -\pi,$$

$$\int_{C_4} \frac{-y\,dx + x\,dy}{x^2 + y^2} = \int_{-\pi/2}^{\pi/2} dt = \pi.$$

The notion of being simply connected can be extended to regions in any number of dimensions. The technical definition of being simply connected amounts to saying that the region can be transformed into a ball by stretching, twisting, or otherwise deforming, but not cutting or pasting. Here, bagels and coffee cups are indistinguishable because they each have one unremovable hole. Neither is simply connected. Henri Poincaré proved the following theorem, which we shall prove for special cases in the next sections.

Theorem 10.4 (Poincaré's Lemma) *If ω is a differential form that is differentiable and closed in a simply connected region, R, then ω is exact in R.*

The reader may have noticed a very appealing duality between differential forms and geometric objects:

$$d(d\omega) = 0 \quad \longleftrightarrow \quad \partial(\partial M) = \emptyset$$

$$\left.\begin{array}{c} \text{a closed form} \\ \text{in a simply connected} \\ \text{region is the differential} \\ \text{of a form with} \\ \text{one less degree} \end{array}\right\} \quad \longleftrightarrow \quad \left\{\begin{array}{c} \text{a closed manifold} \\ \text{in a simply connected} \\ \text{region is the boundary} \\ \text{of a manifold with} \\ \text{one more dimension} \end{array}\right.$$

What fuels this duality is the fundamental theorem of calculus:

$$\int_{\partial M} \omega = \int_M d\omega.$$

In the twentieth century, an entire branch of mathematics, algebraic topology, has arisen from the study and exploitation of this connection.

A Contrary Example

It should be kept in mind that Poincaré's lemma gives a sufficient but not necessary condition for ω to be exact. If ω is not closed, then it cannot be exact. But, if it is closed in a region that is not simply connected, then it may or may not be exact. As an example, consider

$$\omega = \frac{(y^2 - xy)\,dx + (x^2 - xy)\,dy}{(x^2 + y^2)^{3/2}}.$$

This form is closed everywhere except $(0,0)$, where it is not defined:

$$d\left(\frac{(y^2 - xy)\,dx + (x^2 - xy)\,dy}{(x^2 + y^2)^{3/2}}\right)$$

$$= \frac{(2yx^2 - y^3 - x^3 + 2xy^2)\,dy\,dx + (2yx^2 - y^3 - x^3 + 2xy^2)\,dx\,dy}{(x^2 + y^2)^{5/2}}$$

$$= 0.$$

If we compare the integrals of this form over C_3 and C_4, given on page 289, we see that

$$\int_{C_3} \omega = \int_{-\pi/2}^{\pi/2} (\cos^2 t - \sin t \cos t)\cos t + (\sin^2 t - \sin t \cos t)(-\sin t)\,dt$$

$$= \int_{-\pi/2}^{\pi/2} \cos t - \sin t\,dt = \sin t + \cos t\Big|_{-\pi/2}^{\pi/2} = 2,$$

$$\int_{C_4} \omega = \int_{-\pi/2}^{\pi/2} (\cos^2 t + \sin t \cos t)\cos t + (\sin^2 t + \sin t \cos t)\sin t\,dt$$

$$= \int_{-\pi/2}^{\pi/2} \cos t + \sin t \; dt = \sin t - \cos t \Big|_{-\pi/2}^{\pi/2} = 2.$$

For this form, the integral over C_3 is the same as the integral over C_4, even though they pass on opposite sides of our singularity at $(0,0)$. It is not just these two integrals that must be the same. We have found one path that encircles our singularity, namely, from $(-1,0)$ to $(1,0)$ over C_4 and back over C_3 in the opposite direction, over which the total integral is zero. As shown in the exercises of Section 10.3, for a closed form with one singularity, once we can find one path encircling the singularity for which the integral is zero, then the integral is zero over *any* path encircling the singularity. From this, it follows that the integral of our differential form between any two points is independent of the path, and thus the differential form must be exact.

Finding Antidifferentials

The remaining problem is to find a scalar field F for which $\omega = dF$ when ω is exact. For example, let

$$\omega = \left(2xy + \sqrt{1+z^2}\right) dx + (x^2 + \sin z) \, dy + \left(y \cos z + \frac{xz}{\sqrt{1+z^2}}\right) dz.$$

We observe that

$$\begin{aligned}
d\omega &= \left(2x \, dy + \frac{z}{\sqrt{1+z^2}} dz\right) dx \\
&\quad + (2x \, dx + \cos z \, dz) \, dy \\
&\quad + \left(\cos z \, dy + \frac{z}{\sqrt{1+z^2}} dx\right) dz \\
&= 0.
\end{aligned}$$

Since ω is closed and differentiable everywhere, it must be exact. To find a scalar field F for which

$$\omega = dF,$$

we note that

$$\frac{\partial F}{\partial x} = 2xy + \sqrt{1+z^2}.$$

Integrating the right-hand side with respect to x shows us that

$$F(x,y,z) = x^2 y + x\sqrt{1+z^2} + f(y,z),$$

where f is a still unknown function that does not depend on x but may depend on y and z. Integrating the second coefficient with respect to y yields

$$F(x,y,z) = x^2 y + y \sin z + g(x,z).$$

Similarly, by integrating the third coefficient with respect to z, we get

$$F(x, y, z) = y \sin z + x\sqrt{1 + z^2} + h(x, y).$$

We see that these three equations can be simultaneously satisfied if we set

$$F(x, y, z) = x^2 y + x\sqrt{1 + z^2} + y \sin z.$$

Of course, adding any constant gives us an equally valid antidifferential.

We can now use this antidifferential to integrate ω. Choose any path C from $(1,2,0)$ to $(2, -1, \pi/2)$, then

$$
\begin{aligned}
\int_C \omega &= \left[x^2 y + x\sqrt{1 + z^2} + y \sin z \right]_{(1,2,0)}^{(2,-1,\pi/2)} \\
&= \left(-4 - 1 + \sqrt{4 + \pi^2} \right) - (2 + 0 + 1) \\
&= -8 + \sqrt{4 + \pi^2}.
\end{aligned}
$$

Loop Integrals

Integrals around closed curves arise so frequently that they have a special notation:

$$\oint_C f \, dx + g \, dy.$$

The circle on the integral emphasizes that C is a closed curve. Sometimes an arrow will appear on this circle, specifying whether the path is to be traveled counterclockwise or clockwise:

If no arrow is given, it is assumed that the curve is followed in the counterclockwise direction.

Since the starting and ending points of a closed curve are the same point, the loop integral of an exact form is always zero:

$$\oint_C \frac{\partial F}{\partial x} \, dx + \frac{\partial F}{\partial y} \, dy = 0. \tag{10.13}$$

An Alternate Notation

The force field that corresponds to the differential of a scalar field can also be described as the gradient of that scalar field:

$$dF = \frac{\partial F}{\partial x}\,dx + \frac{\partial F}{\partial y}\,dy + \frac{\partial F}{\partial z}\,dz,$$

$$\updownarrow$$

$$\nabla F = \left(\frac{\partial F}{\partial x}, \frac{\partial F}{\partial y}, \frac{\partial F}{\partial z}\right).$$

The scalar field, F, is usually called the *potential field* because it describes the potential energy at any given point. Using the alternate notation given in Equation (5.5), if C is a curve from $\vec{a}$ to $\vec{b}$, then Theorem 10.2 can be written as

$$\int_C \nabla F \cdot d\vec{r} = F(\vec{b}) - F(\vec{a}). \tag{10.14}$$

This amounts to saying that the work done by the field as it moves a particle from $\vec{a}$ to $\vec{b}$ equals the change in potential energy.

10.3   Exercises

1. Let
$$\omega_1 = f\,dx + g\,dy + h\,dz, \quad \omega_2 = r\,dx + s\,dy + t\,dz,$$
be 1-forms defined in $\mathbf{R}^3$. Prove that
$$d(\omega_1\omega_2) = (d\omega_1)\omega_2 - \omega_1(d\omega_2).$$

2. Let R be a solid region in $\mathbf{R}^3$ and let ω_1, ω_2 be 1-forms. Use the fundamental theorem of calculus and the preceding exercise to prove the following analog of the integration by parts formula:
$$\int_R (d\omega_1)\omega_2 = \int_{\partial R} \omega_1\omega_2 + \int_R \omega_1(d\omega_2). \tag{10.15}$$

3. Show that if ω_1 is a 1-form and ω_2 is any differential form, then
$$d(\omega_1\omega_2) = (d\omega_1)\omega_2 - \omega_1(d\omega_2).$$

4. Show that if ω_1 is a k-form and ω_2 is any differential form, then
$$d(\omega_1\omega_2) = (d\omega_1)\omega_2 + (-1)^k\omega_1(d\omega_2).$$

5. Let ω_1 be a p-form and ω_2 a q-form and let M be a $(p + q + 1)$-dimensional manifold. Use the fundamental theorem of calculus and the preceding exercise to prove the following generalization of the integration by parts formula:
$$\int_M (d\omega_1)\omega_2 = \int_{\partial M} \omega_1\omega_2 - (-1)^p \int_M \omega_1(d\omega_2). \tag{10.16}$$

6. Show that
$$\frac{-y\,dx + x\,dy}{x^2 + y^2} = d(\arctan(y/x)).$$

Why is the following argument false? Let C be any path from $(-1,0)$ to $(1,0)$ that avoids the origin, then

$$\int_C \frac{-y\,dx + x\,dy}{x^2 + y^2} = \arctan\left(\frac{y}{x}\right)\Big|_{(-1,0)}^{(1,0)}$$
$$= \arctan(0) - \arctan(0) = 0.$$

7. Let ω be a closed 1-form in $\mathbf{R}^2$ with exactly one singularity at $(0,0)$. Show that if C_1 and C_2 are any pair of closed curves encircling the origin, then

$$\oint_{C_1} \omega = \oint_{C_2} \omega.$$

8. Show that if ω is a differential 1-form in $\mathbf{R}^2$ for which the integral around any closed curve is 0, then the integral of ω between any two points is independent of the path.

9. Show that if ω is a differential 1-form in $\mathbf{R}^2$ for which the integral around some closed curve is not 0, then the integral of ω between any two points always depends on the path. (We assume that our 1-form is defined in a *connected* region. That is, it is always possible to get from any point in the domain to any other by means of a path that remains in the domain.)

10. Show that
$$\frac{(x+1)\,dx + y\,dy}{\sqrt{(x+1)^2 + y^2}}$$

is closed. Where is this 1-form not defined? Integrate it over each of the following curves from $(-2,0)$ to $(2,0)$:

$$C_1 = \left\{(2\sin t, 2\cos t \mid -\frac{\pi}{2} \le t \le \frac{\pi}{2}\right\},$$

and

$$C_2 = \left\{(2\sin t, -2\cos t \mid -\frac{\pi}{2} \le t \le \frac{\pi}{2}\right\}.$$

Is the integral of this 1-form independent of the path?

11. For each of the following 1-forms determine whether or not it is closed:

   (a) $(2xy + y^2)\,dx + (2xy + x^2)\,dy$,
   (b) $(-y\,dx + x\,dy)/\sqrt{x^2 + y^2}$,
   (c) $e^{xy}\,dx + (x/y)e^{xy}\,dy$,

(d) $y\cos(xy)\,dx + x\cos(xy)\,dy$,

(e) $(y+z)\,dx + (z+x)\,dy + (x+y)\,dz$,

(f) $y^2\,dx + (2xy + z^2)\,dy + 2yz\,dz$,

(g) $(x^2 + 2y)\,dx + (2x + 2xy)\,dy + (y^2 + 2xz)\,dz$,

(h) $(x\,dx + y\,dy + z\,dz)/(x^2 + y^2 + z^2)^2$,

(i) $(x\,dx + y\,dy + z\,dz)/(x^2 + y^2 + z^2)$.

12. Which 1-forms in Exercise 11 are exact? For those that are, find a scalar field of which it is the differential.

13. Which 1-forms in Exercise 11 are not exact? For each of these, find two paths with the same endpoints for which the values of the line integrals are different.

14. A 1-form in $\mathbf{R}^2$, $f\,dx + g\,dy$, is closed if and only if $\partial f/\partial y = \partial g/\partial x$. State the partial differential equations that must hold if $f\,dx + g\,dy + h\,dz$ is to be closed.

15. State the partial differential equations that must hold if $f\,dx + g\,dy + h\,dz + k\,dt$ is to be closed.

16. Consider a force field described by the 1-form

$$(2x + y)\,dx + x\,dy.$$

Find the amount of work done by this field in moving a particle from $(1, -2)$ to $(2, 1)$ along any curve.

17. Find the rate at which fluid crosses the curve

$$y = 1 - x^2$$

from $(-1, 0)$ to $(2, -3)$, directed from left to right, if the fluid velocity at (x, y) is described by the vector $(3y^2, -2x)$.

18. Show that if we replace the curve of Exercise 17 by any other curve from $(-1, 0)$ to $(1, 0)$ then the rate at which fluid crosses the curve does not change.

19. Show that $(2x + y)\,dx + (x + zt)\,dy + (yt - t)\,dz + (yz - z)\,dt$ is closed. Find a scalar field of which it is the differential.

20. Let $f(u)$ be a differentiable function of u for all real u. Prove that the 1-form $f(|\vec{x}|)\,(x_1\,dx_1 + x_2\,dx_2 + \cdots + x_n\,dx_n)$ is exact.

21. Let f and g be differentiable scalar fields from $\mathbf{R}^2$ to $\mathbf{R}$ that satisfy

$$\frac{\partial f}{\partial x} = \frac{\partial g}{\partial y}, \quad \frac{\partial f}{\partial y} = -\frac{\partial g}{\partial x}$$

(called the Cauchy-Riemann equations). Show that for any closed curve, C,

$$\oint_C f\, dx - g\, dy = 0, \quad \oint_C g\, dx + f\, dy = 0.$$

10.4   The Divergence Theorems

Proof of Green's Theorem, Equation (10.6)

Let ω be a continuously differentiable 1-form in $\mathbf{R}^2$:

$$\omega = f\, dx + g\, dy,$$

and let R be a bounded simply connected region of $\mathbf{R}^2$ with the boundary

$$C = \partial R.$$

We want to prove that

$$\oint_C f\, dx + g\, dy = \int_R \left(\frac{\partial g}{\partial x} - \frac{\partial f}{\partial y} \right) dx\, dy. \qquad (10.17)$$

Since $\partial g/\partial x - \partial f/\partial y$ stays bounded inside R, we can find a union of nonoverlapping rectangles in R with sides parallel to the x and y axes (Figure 10.5),

$$\cup R_i \subseteq R,$$

such that the integral over the region in R excluded from this union can be made as small as we please:

$$\int_{R - \cup R_i} \left(\frac{\partial g}{\partial x} - \frac{\partial f}{\partial y} \right) dx\, dy < \epsilon,$$

for any prespecified ϵ. The integral over this union of rectangles is the sum of the integrals over the individual rectangles:

$$\int_{\cup R_i} \left(\frac{\partial g}{\partial x} - \frac{\partial f}{\partial y} \right) dx\, dy = \sum \int_{R_i} \left(\frac{\partial g}{\partial x} - \frac{\partial f}{\partial y} \right) dx\, dy. \qquad (10.18)$$

On the other side of Equation (10.17), the continuity of f and g guarantees that small deviations in the path of integration result in small changes

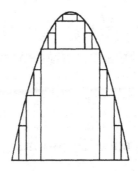

FIGURE 10.5. $\cup R_i \subseteq R$.

in the value of the integral. If the deviations are small enough, the change in the value of the integral can be made arbitrarily small. If we can prove that

$$\int_{\partial(\cup R_i)} f \, dx + g \, dy = \int_{\cup R_i} \left(\frac{\partial g}{\partial x} - \frac{\partial f}{\partial y} \right) dx \, dy \qquad (10.19)$$

for any union of rectangles with sides parallel to the x and y axes, then continuity arguments can be used to prove Equation (10.17).

We next observe that the integral over the boundary of $\cup R_i$ equals the sum of the integrals over the boundaries of the individual rectangles R_i (see Figure 10.6). Each boundary of an R_i that is also part of the outer boundary is integrated once in the proper direction. If some boundary of an R_i is not part of the outer boundary, then it is the common boundary of two rectangles, and so we integrate over it twice, once in each direction. The net effect is that these interior boundaries contribute 0:

$$\oint_{\partial(\cup R_i)} f \, dx + g \, dy = \sum \oint_{\partial R_i} f \, dx + g \, dy. \qquad (10.20)$$

Equation (10.19) is thus equivalent to

$$\sum \oint_{\partial R_i} f \, dx + g \, dy = \sum \int_{R_i} \left(\frac{\partial g}{\partial x} - \frac{\partial f}{\partial y} \right) dx \, dy, \qquad (10.21)$$

which will be true if we can prove that

$$\oint_{\partial R_i} f \, dx + g \, dy = \int_{R_i} \left(\frac{\partial g}{\partial x} - \frac{\partial f}{\partial y} \right) dx \, dy, \qquad (10.22)$$

for any rectangle, R_i, with sides parallel to the x and y axes.

Let R_i be the rectangle with vertices at

$$(a, c), \ (a, d), \ (b, c), \ \text{and} \ (b, d).$$

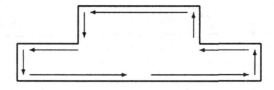

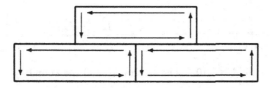

FIGURE 10.6. The integral over the boundary of the union equals the sum of the integrals over the boundaries.

In the following argument, we use the 1-dimensional fundamental theorem of calculus, which tells us that

$$g(b, y) - g(a, y) = \int_a^b \frac{\partial g}{\partial x}\, dx,$$

$$f(x, d) - f(x, c) = \int_c^d \frac{\partial f}{\partial y}\, dy.$$

We begin with the integral over the boundary of R_i and split it into its four pieces (Figure 10.7), two horizontal segments, where $dy = 0$, and two vertical segments, where $dx = 0$:

$$
\begin{aligned}
\oint_{\partial R_i} f\, dx + g\, dy &= \int_a^b f(x, c)\, dx + \int_c^d g(b, y)\, dy \\
&\quad + \int_b^a f(x, d)\, dx + \int_d^c g(a, y)\, dy \\
&= \int_c^d (g(b, y) - g(a, y))\, dy + \int_a^b (f(x, c) - f(x, d))\, dx \\
&= \int_c^d \int_a^b \frac{\partial g}{\partial x}\, dx\, dy - \int_a^b \int_c^d \frac{\partial f}{\partial x}\, dy\, dx \\
&= \int_{R_i} \left(\frac{\partial g}{\partial x} - \frac{\partial f}{\partial y} \right) dx\, dy.
\end{aligned}
$$

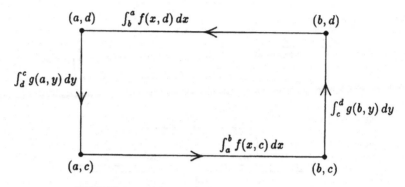

FIGURE 10.7. The integral around the boundary.

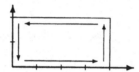

FIGURE 10.8. Rectangle with vertices at $(0,0),(4,0),(4,2),(0,2)$.

Recall that iterated integrals can be taken in either order:

$$\int_a^b \int_c^d \frac{\partial f}{\partial x}\, dy\, dx = \int_c^d \int_a^b \frac{\partial f}{\partial x}\, dx\, dy.$$

$$\textbf{Q.E.D.}$$

Examples

Let R be the rectangle with vertices at $(0,0)$, $(4,0)$, $(4,2)$, $(0,2)$ (Figure 10.8). We can use Green's theorem to evaluate the work done by the force field

$$(x^2 - y^2)\, dx + 2xy\, dy$$

in moving a particle counterclockwise around the boundary of R:

$$\text{Work} = \oint_{\partial R} (x^2 - y^2)\, dx + 2xy\, dy$$

$$= \int_R (2y + 2y)\, dx\, dy$$

$$= \int_0^2 \int_0^4 4y \, dx \, dy$$

$$= \int_0^2 16y \, dy \; = \; 32.$$

As a second example, consider computing the rate at which the flow (x^3, y^3) leaves the unit disc $D = \{(x,y) \mid x^2 + y^2 \leq 1\}$:

$$\oint_{\partial D} -y^3 \, dx + x^3 \, dy \; = \; \int_D (3y^2 + 3x^2) dx \, dy$$

$$= \int_0^{2\pi} \int_0^1 3r^2 \cdot r \, dr \, d\theta$$

$$= \int_0^{2\pi} \frac{3}{4} \, d\theta = \frac{3\pi}{2}.$$

Poincaré's Lemma for 1-Forms in $\mathbf{R}^2$

Green's theorem implies the following corollary.

Corollary 10.1 *If $\omega = f \, dx + g \, dy$ is a continuously differentiable 1-form in $\mathbf{R}^2$ and if*

$$\frac{\partial g}{\partial x} = \frac{\partial f}{\partial y}$$

everywhere inside a bounded, simply connected region, R, then

$$\oint_{\partial R} \omega = 0.$$

This in turn implies that given any curve C inside R the value of the integral of ω over C,

$$\int_C \omega,$$

is independent of the path.

Proof: The first part is simply the observation that under these conditions the right-hand side of Equation (10.17) is zero. For the second part, let C_1 and C_2 be two paths inside R that both start at $\vec{a}$ and end at $\vec{b}$. Let C be the curve obtained by following C_1 from $\vec{a}$ to $\vec{b}$ and then going back from $\vec{b}$ to $\vec{a}$ along C_2 in the reverse direction (Figure 10.9). The curve C is closed and therefore

$$0 = \oint_C \omega = \int_{C_1} \omega - \int_{C_2} \omega, \qquad \int_{C_1} \omega = \int_{C_2} \omega.$$

Q.E.D.

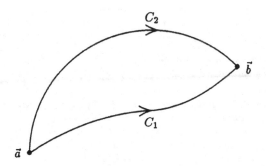

FIGURE 10.9. $C = C_1 - C_2$.

We have proven that a continuously differentiable closed 1-form,

$$\omega = f\,dx + g\,dy,$$

in a simply connected region, $R \subseteq \mathbf{R}^2$, has line integrals in R that are independent of path. To prove Poincaré's lemma for this case, we exhibit the obvious candidate for the antidifferential and check that it works. Choose any fixed point $(a, b) \in R$ and define

$$F(r, s) = \int_{(a,b)}^{(r,s)} f\,dx + g\,dy.$$

This integral is well defined because we have shown it does not depend on which curve we follow to get from (a, b) to (r, s).

   To show that

$$\frac{\partial F}{\partial r} = f(r, s) \quad \text{and} \quad \frac{\partial F}{\partial s} = g(r, s),$$

we consider two curves: C_1 consists of the straight lines from (a, b) to (a, s) to (r, s); C_2 consists of the straight lines from (a, b) to (r, b) to (r, s) (Figure 10.10). By integrating over C_1, we obtain

$$F(r, s) = \int_b^s g(a, y)\,dy + \int_a^r f(x, s)\,dx,$$

while by integrating over C_2 we see that

$$F(r, s) = \int_a^r f(x, b)\,dx + \int_b^s g(r, y)\,dy.$$

We compute $\partial F/\partial r$ using the first representation of F, $\partial F/\partial s$ using the second representation of F:

$$\frac{\partial F}{\partial r} = \frac{\partial}{\partial r} \int_b^s g(a, y)\,dy + \frac{\partial}{\partial r} \int_a^r f(x, s)\,dx = f(r, s),$$

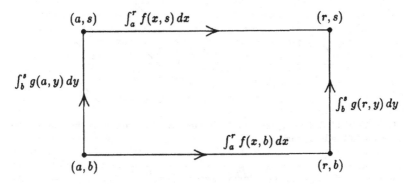

FIGURE 10.10. Two paths from (a, b) to (r, s).

$$\frac{\partial F}{\partial s} = \frac{\partial}{\partial s} \int_a^r f(x, b)\, dx + \frac{\partial}{\partial s} \int_b^s g(r, b)\, dy = g(r, s).$$

Thus, $dF = f\, dx + g\, dy$.

Proof of Gauss's Theorem, Equation (10.7)

Let ω be a continuously differentiable 2-form in $\mathbf{R}^3$:

$$\omega = F\, dy\, dz + G\, dz\, dx + H\, dx\, dy,$$

and let R be a simply connected bounded region of $\mathbf{R}^3$ with the boundary

$$S = \partial R,$$

a closed surface in $\mathbf{R}^3$ oriented so that the positive direction is outward. We want to prove that

$$\int_S F\, dy\, dz + G\, dz\, dx + H\, dx\, dy = \int_R \left(\frac{\partial F}{\partial x} + \frac{\partial G}{\partial y} + \frac{\partial H}{\partial z} \right) dx\, dy\, dz.$$
$$(10.23)$$

As in the proof of Green's theorem, we can approximate R by a union of rectangular blocks whose sides are parallel to the y, z; z, x; and x, y planes. The integral over the boundary of this union is the sum of the integrals over the individual boundaries (Figure 10.11), and so it is enough to prove this theorem when R is a rectangular block with sides parallel to the coordinate planes.

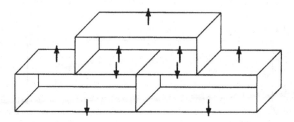

FIGURE 10.11. The integral over the boundary of the union is the sum of the integrals of the boundaries.

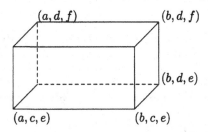

FIGURE 10.12. The six sides of the rectangular block.

The integral on the left breaks into a sum of six integrals, one for each side of the rectangular block (Figure 10.12). The sign on the integral is determined by the direction of the outward normal:

$$\int_S F\, dy\, dz + G\, dz\, dx + H\, dx\, dy$$

$$= -\int_e^f \int_c^d F(a, y, z)\, dy\, dz + \int_e^f \int_c^d F(b, y, z)\, dy\, dz$$

$$-\int_a^b \int_e^f G(x, c, z)\, dz\, dx + \int_a^b \int_e^f G(x, d, z)\, dz\, dx$$

$$-\int_c^d \int_a^b H(x, y, e)\, dx\, dy + \int_c^d \int_a^b H(x, y, f)\, dx\, dy$$

$$= \int_e^f \int_c^d \int_a^b \frac{\partial F}{\partial x}\, dx\, dy\, dz + \int_a^b \int_e^f \int_c^d \frac{\partial G}{\partial y}\, dy\, dz\, dx$$

$$+ \int_c^d \int_a^b \int_e^f \frac{\partial H}{\partial z}\, dz\, dx\, dy$$

$$= \int_R \left(\frac{\partial F}{\partial x} + \frac{\partial G}{\partial y} + \frac{\partial H}{\partial z} \right) dx\, dy\, dz.$$

Q.E.D.

Example

Let $B = B(\vec{0}, 1)$ be the unit ball with its center at the origin. If we have a flow described by the 2-form

$$\omega = x^2 y\, dy\, dz + (y^3 - z^2)\, dz\, dx + xz^3\, dx\, dy,$$

then the rate at which this flow crosses the boundary of B can be evaluated using Gauss's theorem and then changing to spherical coordinates. We integrate first with respect to θ because two of the terms are immediately seen to integrate to 0:

$$
\begin{aligned}
\int_{\partial B} \omega &= \int_B (2xy + 3y^2 + 3xz^2)\, dx\, dy\, dz \\
&= \int_{-\pi/2}^{\pi/2} \int_0^{2\pi} \int_0^1 (2\rho^2 \cos\theta \sin\theta \cos^2\phi + 3\rho^2 \sin^2\theta \cos^2\phi \\
&\qquad\qquad + 3\rho^3 \cos\theta \cos\phi \sin^2\phi)\rho^2 \cos\phi\, d\rho\, d\theta\, d\phi \\
&= \int_{-\pi/2}^{\pi/2} \int_0^1 3\pi\rho^4 \cos^3\phi\, d\rho\, d\phi \\
&= \int_{-\pi/2}^{\pi/2} \frac{3\pi}{5} \cos^3\phi\, d\phi = \frac{4\pi}{5}.
\end{aligned}
$$

Divergence

The theorems of Green and Gauss are more similar than may be apparent at first glance. Let $\vec{F} = (f, g)$ describe a 2-dimensional fluid flow. As we have seen, the corresponding 1-form is

$$-g\, dx + f\, dy.$$

Green's theorem states that if $C = \partial R$, then

$$\oint_C -g\, dx + f\, dy = \int_R \left(\frac{\partial f}{\partial x} + \frac{\partial g}{\partial y} \right) dx\, dy; \qquad (10.24)$$

the rate at which our fluid crosses the closed curve C is equal to the integral of

$$\frac{\partial f}{\partial x} + \frac{\partial g}{\partial y}$$

over the region bounded by C. Since ∇ denotes the operator

$$\nabla = \left(\frac{\partial}{\partial x}, \frac{\partial}{\partial y} \right)$$

in $\mathbf{R}^2$ and

$$\nabla = \left(\frac{\partial}{\partial x}, \frac{\partial}{\partial y}, \frac{\partial}{\partial z} \right)$$

in $\mathbf{R}^3$, it is natural to denote the sum of the partial derivatives arising in the divergence theorems as

$$\nabla \cdot \vec{F} = \frac{\partial f}{\partial x} + \frac{\partial g}{\partial y} \quad \text{or} \quad \nabla \cdot \vec{F} = \frac{\partial f}{\partial x} + \frac{\partial g}{\partial y} + \frac{\partial h}{\partial z},$$

depending on whether $\vec{F} = (f, g)$ or $= (f, g, h)$. This is called the *divergence* of $\vec{F}$, also denoted by "div $\vec{F}$." It is a measure of the incompressibility of the flow described by $\vec{F}$. In this notation, Gauss's theorem becomes

$$\int_{\partial R} \vec{F} \cdot \vec{n} \, d\sigma = \int_R \nabla \cdot \vec{F} \, dx \, dy \, dz. \qquad (10.25)$$

Incompressibility

If the divergence of $\vec{F}$ is identically 0 in any simply connected region, then we say that $\vec{F}$ is *incompressible*. As will now be demonstrated, incompressibility says that in each region, no matter how small, the net result of flow out minus flow in must be zero. In other words, whatever is flowing is not permitted to accumulate. This is true of the flow of most fluids under normal circumstances. It is also true of the flow lines of magnetic force, which is why this property is sometimes called *solenoidal*, referring to the magnetic field created by a solenoid or coil of wire carrying a current.

If the divergence of the flow $\vec{F}$ is 0, then Gauss's theorem tells us that for any closed, bounded region R the net flow across the boundary of R is

$$\int_{\partial R} \vec{F} \cdot \vec{n} \, d\sigma = \int_R \nabla \cdot \vec{F} \, dV = 0,$$

where dV denotes $dx \, dy \, dz$. This means that if the divergence is 0, then the flow cannot accumulate. The following lemma gives us the first step in proving that $\nabla \cdot \vec{F} = 0$ is also a consequence of the physical statement that flow cannot accumulate. The lemma states that for any continuous scalar field, f, the average value of f over a ball with its center at $\vec{a}$ approaches $f(\vec{a})$ as the radius of the ball approaches 0.

Lemma 10.1 *Let f be a scalar field that is continuous at $\vec{a}$ and let $B = B(\vec{a}, r)$ be the solid ball in $\mathbf{R}^3$ with its center at $\vec{a}$ and radius r. We have*

$$f(\vec{a}) = \lim_{r \to 0} \frac{\displaystyle\int_{B(\vec{a}, r)} f(\vec{x}) \, dV}{\displaystyle\int_B dV}. \qquad (10.26)$$

Proof: We begin by noting that

$$\int_B f(\vec{x})\, dV = \int_B f(\vec{a})\, dV + \int_B (f(\vec{x}) - f(\vec{a}))\, dV$$

$$= f(\vec{a})\int_B dV + \int_B (f(\vec{x}) - f(\vec{a}))\, dV,$$

$$\frac{\int_B f(\vec{x})\, dV}{\int_B dV} = f(\vec{a}) + \frac{\int_B [f(\vec{x}) - f(\vec{a})]\, dV}{\int_B dV}.$$

Now, we know that

$$\left| \int_B (f(\vec{x}) - f(\vec{a}))\, dV \right| \le \max_{\vec{x} \in B} |f(\vec{x}) - f(\vec{a})| \int_B dV.$$

Since f is continuous at $\vec{a}$, we can find an r such that

$$\left| \int_{B(\vec{a},r)} (f(\vec{x}) - f(\vec{a}))\, dV \right| < \epsilon \int_{B(\vec{a},r)} dV$$

for any prespecified positive ϵ. It follows that

$$\lim_{r \to 0} \frac{\int_B (f(\vec{x}) - f(\vec{a}))\, dV}{\int_B dV} = 0,$$

$$\lim_{r \to 0} \frac{\int_B f(\vec{x})\, dV}{\int_B dV} = f(\vec{a}) + \lim_{r \to 0} \frac{\int_B (f(\vec{x}) - f(\vec{a}))\, dV}{\int_B dV} = f(\vec{a}).$$

Q.E.D.

Given a flow, $\vec{F}$, that does not accumulate, we have

$$\int_{\partial B} \vec{F} \cdot \vec{n}\, d\sigma = 0$$

across the surface of any ball in our region. Combining this with our lemma, we see that

$$\nabla \cdot \vec{F}(\vec{a}) = \lim_{r \to 0} \frac{\int_B \nabla \cdot \vec{F}\, dV}{\int_B dV} = \lim_{r \to 0} \frac{\int_{\partial B} \vec{F} \cdot \vec{n}\, d\sigma}{\int_B dV} = \lim_{r \to 0} 0 = 0.$$

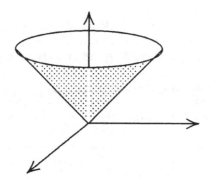

FIGURE 10.13. $z^2 = x^2 + y^2$, $0 \le z \le 1$.

Integrating an Incompressible Flow

Any two surfaces sharing a common boundary can be glued together along that boundary to form a closed surface. For example (Figure 10.13), the cone,

$$S_1 = \{(x, y, z) \mid z^2 = x^2 + y^2, \ 0 \le z \le 1\},$$

and the disc,

$$S_2 = \{(x, y, z) \mid 0 \le x^2 + y^2 \le 1, \ z = 1\},$$

both have the same boundary, namely, the circle,

$$x^2 + y^2 = 1, \quad z = 1.$$

The union of these two surfaces is the closed surface of the solid cone. Let us take a counterclockwise flow (as seen from above) around the boundary circle to determine the orientation of both surfaces. The right-hand rule tells us that for both S_1 and S_2 flow in the direction of $(0, 0, 1)$ is positive flow. If we let $-S_1$ denote the surface S_1 with opposite orientation, then $-S_1 \cup S_2$ is the surface of our solid cone where the positive orientation is now determined by outward flow.

If ω is any closed form that is well defined everywhere in and on our solid cone, then the divergence theorem tells us that

$$\int_{-S_1 \cup S_2} \omega = 0.$$

But, the integral over $-S_1 \cup S_2$ is equal to the integral over S_2 minus the integral over S_1, and so we have

$$0 = -\int_{S_1} \omega + \int_{S_2} \omega,$$

$$\int_{S_1} \omega = \int_{S_2} \omega.$$

We have demonstrated the following corollary to Theorem 10.7.

Corollary 10.2 *If S_1 and S_2 are any two oriented surfaces with a common oriented boundary, such that the orientation on the common boundary agrees with the orientations on both surfaces, and if ω is a closed 2-form that is well defined everywhere in the region enclosed by the union of S_1 and S_2, then*

$$\int_{S_1} \omega = \int_{S_2} \omega. \tag{10.27}$$

Example

Let us continue with the specific surfaces given above: $S_1 = \{(x,y,z) \mid z^2 = x^2 + y^2, 0 \le z \le 1\}$ and $S_2 = \{(x,y,z) \mid 0 \le x^2 + y^2 \le 1, z = 1\}$, and let us take

$$\omega = (x^2 y - 2xy^2 z)\, dy\, dz - (x^2 y + xy^2)\, dz\, dx + (y^2 z^2 + x^2 z)\, dx\, dy.$$

We first verify that ω is closed:

$$d\omega = (2xy - 2y^2 z - x^2 - 2xy + 2y^2 z + x^2)dx\, dy\, dz = 0.$$

If we parametrize S_1 by

$$x = r\cos\theta,$$
$$y = r\sin\theta,$$
$$z = r,$$
$$0 \le r \le 1, \quad 0 \le \theta \le 2\pi,$$

then the integral over S_1 is

$$\begin{aligned}
\int_{S_1} &= \int_0^{2\pi}\int_0^1 \big[(r^3\cos^2\theta\sin\theta - 2r^4\cos\theta\sin^2\theta)(-r\cos\theta) \\
&\quad - (r^3\cos^2\theta\sin\theta + r^3\cos\theta\sin^2\theta)(-r\sin\theta) \\
&\quad + (r^4\sin^2\theta + r^3\cos^2\theta)(r)\big]\, dr\, d\theta \\
&= \int_0^{2\pi}\int_0^1 \big[r^4(-\cos\theta\sin\theta + \cos^2\theta\sin^2\theta + \cos^2\theta) \\
&\quad + r^5(2\cos^2\theta\sin^2\theta + \sin^2\theta)\big]\, dr\, d\theta \\
&= 0 + \frac{\pi}{20} + \frac{\pi}{5} + \frac{\pi}{12} + \frac{\pi}{6} = \frac{\pi}{2}.
\end{aligned}$$

Integrating over S_2 is considerably simpler. We begin with the parametrization

$$
\begin{aligned}
x &= r\cos\theta, \\
y &= r\sin\theta, \\
z &= 1, \\
0 \le r \le 1, &\quad 0 \le \theta \le 2\pi.
\end{aligned}
$$

The integral over S_2 is

$$
\int_{S_2} \omega = \int_0^{2\pi} \int_0^1 (r^2\cos^2\theta + r^2\sin^2\theta) r\, dr\, d\theta = \frac{\pi}{2}.
$$

The moral of this example is that when we are integrating a *closed* 2-form over a surface, we should look for a simpler surface with the same boundary.

10.5   Exercises

1. Use Green's theorem to evaluate

$$
\oint_C x^3 y\, dx + xy\, dy,
$$

where C is the square with vertices at $(0,0)$, $(2,0)$, $(2,2)$, $(0,2)$.

2. Use Green's theorem to evaluate

$$
\oint_C x\sin y\, dx - x\cos y\, dy,
$$

where C is the rectangle with vertices at

$$
(1,\pi/2),(3,\pi/2),(3,\pi),(1,\pi).
$$

3. Use Green's theorem to evaluate

$$
\oint_C (x^2 + y)\, dx - (3x + y^3)\, dy,
$$

where C is the ellipse

$$
x^2 + 4y^2 = 4.
$$

4. Use Green's theorem to evaluate

$$
\oint_C -x^2 y\, dx + xy^2\, dy,
$$

where C is the circle

$$
x^2 + y^2 = 1.
$$

5. Use Green's theorem to evaluate

$$\oint_C y\,dx + x^2\,dy,$$

where C follows the parabola $y = x^2$ from $(-1,1)$ to $(1,1)$ and then returns on the straight line from $(1,1)$ to $(-1,1)$.

6. Find the work done by the force field

$$\left(y^2 + \frac{x^3}{\sqrt{1+x^2}}\right) dx + \left(x^3 + \frac{y^2}{\sqrt{1+y^2}}\right) dy$$

in pushing a particle counterclockwise once around the circle

$$x^2 + y^2 = 4.$$

7. If a fluid has velocity vector $(3x^2 - y^2, x^2 + 3y^2)$ at (x,y), find the rate at which the fluid is flowing out of the rectangle with vertices at $(1,0)$, $(3,0)$, $(3,2)$, $(1,2)$.

8. Let $\vec{n}$ be the principal normal to the curve C. Show that if $\vec{F} = (f,g)$, then

$$\int_C \vec{F}\cdot\vec{n}\,ds = \int_C -g\,dx + f\,dy.$$

9. Let

$$\frac{\partial f}{\partial n} = \nabla f \cdot \vec{n},$$
$$\nabla^2 f = \nabla\cdot(\nabla f) = \frac{\partial^2 f}{\partial x^2} + \frac{\partial^2 f}{\partial y^2}.$$

(The operator $\nabla^2 = \nabla\cdot\nabla$ is called the *Laplacian*.) Translate each statement into the language of differential forms and then explain why it is true:

(a)

$$\oint_{\partial R} \frac{\partial g}{\partial n}\,ds = \int_R \nabla^2 g\,dx\,dy,$$

(b)

$$\oint_{\partial R} f\frac{\partial g}{\partial n}\,ds = \int_R (f\nabla^2 g + \nabla f\cdot\nabla g)\,dx\,dy.$$

10. Let

$$\begin{aligned} \omega_1 &= -x^2 y \, dx + xy^2 \, dy, \\ \omega_2 &= -\frac{y^3}{3} \, dx + \frac{x^3}{3} \, dy, \\ \omega_3 &= (2x^2 y - y^3) \, dx + (x^3 - 2xy^2) \, dy. \end{aligned}$$

Show that in each case,

$$d\omega_i = (x^2 + y^2) \, dx \, dy.$$

Let C be the boundary of the rectangle with vertices at

$$(0,0), (2,0), (2,1), (0,1).$$

Verify by computing the line integrals that

$$\oint_C \omega_1 = \oint_C \omega_2 = \oint_C \omega_3.$$

11. Use Gauss's theorem to evaluate

$$\int_S x^2 y \, dy \, dz + 3y^2 \, dz \, dx - 2xz^2 \, dx \, dy,$$

where S is the surface of the unit cube

$$0 \leq x \leq 1, \ 0 \leq y \leq 1, \ 0 \leq z \leq 1,$$

and the positive direction is the outward normal.

12. Use Gauss's theorem to evaluate

$$\int_S (x - y) \, dy \, dz + (y^2 + z^2) \, dz \, dx + (y - x^2) \, dx \, dy,$$

where S is the surface of the unit sphere with its center at the origin and the positive direction is the outward normal.

13. Use Gauss's theorem to evaluate

$$\begin{aligned} \int_S (x^3 + yz^2 \sin x) \, dy \, dz &+ (4y^3 + y^2 z^2 \cos x) \, dz \, dx \\ &+ (9z^3 - yz^3 \cos x) \, dx \, dy, \end{aligned}$$

where S is the ellipsoid

$$S = \{(x, y, z) \mid x^2 + 4y^2 + 9z^2 = 36\}$$

and the positive direction is the outward normal.

14. Sketch the torus given by the equation $(r - 4)^2 + z^2 = 1$, where $x = r \cos \theta$ and $y = r \sin \theta$. Use Gauss's theorem to determine the rate at which the flow described by $x \, dy \, dz + y \, dz \, dx + z \, dx \, dy$ crosses the surface of this torus.

15. Let

$$S_1 = \{(x, y, z) \mid x^2 + y^2 + z^2 = 1, \ z \geq 0\},$$
$$S_2 = \{(x, y, z) \mid x^2 + y^2 = 1, \ z = 0\},$$

where S_1 and S_2 are both oriented so that $(0, 0, 1)$ points in the positive direction. Show by explicit calculation over the surfaces that

$$\int_{S_1} xz^2 \, dy \, dz - yz^2 \, dz \, dx + (y - x^2) \, dx \, dy$$
$$= \int_{S_2} xz^2 \, dy \, dz - yz^2 \, dz \, dx + (y - x^2) \, dx \, dy.$$

16. Find f, g, and h such that

$$\frac{\partial f}{\partial x} = x^2 + y^2 + z^2,$$
$$\frac{\partial g}{\partial y} = x^2 + y^2 + z^2,$$
$$\frac{\partial h}{\partial z} = x^2 + y^2 + z^2.$$

Show that for any closed surface S

$$\int_S f \, dy \, dz + dz \, dx + dx \, dy = \int_S dy \, dz + g \, dz \, dx + dx \, dy$$
$$= \int_S dy \, dz + dz \, dx + h \, dx \, dy$$
$$= \frac{1}{3} \int_S f \, dy \, dz + g \, dz \, dx + h \, dx \, dy.$$

17. Show that an integrable 2-form in $\mathbf{R}^2$ or an integrable 3-form in $\mathbf{R}^3$ is always exact.

18. For each of the following integrals, verify that $d\omega = 0$. Sketch the surface S and then evaluate the integral by finding a simpler surface with the same boundary.

(a)

$$\int_S xz^2 \, dy \, dz - yz^2 \, dz \, dx - x^2 y^2 \, dx \, dy,$$

$S = \{(x, y, z) \mid z = 1 - x^2 - y^2, \ z \geq 0\}$, $(0,0,1)$ points in the positive direction.

(b)

$$\int_S (x^3 + z^3)\, dy\, dz - 2x^2 y\, dz\, dx - x^2 z\, dx\, dy,$$

$S = \{(x, y, z) \mid x^2 + y^2 + z^2 = 4,\ y \geq 1\}$, $(0,1,0)$ points in the positive direction.

(c)

$$\int_S \frac{x}{1 + y^2}\, dy\, dz - \frac{x}{1 + z^2}\, dz\, dx - \frac{z}{1 + y^2}\, dx\, dy,$$

S consists of the five faces of the unit cube $0 \leq x, y, z \leq 1$ that are not in the z, x plane. The outer normal is the positive direction.

10.6   Stokes' Theorem

Proof of Equation (10.10)

Let ω be a continuously differentiable 1-form in $\mathbf{R}^3$:

$$\omega = f\, dx + g\, dy + h\, dz,$$

and let S be a surface with a twice continuously differentiable parametrization

$$S = \{(x(u, v), y(u, v), z(u, v)) \mid (u, v) \in R\}.$$

The fact that x, y, and z are twice continuously differentiable with respect to u and v guarantees that the second mixed partial derivatives will be equal:

$$\frac{\partial^2 x}{\partial u \partial v} = \frac{\partial^2 x}{\partial v \partial u},$$

$$\frac{\partial^2 y}{\partial u \partial v} = \frac{\partial^2 y}{\partial v \partial u},$$

$$\frac{\partial^2 z}{\partial u \partial v} = \frac{\partial^2 z}{\partial v \partial u}.$$

If we use the chain rule, rearrange terms, and then apply Green's theorem, we have that

$$\int_{\partial S} f\, dx + g\, dy + h\, dz = \int_{\partial R} f\left(\frac{\partial x}{\partial u}\, du + \frac{\partial x}{\partial v}\, dv\right)$$

$$+ g\left(\frac{\partial y}{\partial u}\, du + \frac{\partial y}{\partial v}\, dv\right) + h\left(\frac{\partial z}{\partial u}\, du + \frac{\partial z}{\partial v}\, dv\right)$$

$$= \int_{\partial R} \left(f\frac{\partial x}{\partial u} + g\frac{\partial y}{\partial u} + h\frac{\partial z}{\partial u} \right) du$$
$$+ \left(f\frac{\partial x}{\partial v} + g\frac{\partial y}{\partial v} + h\frac{\partial z}{\partial v} \right) dv$$

$$= \int_R \left[\frac{\partial}{\partial u} \left(f\frac{\partial x}{\partial v} + g\frac{\partial y}{\partial v} + h\frac{\partial z}{\partial v} \right) \right.$$
$$\left. - \frac{\partial}{\partial v} \left(f\frac{\partial x}{\partial u} + g\frac{\partial y}{\partial u} + h\frac{\partial z}{\partial u} \right) \right] du\, dv.$$

Continuing to manipulate the right side, we use the product rule to evaluate the partial derivatives, rearrange terms, and then use the chain rule a second time:

$$\int_{\partial S} f\, dx + g\, dy + h\, dz = \int_R \left(\frac{\partial f}{\partial u}\frac{\partial x}{\partial v} + f\frac{\partial^2 x}{\partial u \partial v} + \frac{\partial g}{\partial u}\frac{\partial y}{\partial v} + g\frac{\partial^2 y}{\partial u \partial v} \right.$$
$$+ \frac{\partial h}{\partial u}\frac{\partial z}{\partial v} + h\frac{\partial^2 z}{\partial u \partial v} - \frac{\partial f}{\partial v}\frac{\partial x}{\partial u} - f\frac{\partial^2 x}{\partial v \partial u}$$
$$\left. - \frac{\partial g}{\partial v}\frac{\partial y}{\partial u} - g\frac{\partial^2 y}{\partial v \partial u} - \frac{\partial h}{\partial v}\frac{\partial z}{\partial u} - h\frac{\partial^2 z}{\partial v \partial u} \right) du\, dv$$

$$= \int_R \left(\frac{\partial f}{\partial u}\frac{\partial x}{\partial v} + \frac{\partial g}{\partial u}\frac{\partial y}{\partial v} + \frac{\partial h}{\partial u}\frac{\partial z}{\partial v} \right.$$
$$\left. - \frac{\partial f}{\partial v}\frac{\partial x}{\partial u} - \frac{\partial g}{\partial v}\frac{\partial y}{\partial u} - \frac{\partial h}{\partial v}\frac{\partial z}{\partial u} \right) du\, dv$$

$$= \int_R \left[\left(\frac{\partial f}{\partial x}\frac{\partial x}{\partial u} + \frac{\partial f}{\partial y}\frac{\partial y}{\partial u} + \frac{\partial f}{\partial z}\frac{\partial z}{\partial u} \right)\frac{\partial x}{\partial v} \right.$$
$$+ \left(\frac{\partial g}{\partial x}\frac{\partial x}{\partial u} + \frac{\partial g}{\partial y}\frac{\partial y}{\partial u} + \frac{\partial g}{\partial z}\frac{\partial z}{\partial u} \right)\frac{\partial y}{\partial v}$$
$$+ \left(\frac{\partial h}{\partial x}\frac{\partial x}{\partial u} + \frac{\partial h}{\partial y}\frac{\partial y}{\partial u} + \frac{\partial h}{\partial z}\frac{\partial z}{\partial u} \right)\frac{\partial z}{\partial v}$$
$$- \left(\frac{\partial f}{\partial x}\frac{\partial x}{\partial v} - \frac{\partial f}{\partial y}\frac{\partial y}{\partial v} - \frac{\partial f}{\partial z}\frac{\partial z}{\partial v} \right)\frac{\partial x}{\partial u}$$
$$- \left(\frac{\partial g}{\partial x}\frac{\partial x}{\partial v} - \frac{\partial g}{\partial y}\frac{\partial y}{\partial v} - \frac{\partial g}{\partial z}\frac{\partial z}{\partial v} \right)\frac{\partial y}{\partial u}$$
$$\left. - \left(\frac{\partial h}{\partial x}\frac{\partial x}{\partial v} - \frac{\partial h}{\partial y}\frac{\partial y}{\partial v} - \frac{\partial h}{\partial z}\frac{\partial z}{\partial v} \right)\frac{\partial z}{\partial u} \right] du\, dv.$$

We complete the proof of Stokes' theorem by again rearranging terms and then applying the chain rule for the third time, returning us to an integral in x, y, z space:

$$
\begin{aligned}
\int_{\partial S} f\, dx + g\, dy + h\, dz \ = \ & \int_R \left(\frac{\partial h}{\partial y} - \frac{\partial g}{\partial z} \right) \left(\frac{\partial y}{\partial u} \frac{\partial z}{\partial v} - \frac{\partial z}{\partial u} \frac{\partial y}{\partial v} \right) du\, dv \\
& + \left(\frac{\partial f}{\partial z} - \frac{\partial h}{\partial x} \right) \left(\frac{\partial z}{\partial u} \frac{\partial x}{\partial v} - \frac{\partial x}{\partial u} \frac{\partial z}{\partial v} \right) du\, dv \\
& + \left(\frac{\partial g}{\partial x} - \frac{\partial f}{\partial y} \right) \left(\frac{\partial x}{\partial u} \frac{\partial y}{\partial v} - \frac{\partial y}{\partial u} \frac{\partial x}{\partial v} \right) du\, dv \\
= \ & \int_S \left(\frac{\partial h}{\partial y} - \frac{\partial g}{\partial z} \right) dy\, dz + \left(\frac{\partial f}{\partial z} - \frac{\partial h}{\partial x} \right) dz\, dx \\
& + \left(\frac{\partial g}{\partial x} - \frac{\partial f}{\partial y} \right) dx\, dy.
\end{aligned}
$$

Q.E.D.

A Second Look at the Proof

It is worth pointing out that if $\vec{T}$ is the mapping from R in u, v space to S in x, y, z space, and if we use $\vec{T}^{-1}\omega$ to denote the pullback of ω, then the bulk of the manipulations given in this proof amount to demonstrating that the differential operator d commutes with the pullback operation:

$$d(\vec{T}^{-1}\omega) = \vec{T}^{-1}(d\omega). \tag{10.28}$$

We have assumed that the inverse mapping commutes with the boundary operator:

$$\partial(\vec{T}^{-1}S) = \vec{T}^{-1}(\partial S). \tag{10.29}$$

The proof given above can be summarized as follows:

$$
\int_{\partial S} \omega \ = \ \int_{\vec{T}^{-1}(\partial S)} \vec{T}^{-1}\omega \ = \ \int_{\partial(\vec{T}^{-1}S)} \vec{T}^{-1}\omega \ = \ \int_{\vec{T}^{-1}S} d(\vec{T}^{-1}\omega)
$$

$$
= \ \int_{\vec{T}^{-1}S} \vec{T}^{-1}(d\omega) \ = \ \int_S d\omega.
$$

The Curl

Stokes' theorem is less useful as a tool for simplifying integrals than for what it says about when a 1-form in $\mathbf{R}^3$ is exact. Given a differentiable vector field in $\mathbf{R}^3$,

$$\vec{F} = (f, g, h),$$

we define the *curl* of $\vec{F}$ to be the vector field

$$\nabla \times \vec{F} = \left(\frac{\partial h}{\partial y} - \frac{\partial g}{\partial z}, \frac{\partial f}{\partial z} - \frac{\partial h}{\partial x}, \frac{\partial g}{\partial x} - \frac{\partial f}{\partial y} \right). \qquad (10.30)$$

This is also sometimes denoted by "curl $\vec{F}$." In our alternate notation, Stokes' theorem can be written in the form

$$\int_{\partial S} \vec{F} \cdot d\vec{r} = \int_S \nabla \times \vec{F} \cdot \vec{n} \, d\sigma. \qquad (10.31)$$

If we have a vector field $\vec{F}$ whose curl is identically $\vec{0}$ in a simply connected region, then we say that $\vec{F}$ is *irrotational* in that region. It follows from Stokes' theorem that if $\vec{F}$ is irrotational, then

$$\oint_C \vec{F} \cdot d\vec{r} = 0$$

for any closed curve C in the region. This says that the amount of work done by $\vec{F}$ in moving a particle around a closed curve is always zero. An example of such a field is the gravitational field generated by a mass at the origin:

$$\vec{F} = -r^{-3}(x, y, z), \quad \text{where} \quad r = \sqrt{x^2 + y^2 + z^2},$$

whose corresponding 1-form is

$$\omega = -r^{-3}(x \, dx + y \, dy + z \, dz).$$

We see that

$$
\begin{aligned}
d\omega &= 3r^{-4} \frac{2x \, dx + 2y \, dy + 2z \, dz}{(x^2 + y^2 + z^2)^{1/2}} (x \, dx + y \, dy + z \, dz) \\
&= 3r^{-5} \left[(2yz - 2zy) \, dy \, dz \right. \\
&\qquad \left. + (2zx - 2xz) \, dz \, dx + (2xy - 2yx) \, dx \, dy \right] \\
&= 0,
\end{aligned}
$$

or equivalently,

$$\nabla \times \vec{F} = \vec{0}.$$

If the curl is zero, then our force field has no eddies or vortices. A particle drifting with the field is never returned to the same position. As illustrated in Exercise 6, even more is true when the curl is zero. It is not just that the force field never curls back on itself, the force field has no tendency to rotate any object that is propelled by it.

To say that the curl of the vector field is zero is the same as saying that the corresponding 1-form is closed. We can use Stokes' theorem to prove the following stronger version of Poincaré's lemma for 1-forms in $\mathbf{R}^3$. We do not insist that our region, R, be simply connected, but only that it be

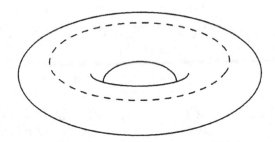

FIGURE 10.14. A doughnut is not simply loop connected.

simply loop-connected. That is, any closed curve in R can be contracted down to a single point without leaving R. Equivalently, any closed curve in R is the boundary of a surface contained entirely in R. A ring or doughnut (Figure 10.14) in $\mathbf{R}^3$ is *not* simply loop connected because a loop that encircles the hole cannot be contracted to a single point without leaving the doughnut. On the other hand, a hollow sphere is not simply connected, but it is simply loop connected.

Proposition 10.2 *Let*

$$\omega = f \, dx + g \, dy + h \, dz$$

be a continuously differentiable 1-form in $\mathbf{R}^3$ that is closed ($d\omega = 0$) in some simply loop-connected region, R, then ω is exact in R.

In the language of curls and gradients, this says that if $\vec{F}$ is a continuously differentiable vector field,

$$\vec{F} : \mathbf{R}^3 \longrightarrow \mathbf{R}^3,$$

and if $\vec{F}$ is irrotational,

$$\nabla \times \vec{F} = \vec{0},$$

in some simply loop-connected region, R, then there exists a scalar field, ϕ, called the *potential field*, such that within R,

$$\vec{F} = \nabla \phi.$$

As an example, the gravitational field generated by a point mass at the origin,

$$\omega = \frac{-k}{r^3} (x \, dx + y \, dy + z \, dz),$$

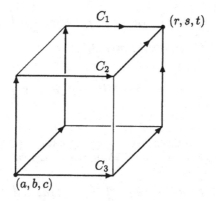

FIGURE 10.15. Three paths from (a, b, c) to (r, s, t).

is the differential of the gravitational potential,

$$\phi = kr^{-1}.$$

Proof: We first observe that if C is any closed curve in R, then we can find a surface, S, contained entirely in R and for which

$$C = \partial S.$$

It now follows from Stokes' theorem that

$$\oint_C \omega = \int_S d\omega = 0.$$

As in Corollary 10.1, if C_1 and C_2 are any two curves in R with the same initial points and the same terminal points, then

$$\int_{C_1} \omega = \int_{C_2} \omega.$$

We now choose a fixed point $(a, b, c) \in R$ and define (Figure 10.15)

$$\phi(r, s, t) = \int_{(a,b,c)}^{(r,s,t)} \omega$$

$$= \int_b^s g(a, y, c)\, dy + \int_c^t h(a, s, z)\, dz + \int_a^r f(x, s, t)\, dx,$$

integrating over C_1,

$$= \int_c^t h(a,b,z)\,dz + \int_a^r f(x,b,t)\,dx + \int_b^s g(r,y,t)\,dy,$$

integrating over C_2,

$$= \int_a^r f(x,b,c)\,dx + \int_b^s g(r,y,c)\,dy + \int_c^t h(r,s,z)\,dz,$$

integrating over C_3.

The fact that we have three distinct representations of ϕ is a consequence of the independence of the path. Taking partial derivatives yields

$$
\begin{aligned}
\frac{\partial \phi}{\partial r} &= \frac{\partial}{\partial r}\left(\int_b^s g(a,y,c)\,dy + \int_c^t h(a,s,z)\,dz + \int_a^r f(x,s,t)\,dx \right) \\
&= f(r,s,t),
\end{aligned}
$$

$$
\begin{aligned}
\frac{\partial \phi}{\partial s} &= \frac{\partial}{\partial s}\left(\int_c^t h(a,b,z)\,dz + \int_a^r f(x,b,t)\,dx + \int_b^s g(r,y,t)\,dy \right) \\
&= g(r,s,t),
\end{aligned}
$$

$$
\begin{aligned}
\frac{\partial \phi}{\partial t} &= \frac{\partial}{\partial t}\left(\int_a^r f(x,b,c)\,dx + \int_b^s g(r,y,c)\,dy + \int_c^t h(r,s,z)\,dz \right) \\
&= h(r,s,t).
\end{aligned}
$$

Q.E.D.

Example

Let

$$\omega = 2xyz\,dx + x^2 z\,dy + x^2 y\,dz.$$

This is continuously differentiable and closed everywhere since

$$d\omega = (x^2 - x^2)\,dy\,dz + (2xy - 2xy)\,dz\,dx + (2xz - 2xz)\,dx\,dy = 0.$$

We can let $(0,0,0)$ be the fixed point and use the straight line from the origin to (r,s,t) as the curve of integration:

$$C = \{(ru, su, tu)\,|\,0 \le u \le 1\}.$$

We then have

$$\phi(r,s,t) = \int_C \omega = \int_0^1 (2r^2 stu^3 + r^2 stu^3 + r^2 stu^3)\,du = r^2 st.$$

10.7   Summary for $\mathbf{R}^3$

It may be helpful to pause and summarize what we have proven for 3-dimensional space. This is also an opportunity to lay out the translation from vector to differential notation. I hope that the reader will see the clear advantages of differential notation. In any event, in the next chapter, we shall move on to 4-dimensional space–time, and it is only differential notation that carries into higher dimensions.

Basic Objects

Scalar fields	$\longleftrightarrow$	$\left\{\begin{array}{l}\text{0-forms defined on points}\\\text{3-forms defined on solids}\end{array}\right.$
$f : \mathbf{R}^3 \longrightarrow \mathbf{R}$		
Vector fields	$\longleftrightarrow$	$\left\{\begin{array}{l}\text{1-forms defined on curves}\\\text{2-forms defined on surfaces}\end{array}\right.$
$\vec{F} : \mathbf{R}^3 \longrightarrow \mathbf{R}^3$		

Note that a vector field can describe either a force field, in which case it corresponds to a 1-form, or a fluid flow, in which case it corresponds to a 2-form. In differential notation, it is useful to be able to pass back and forth between 1-forms and 2-forms:

$$f\,dx + g\,dy + h\,dz \qquad \longleftrightarrow \qquad f\,dy\,dz + g\,dz\,dx + h\,dx\,dy.$$

Multiplication

Dot product	$\longleftrightarrow$	1-form times 2-form
Cross product	$\longleftrightarrow$	1-form times 1-form

Basic Operators

Gradient
$$\nabla f = \left(\frac{\partial f}{\partial x}, \frac{\partial f}{\partial y}, \frac{\partial f}{\partial z}\right) \qquad \longleftrightarrow \qquad \text{differential of a 0-form}$$

Curl
$$\nabla \times \vec{F} \qquad \longleftrightarrow \qquad \text{differential of a 1-form}$$
$$= \left(\frac{\partial h}{\partial y} - \frac{\partial g}{\partial z}, \frac{\partial f}{\partial z} - \frac{\partial h}{\partial x}, \frac{\partial g}{\partial x} - \frac{\partial f}{\partial y}\right)$$

Divergence

$$\nabla \cdot (f, g, h) = \frac{\partial f}{\partial x} + \frac{\partial g}{\partial y} + \frac{\partial h}{\partial z} \qquad \longleftrightarrow \qquad \text{differential of a 2-form}$$

Laplacian

$$\nabla^2 f = \frac{\partial^2 f}{\partial x^2} + \frac{\partial^2 g}{\partial y^2} + \frac{\partial^2 h}{\partial z^2} \qquad \longleftrightarrow \qquad \begin{array}{l} \text{rewrite } df \text{ as a 2-form and} \\ \text{then take its differential} \end{array}$$

These operators can be combined as long as they are applied to the correct type of field, for example,

$$\nabla \times (\nabla f) = \vec{0} \qquad \longleftrightarrow \qquad \begin{array}{l} d(d\omega) = 0, \\ \omega \text{ is a 0-form,} \end{array}$$

$$\nabla \cdot (\nabla \times \vec{F}) = 0 \qquad \longleftrightarrow \qquad \begin{array}{l} d(d\omega) = 0, \\ \omega \text{ is a 1-form.} \end{array}$$

Integral Notation

If $\vec{F} = (f, g, h)$, then we have the correspondence

$$\int_C \vec{F} \cdot d\vec{r} \qquad \longleftrightarrow \qquad \int_C f\, dx + g\, dy + h\, dz,$$

$$\int_S \vec{F} \cdot \vec{n}\, d\sigma \qquad \longleftrightarrow \qquad \int_S f\, dy\, dz + g\, dz\, dx + h\, dx\, dy,$$

$$\int_S \frac{\partial f}{\partial n}\, d\sigma \qquad \longleftrightarrow \qquad \int_S \frac{\partial f}{\partial x}\, dy\, dz + \frac{\partial f}{\partial y}\, dz\, dx + \frac{\partial f}{\partial z}\, dx\, dy,$$

$$\int_R f\, dV \qquad \longleftrightarrow \qquad \int_R f\, dx\, dy\, dz.$$

The Fundamental Theorem of Calculus

Integration in a potential field

$$\int_{\vec{a}}^{\vec{b}} \nabla f \cdot d\vec{r} = f(\vec{b}) - f(\vec{a}) \qquad \longleftrightarrow \qquad \int_C df = \int_{\partial C} f.$$

Stokes' theorem

$$\int_S \nabla \times \vec{F} \cdot \vec{n}\, d\sigma = \int_{\partial S} \vec{F} \cdot d\vec{r} \qquad \longleftrightarrow \qquad \int_S d\omega = \int_{\partial S} \omega.$$

Gauss's theorem

$$\int_R \nabla \cdot \vec{F} \, dV = \int_{\partial R} \vec{F} \cdot \vec{n} \, d\sigma \qquad \longleftrightarrow \qquad \int_R d\omega = \int_{\partial R} \omega.$$

Poincaré's Lemma

In a simply connected region

$$\nabla \times \vec{F} = \vec{0}$$
implies a ϕ such that $\longleftrightarrow$
$$\vec{F} = \nabla \phi;$$

$d\omega = 0$, ω is a 1-form,
implies a ϕ such that
$$\omega = d\phi;$$

$$\nabla \cdot \vec{F} = 0$$
implies a $\vec{G}$ such that $\longleftrightarrow$
$$\vec{F} = \nabla \times \vec{G}.$$

$d\omega = 0$, ω is a 2-form,
implies a 1-form ν such that
$$\omega = d\nu.$$

10.8 Exercises

Use Stokes' theorem to evaluate the integrals given in Exercises 1 through 4. In each case, determine the direction in which we need to traverse C in order to obtain a positive value.

1.

$$\oint_C y \, dx + (2x - z) \, dy + (z - x) \, dz,$$

where C is the intersection of $x^2 + y^2 + z^2 = 4$ and $z = 1$.

2.

$$\oint_C (y - z) \, dx + (3x + z) \, dy + (x + 2y) \, dz,$$

where C is the intersection of $z = 4 - x^2 - y^2$ and $x + y + z = 0$.

3.

$$\oint_C (y^2 - z^2) \, dx + (z^2 - x^2) \, dy + (x^2 - y^2) \, dz,$$

where C is formed by the intersection of $x + 2y + 3z = 4$ with the planes $x = 3$, $y = 2$, and $z = 1$, respectively. Hint: C is the boundary of a triangle. Find a pullback from this triangle to the fundamental triangle in u, v space.

4.

$$\oint_C (y + z^2)\, dx + (x^2 + z)\, dy + (x + y)\, dz,$$

where C is the intersection of $z = 4 - x^2 - y^2$ and $x + y + z = 0$. Hint: parametrize the surface in terms of x and y and pull back to an integral in the x, y plane. Then, change variables:

$$x = \frac{1}{2} + r\cos\theta, \quad y = \frac{1}{2} + r\sin\theta.$$

5. Let $(x, y, z) = \vec{T}(u, v)$, $\omega = f\, dx + g\, dy + h\, dz$. Express the equality $d(\vec{T}^{-1}\omega) = \vec{T}^{-1}(d\omega)$ in terms of partial derivatives.

6. Verify that all image vectors of the field $\vec{F} = \left(0, e^{-x^2}, 0\right)$ are parallel, but that the curl of this field is not $\vec{0}$. Explain why a leaf, pushed by a stream whose flow was described by this vector field, would rotate as it moved.

7. Let
$$\omega = \frac{-y\, dx}{x^2 + y^2} + \frac{x\, dy}{x^2 + y^2} + dz.$$

Verify that $d\omega = 0$, but show that ω is not exact by evaluating

$$\oint_C \frac{-y\, dx}{x^2 + y^2} + \frac{x\, dy}{x^2 + y^2} + dz,$$

where $C = \{(\cos\theta, \sin\theta, 0) \mid 0 \le \theta \le 2\pi\}$.

8. Let $r = \sqrt{x^2 + y^2 + z^2}$ and set

$$\omega = f(r)\, (x\, dx + y\, dy + z\, dz),$$

where $f(r)$ is a differentiable function of r. Show that ω is closed except possibly at the origin, where it may not be defined. It follows that any radial force field whose strength is a function of the distance from the origin has a potential field.

9. Find a potential field for

$$\omega = r^a(x\, dx + y\, dy + z\, dz), \quad a \ne -2.$$

10. Find a potential field for

$$\omega = \frac{x\, dx + y\, dy + z\, dz}{x^2 + y^2 + z^2}.$$

11. Justify each of the following identities. Where possible, translate them into the language of differential forms. We use the notation

$$\frac{\partial f}{\partial n} = \nabla f \cdot \vec{n}, \quad dV = dx\,dy\,dz.$$

(a)

$$\int_{\partial R} \frac{\partial f}{\partial n}\,d\sigma = \int_R \nabla^2 f\,dV,$$

(b)

$$\int_{\partial R} f \frac{\partial g}{\partial n}\,d\sigma = \int_R f \nabla^2 g\,dV + \int_R \nabla f \cdot \nabla g\,dV,$$

(c)

$$\int_R f \nabla^2 g\,dV + \int_{\partial R} g \frac{\partial f}{\partial n}\,d\sigma = \int_R g \nabla^2 f\,dV + \int_{\partial R} f \frac{\partial g}{\partial n}\,d\sigma,$$

(d)

$$\int_S (\nabla f) \times \vec{g} \cdot \vec{n}\,d\sigma = \oint_{\partial S} f \vec{g} \cdot d\vec{r} - \int_S f(\nabla \times \vec{g}) \cdot \vec{n}\,d\sigma,$$

(e)

$$\int_R (\nabla f) \cdot \vec{g}\,dV = \int_{\partial R} f \vec{g} \cdot \vec{n}\,d\sigma - \int_R f(\nabla \cdot \vec{g})\,dV,$$

(f)

$$\int_R (\nabla \times \vec{f}) \cdot \vec{g}\,dV = \int_{\partial R} \vec{f} \times \vec{g} \cdot \vec{n}\,d\sigma + \int_R \vec{f} \cdot (\nabla \times \vec{g})\,dV.$$

12. Let $\vec{F} = (x^2 - y^2 + 3z, y^2 + z^2 - 2x, z^2 - 2x^2 + y)$. Using Stokes' theorem, evaluate

$$\int_S \nabla \times \vec{F} \cdot \vec{n}\,d\sigma,$$

where S is that portion of the paraboloid $z = 4 - x^2 - y^2$ that lies above the plane $z = 2$, oriented so that the vector $(0, 0, 1)$ points in the direction of positive flow.

13. Let $\vec{F}$ and $\vec{G}$ be vector fields from $\mathbf{R}^3$ to $\mathbf{R}^3$. Prove that

$$\nabla \cdot (\vec{F} \times \vec{G}) = (\nabla \times \vec{F}) \cdot \vec{G} - \vec{F} \cdot (\nabla \times \vec{G}). \tag{10.32}$$

14. Translate the equality of the previous exercise into the language of differential forms. Show that this equation is a special case of Exercise 3 of Section 10.3.

10.9   Potential Theory

By the early nineteenth century, many different physical phenomena were recognized to have models as vector fields $(\vec{F})$ that were both irrotational $(\nabla \times \vec{F} = \vec{0})$ and incompressible $(\nabla \cdot \vec{F} = 0)$. As we have seen, the first attribute implies the existence of a potential field, usually denoted by ϕ. The second states that the Laplacian of ϕ is zero, establishing Laplace's equation, which we first encountered in Section 10.1 (Equation 10.9):

$$\frac{\partial^2 \phi}{\partial x^2} + \frac{\partial^2 \phi}{\partial y^2} + \frac{\partial^2 \phi}{\partial z^2} = 0 \qquad (10.33)$$

$(\nabla^2 \phi = 0)$. It was the 2-dimensional version of this equation that Fourier discovered in 1807 in the course of modeling the flow of heat in a lamina. The problem of solving this equation led him to introduce his infinite series of trigonometric functions, the Fourier series. A twice differentiable scalar field ϕ satisfying Equation (10.33) is said to be *harmonic*. This term arises from the fact that such functions first appeared in the study of vibrating strings, where they explained the occurrence of overtones or harmonics.

Laplace was the first to exhibit the power of working with the potential field instead of the vector field. This was demonstrated in his article of 1782, "Théorie des attractions des sphéroïdes et de la figure des planètes" ("Theory of the attractions of spheroids and the shape of planets"), and more fully in his five-volume *Mécanique céleste* (*Celestial Mechanics*), published over the period 1799–1825. As a historical sidelight, Napoleon is reported to have commented to Laplace on the lack of any mention of God in this work to which Laplace replied, "I have no need for that hypothesis." On hearing of this, Lagrange is said to have retorted, "Ah, but it is a beautiful hypothesis."

Harmonic Functions

It follows from Exercise 11b of Section 10.8 that if ϕ is harmonic $(\nabla^2 \phi = 0)$ then

$$\int_{\partial R} \phi \frac{\partial \phi}{\partial n}\, d\sigma = \int_R |\nabla \phi|^2\, dV. \qquad (10.34)$$

This implies the following theorem.

Theorem 10.5 *If ϕ is a harmonic scalar field in a bounded, simply connected region, $R \subset \mathbf{R}^3$, with a twice continuously differentiable parametrization, and if $\phi(\vec{x}) = 0$ for all $\vec{x} \in \partial R$, then $\phi(\vec{x}) = 0$ for all $\vec{x} \in R$.*

Proof: If $\phi(\vec{x}) = 0$ for all $\vec{x} \in \partial R$, then the left side of Equation (10.34) is zero. Since ϕ is twice differentiable, $\nabla \phi$ is continuous. If we could find a point $\vec{a} \in R$ for which $\nabla \phi(\vec{a}) \neq \vec{0}$, then we could find a small ball around $\vec{a}$, $B(\vec{a}, \epsilon) \subseteq R$, where $|\nabla \phi|$ was at least $|\nabla \phi(\vec{a})|/2$. It would then follow that

$$\int_R |\nabla \phi|^2 \, dV \geq \int_{B(\vec{a}, \epsilon)} |\nabla \phi|^2 \, dV \geq \int_B \frac{|\nabla \phi(\vec{a})|^2}{4} \, dV = \frac{|\nabla \phi(\vec{a})|^2}{4} \frac{4}{3} \pi \epsilon^3 > 0,$$

contradicting the fact that

$$0 = \int_{\partial R} \phi \frac{\partial \phi}{\partial n} \, d\sigma = \int_R |\nabla \phi|^2 \, dV.$$

Therefore, the gradient of ϕ is $\vec{0}$ everywhere, and so ϕ is constant. Since ϕ is 0 on ∂R and it is continuous, it must be 0 everywhere.

Q.E.D.

Corollary 10.3 *If ϕ and ψ are harmonic scalar fields in a bounded, simply connected region $R \subset \mathbf{R}^3$, and if $\phi(\vec{x}) = \psi(\vec{x})$ for all $\vec{x} \in \partial R$, then $\phi(\vec{x}) = \psi(\vec{x})$ for all $\vec{x} \in R$.*

Proof: Apply Theorem 10.5 to $\phi(\vec{x}) - \psi(\vec{x})$.

Q.E.D.

This corollary implies that if we know the values of a harmonic function on the boundary of a suitable region, then these values uniquely determine the function inside the region. A harmonic potential field is uniquely determined by its values on the boundary of the region on which it is defined. The problem of determining when a given set of boundary values corresponds to a harmonic function, and finding that harmonic function when it exists, became known as the *Dirichlet problem*, named for Peter Gustav Lejeune-Dirichlet (1805–1859). Techniques for finding solutions were developed by Dirichlet, George Green, and William Thomson (Lord Kelvin). Proofs that solutions always exist provided the boundary values are continuous and R is suitably "nice" were discovered in the late 1800s by Carl G. Neumann (1832–1925), Henri Poincaré, and David Hilbert (1862–1943).

Gravitational Potential

If we assume that the gravitational field created by a mass at the origin acts radially with a strength depending only on the distance from the origin, then it must be of the form

$$\omega = f(r) \, (x \, dx + y \, dy + z \, dz), \quad r = \sqrt{x^2 + y^2 + z^2}.$$

Exercise 8 of Section 10.8 showed that such a vector field is always irrotational, and thus there must be a potential field of the form

$$\phi(x, y, z) = F(r) \quad \text{where} \quad \frac{dF}{dr} = r\,f(r).$$

If we now assume that the gravitational field can be viewed as a fluid flow that has reached a steady state, then ω is incompressible and ϕ must satisfy Equation (10.33). We observe that

$$\frac{\partial \phi}{\partial x} = \frac{dF}{dr}\frac{\partial r}{\partial x} = r\,f(r)\frac{x}{r} = f(r)\,x,$$

$$\frac{\partial^2 \phi}{\partial x^2} = f(r) + f'(r)\frac{x^2}{r},$$

and thus Equation (10.33) becomes

$$3f(r) + f'(r)\frac{x^2 + y^2 + z^2}{r} = 0,$$

$$3f(r) = -f'(r)\,r,$$

$$\frac{f'(r)}{f(r)} = -\frac{3}{r}.$$

We therefore have that

$$f(r) = -kr^{-3},$$

$$\frac{dF}{dr} = -kr^{-2},$$

$$\phi(x, y, z) = \frac{k}{r}.$$

This implies that that if $k \neq 0$, then the gravitational potential is inversely proportional to the distance, and the gravitational force is inversely proportional to the square of the distance.

If our gravitational field is generated by a single particle of mass m_1 located at $\vec{c}$, then the constant k equals γm_1, where γ is the gravitational constant. The potential field is given by

$$\phi(\vec{x}) = \frac{\gamma m_1}{|\vec{x} - \vec{c}|}. \tag{10.35}$$

Since forces add componentwise and the differential operator preserves sums, the potential field generated by several point masses, $m_1, m_2, \ldots, m_n$, located at $\vec{c}_1, \vec{c}_2, \ldots, \vec{c}_n$, is the sum of the individual potential fields:

$$\phi(\vec{x}) = \gamma \sum_{i=1}^{n} \frac{m_i}{|\vec{x} - \vec{c}_i|}. \tag{10.36}$$

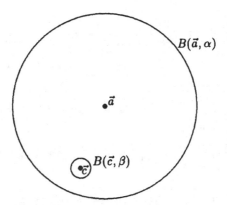

FIGURE 10.16. $B(\vec{c}, \beta) \subset B(\vec{a}, \alpha)$.

If instead of discrete masses we have a solid R, whose density at $\vec{c} = (c_1, c_2, c_3)$ is given by the scalar field $\rho(\vec{c})$, then the corresponding potential field is

$$\phi(\vec{x}) = \gamma \int_R \frac{\rho(\vec{c})\, dc_1\, dc_2\, dc_3}{|\vec{x} - \vec{c}|}. \tag{10.37}$$

Poisson's Equation

What happens to the gravitational potential *inside* the solid R? Laplace had assumed that it was still harmonic. In 1813, Siméon-Denis Poisson (1781–1840) showed that it was not, though it was not until 1839 that Gauss firmly established Poisson's observation, deriving his theorem, Equation (10.7), specifically for this purpose.

Let $B = B(\vec{a}, \alpha)$ be the solid sphere with its center at $\vec{a}$ and radius α. Viewing the gravitational field as a flow, we can ask for the rate at which it crosses the boundary of B, usually referred to as the *flux of the gravitational field across* ∂B. The gravitational field is incompressible. If the field is generated by a point mass outside B, then the flux across ∂B is zero. Because potential fields are added, all of the mass outside B contributes nothing to the flux across ∂B.

If we have a gravitational field generated by a point mass, m, located at some point, $\vec{c} = (c_1, c_2, c_3)$, in the interior of $B(\vec{a}, \alpha)$, then we can find a small positive radius, β, such that (Figure 10.16)

$$B(\vec{c}, \beta) \subset B(\vec{a}, \alpha).$$

Again using the fact that a gravitational field is incompressible, the flux across $\partial B(\vec{c}, \beta)$ is equal to the flux across $\partial B(\vec{a}, \alpha)$:

$$
\int_{\partial B(\vec{a},\alpha)} \frac{\partial \phi}{\partial x}\, dy\, dz + \frac{\partial \phi}{\partial y}\, dz\, dx + \frac{\partial \phi}{\partial z}\, dx\, dy
$$

$$
= -m\gamma \int_{\partial B(\vec{a},\alpha)} \frac{(x-c_1)\, dy\, dz + (y-c_2)\, dz\, dx + (z-c_3)\, dx\, dy}{|\vec{x}-\vec{c}|^3}
$$

$$
= -m\gamma \int_{\partial B(\vec{c},\beta)} \frac{(x-c_1)\, dy\, dz + (y-c_2)\, dz\, dx + (z-c_3)\, dx\, dy}{|\vec{x}-\vec{c}|^3}
$$

$$
= \frac{-m\gamma}{\beta^3} \int_{\partial B(\vec{c},\beta)} (x-c_1)\, dy\, dz + (y-c_2)\, dz\, dx + (z-c_3)\, dx\, dy
$$

$$
= \frac{-m\gamma}{\beta^3} \int_{-\pi/2}^{\pi/2} \int_0^{2\pi} (\beta^3 \cos^2\theta \cos^3\phi + \beta^3 \sin^2\theta \cos^3\phi
$$

$$
+ \beta^3 \cos\phi \sin^2\phi)d\theta\, d\phi
$$

$$
= -m\gamma \int_{-\pi/2}^{\pi/2} \int_0^{2\pi} \cos\phi\, d\theta\, d\phi = -4\pi m\gamma. \tag{10.38}
$$

If we have several point masses, $m_1, m_2, \ldots, m_n$, located at $\vec{c}_1, \vec{c}_2, \ldots, \vec{c}_n$, all inside $B(\vec{a}, \alpha)$, then the total flux across $\partial B(\vec{a}, \alpha)$ is the sum of the individual fluxes:

$$
\int_{\partial B(\vec{a},\alpha)} \frac{\partial \phi}{\partial x}\, dy\, dz + \frac{\partial \phi}{\partial y}\, dz\, dx + \frac{\partial \phi}{\partial z}\, dx\, dy = -4\pi\gamma \sum_{i=1}^{n} m_i. \tag{10.39}
$$

If we have a solid $R \subset B(\vec{a}, \alpha)$ with density $\rho(\vec{c})$ at $\vec{c}$, then the total flux across $\partial B(\vec{a}, \alpha)$ is

$$
\int_{\partial B(\vec{a},\alpha)} \frac{\partial \phi}{\partial x}\, dy\, dz + \frac{\partial \phi}{\partial y}\, dz\, dx + \frac{\partial \phi}{\partial z}\, dx\, dy = -4\pi\gamma \int_R \rho(\vec{c})\, dc_1\, dc_2\, dc_3. \tag{10.40}
$$

If we define $\rho(\vec{c})$ to be 0 if there is no mass at $\vec{c}$, then we can rewrite this last equation as

$$
\int_{\partial B(\vec{a},\alpha)} \frac{\partial \phi}{\partial x}\, dy\, dz + \frac{\partial \phi}{\partial y}\, dz\, dx + \frac{\partial \phi}{\partial z}\, dx\, dy = -4\pi\gamma \int_{B(\vec{a},\alpha)} \rho(\vec{c})\, dc_1\, dc_2\, dc_3. \tag{10.41}
$$

We can now use use Gauss's theorem to rewrite Equation (10.41) as

$$\int_{B(\bar{a},\alpha)} \left(\frac{\partial^2 \phi}{\partial x^2} + \frac{\partial^2 \phi}{\partial y^2} + \frac{\partial^2 \phi}{\partial z^2} \right) dx\, dy\, dz = -4\pi\gamma \int_{B(\bar{a},\alpha)} \rho(x,y,z)\, dx\, dy\, dz,$$

or equivalently,

$$\int_{B(\bar{a},\alpha)} (\nabla^2 \phi + 4\pi\gamma\rho)\, dx\, dy\, dz = 0. \qquad (10.42)$$

The significance of Equation (10.42) lies in the fact that it holds for *any* ball whether all, some, or none of the mass generating the gravitational field lies inside B. Recalling Lemma 10.1, we have

$$\nabla^2 \phi(\bar{a}) \; + \; 4\pi\gamma\rho(\bar{a}) \; = \; \lim_{\alpha \to 0} \frac{\displaystyle\int_{B(\bar{a},\alpha)} (\nabla^2 \phi + 4\pi\gamma\rho) dx\, dy\, dz}{\displaystyle\int_{B(\bar{a},\alpha)} dx\, dy\, dz}$$

$$= \; \lim_{\alpha \to 0} \frac{0}{\displaystyle\int_{B(\bar{a},\alpha)} dx\, dy\, dz} \; = \; 0.$$

This is Poisson's equation, valid at any point in the gravitational field whether mass is present or not:

$$\nabla^2 \phi \; + \; 4\pi\gamma\rho \; = \; 0. \qquad (10.43)$$

Conclusion

The attraction of working with the potential field is twofold. First, as we have seen, it simplifies the computations. Second, it avoids a problem that had been the one serious criticism leveled at Newton's *Principia*: What is the agent by which the force of gravity acts? If the sun exerts a gravitational force on the earth, how does it accomplish this with no physical connection between the two? Newton's response had been, in effect, that he neither knew nor cared. It was not an important question. By talking of potentials instead of forces, the issue of how gravitational forces act at a distance was sidestepped. Ironically, it was this shift of emphasis that ultimately led to our modern understanding of how gravity works, that a mass distorts space–time and that this distortion causes objects to move as if a force were present.

The search for the potential field became a standard part of the mathematical analysis of physical phenomena in the 1800s, even when it was not at all clear what the potential field measured. The significance of Maxwell's equations lies in that they establish the properties of electromagnetic potential, though to this day we do not know what electromagnetic potential is. But, I am getting ahead of myself. This story belongs in our next and final chapter.

11

$E = mc^2$

11.1 Prelude to Maxwell's *Dynamical Theory*

Much as Newton's contribution to calculus was less in the discovery of
the techniques than in his vision of how they linked together and what
could be done with them, so the equations of James Clerk Maxwell (1831–
1879) were not, with one partial exception, his discovery, yet he placed his
mark upon them in recognizing their basic unity and what they implied.
Maxwell's seminal paper, "A Dynamical Theory of the Electro-Magnetic
Field," read before the Royal Society of London in 1864 and published in its
Philosophical Transactions in 1865, explained the nature of electromagnetic
potential, revealed it to be intimately connected to the propagation of light,
and set the stage for Einstein's discovery of special relativity.

In our investigations of the development of Maxwell's equations, we shall
use the notation of vector algebra and differential forms. It should be kept
in mind that this notation was not available to the researchers of the time.
In fact, as explained in earlier chapters, vector algebra and the language of
differential forms arose in the late nineteenth and early twentieth centuries
precisely because the existing mathematical terminology was inadequate.
From our vantage point of more than a century later, we shall abandon
historical fidelity for the sake of clarity.

The Laws of Coulomb and Gauss

To the scientist of the seventeenth and most of the eighteenth centuries,
electricity was static electricity. The earliest research studied the attractive
and repulsive forces of objects charged with static electricity. It was discovered that objects with like charge repelled each other with a force that is inversely proportional to the square of the distance between them. Known as
Coulomb's law for Charles Augustin Coulomb (1736–1806), this basic property of electrostatic force had been previously observed by Daniel Bernoulli
(1700-1782), son of Jean Bernoulli, and by Henry Cavendish (1731–1810).

The only difference between gravitational force and electrostatic force is
that the constant, γ, changes value and sign. A static charge located at
the origin generates a force directed radially away from itself so that it is

described by the field

$$\mathbf{E} = \frac{m\gamma}{r^3}(x\,dx + y\,dy + z\,dz),$$

where $r = \sqrt{x^2 + y^2 + z^2}$, m is the charge, and γ is a positive constant.

By the early 1800s, electrostatic fields were also being described in terms of the *electrostatic potential*, ϕ. The potential field generated by a point charge at the origin is

$$\phi = \frac{m\gamma}{r}. \tag{11.1}$$

The electrostatic force field then satisfies

$$\vec{E} = -\nabla\phi. \tag{11.2}$$

It was quickly realized that Poisson's equation was equally valid in this case, that any electrostatic potential must satisfy the partial differential equation

$$\nabla^2\phi + 4\pi\gamma\rho = 0, \tag{11.3}$$

where $\rho(\vec{x})$ is the *charge density* at $\vec{x}$. It is now common to define

$$\epsilon = \frac{1}{4\pi\gamma},$$

where ϵ is known as the *dielectric constant* or the *electric permittivity*. Equation (11.3) can be rewritten as

$$\epsilon\,\nabla^2\phi + \rho = 0. \tag{11.4}$$

Using the fact that

$$\nabla^2\phi = \nabla\cdot(\nabla\phi) = -\nabla\cdot\vec{E},$$

Equation (11.4) usually appears in the form

$$\epsilon\,\nabla\cdot\vec{E} = \rho. \tag{11.5}$$

This is known as *Gauss's law*.

There is no time dependency in Equation (11.4). The word "static" implies that there is no change over time. Maxwell found the generalization of this equation to a potential field depending on both space and time, $\Phi(x, y, z, t)$, the electromagnetic potential. When t is constant, the partial differential equation satisfied by the electromagnetic potential will reduce to Poisson's equation.

Ampère's Law

On July 21, 1820, Hans Christian Oersted (1777–1851) announced to the world that he had passed an electric current near a magnetic needle and

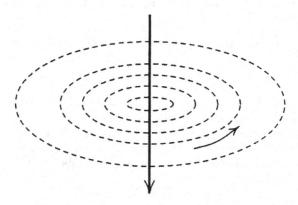

FIGURE 11.1. Magnetic field generated by an infinitely long straight wire.

that the current had deflected the needle. Somehow, electricity and magnetism were related. News of this discovery spread rapidly across Europe. Before the end of the year, both Dominique François Jean Arago (1786–1853) and Humphry Davy (1778–1829) had announced that a current-carrying wire wrapped as a solenoid around a soft iron bar would magnetize the bar. On September 18, 1820, less than two months after Oersted's pronouncement, André Marie Ampère (1775–1836) began a series of lectures before the French Academy of Science explaining the mathematical relationship between the current,

$$\vec{J} = (J_1, J_2, J_3),$$

and the resulting magnetic field, $\vec{B}$. Later the same year, Jean Baptiste Biot (1774–1862) and Félix Savart (1791–1841) announced that a straight wire conducting a constant current generates a magnetic field whose lines of force are circles lying on the plane perpendicular to the wire, centered on the wire, and whose strength is inversely proportional to the distance from the wire (Figure 11.1).

The Biot–Savart law implies that if we assume that our wire is the z axis,

$$\vec{J} = (0, 0, J_3),$$

then the magnetic field is given by

$$\vec{B} = \frac{\alpha J_3}{x^2 + y^2}(y, -x, 0),$$

where α is a constant depending only on the medium through which the magnetic field acts. Let **B** denote the corresponding 1-form,

$$\mathbf{B} = \alpha J_3 \frac{y\, dx - x\, dy}{x^2 + y^2}.$$

If S is any bounded surface in $\mathbf{R}^3$, then the work done by $\mathbf{B}$ in moving a particle around ∂S is

$$\oint_{\partial S} \alpha J_3 \frac{y\, dx - x\, dy}{x^2 + y^2} = \begin{cases} -2\pi\alpha J_3, & \text{if } \partial S \text{ encircles the } z \text{ axis} \\ 0, & \text{otherwise} \end{cases}$$

$$= -2\pi\alpha \times (\text{rate at which current crosses } S).$$

This says that the work done by a magnetic field in moving a particle around a closed loop, ∂S, is equal to a constant $(-2\pi\alpha)$ times the rate at which the current flows across the corresponding surface, S. In the calculations performed above, all of the current flowed through a single point in the surface. If we assume that the current is described by the continuous 2-form

$$J_1\, dy\, dz + J_2\, dz\, dx + J_3\, dx\, dy,$$

then the rate at which the current crosses S is

$$\int_S J_1\, dy\, dz + J_2\, dz\, dx + J_3\, dx\, dy;$$

and the Biot–Savart law implies the relationship

$$\oint_{\partial S} \mathbf{B} = -2\pi\alpha \int_S J_1\, dy\, dz + J_2\, dz\, dx + J_3\, dx\, dy,$$

where $\mathbf{B}$ is the 1-form describing the magnetic field:

$$\mathbf{B} = B_1\, dx + B_2\, dy + B_3\, dz.$$

We can use Stokes' theorem to rewrite the left side of this equality, obtaining

$$\int_S d\mathbf{B} = -2\pi\alpha \int_S J_1\, dy\, dz + J_2\, dz\, dx + J_3\, dx\, dy \qquad (11.6)$$

for any surface S. If we define

$$\mu = -2\pi\alpha,$$

called the *magnetic permeability*, then Equation (11.6) is equivalent to

$$\nabla \times \vec{B} = \mu \vec{J}, \qquad (11.7)$$

provided the current is constant with respect to time. This has become known as *Ampère's law*.

Faraday's Law

We still have not introduced any time dependency. It was Michael Faraday (1791–1867) who, in 1831, discovered that moving a magnetic field around a wire induced an electrical current in that wire. He also explained this phenomenon mathematically. We view the wire as closed loop, ∂S, and consider our magnetic field as a flow:

$$\mathbf{B} = B_1 dy\,dz + B_2 dz\,dx + B_3 dx\,dy,$$

where now each coefficient is a function of both position and time:

$$B_i = B_i(x, y, z, t).$$

A characteristic of the lines of magnetic flow is that they form closed loops. Equivalently, magnetic flow is incompressible:

$$\nabla \cdot \vec{B} = 0. \tag{11.8}$$

A current is only induced if the magnetic field is changing over time. Faraday stated that the work done by the electromotive force,

$$\mathbf{E} = E_1\,dx + E_2\,dy + E_3\,dz,$$

in pushing a particle around the circuit, ∂S, is directly proportional to the rate at which the magnetic flow across S is changing:

$$\oint_{\partial S} \mathbf{E} = \beta \frac{\partial}{\partial t} \int_S \mathbf{B}.$$

It is common to choose units so that $\beta = -1$. Stokes' theorem then implies

$$\int_S d\mathbf{E} = -\int_S \frac{\partial \mathbf{B}}{\partial t},$$

or, equivalently, that

$$\int_S \left(d\mathbf{E} + \frac{\partial \mathbf{B}}{\partial t} \right) = 0, \tag{11.9}$$

where

$$\frac{\partial \mathbf{B}}{\partial t} = \frac{\partial B_1}{\partial t} dy\,dz + \frac{\partial B_2}{\partial t} dz\,dx + \frac{\partial B_3}{\partial t} dx\,dy.$$

In general, one must be careful about moving a partial differential operator across an integral:

$$\frac{\partial}{\partial t} \int_S \longrightarrow \int_S \frac{\partial}{\partial t}.$$

In this particular case, it can be justified (see Exercises 1 through 4, Section 11.4). Since Equation (11.9) holds for any surface, we have

$$d\mathbf{E} + \frac{\partial \mathbf{B}}{\partial t} = 0,$$

or

$$\nabla \times \vec{E} + \frac{\partial \vec{B}}{\partial t} = \vec{0}. \tag{11.10}$$

Equation (11.10) is known as *Faraday's law*. We shall combine it with Equation (11.8). We define the *electromagnetic field* as a 2-form in 4-dimensional space–time:

$$\mathbf{B} + \mathbf{E}\,dt = B_1\,dy\,dz + B_2\,dz\,dx + B_3\,dx\,dy$$
$$+ E_1\,dx\,dt + E_2\,dy\,dt + E_3\,dz\,dt.$$

The differential of this 2-form is

$$d(\mathbf{B} + \mathbf{E}\,dt) = \left(\frac{\partial B_1}{\partial x} + \frac{\partial B_2}{\partial y} + \frac{\partial B_3}{\partial z}\right) dx\,dy\,dz$$
$$+ \left(\frac{\partial B_1}{\partial t} + \frac{\partial E_3}{\partial y} - \frac{\partial E_2}{\partial z}\right) dy\,dz\,dt$$
$$+ \left(\frac{\partial B_2}{\partial t} + \frac{\partial E_1}{\partial z} - \frac{\partial E_3}{\partial x}\right) dz\,dx\,dt$$
$$+ \left(\frac{\partial B_3}{\partial t} + \frac{\partial E_2}{\partial x} - \frac{\partial E_1}{\partial y}\right) dx\,dy\,dt.$$

Equation (11.8) states that the first coefficient is zero. Equation (11.10) states that the next three coefficients are each zero. These two equations are therefore succinctly summed up in the statement that the electromagnetic field, $\mathbf{B} + \mathbf{E}\,dt$, is closed:

$$d(\mathbf{B} + \mathbf{E}\,dt) = 0. \tag{11.11}$$

Faraday's discovery led to a practical means of generating large amounts of electricity, and in 1858, he presided over the installation of the first electric lighthouse in Britain. By 1866, dynamos based on Faraday's law were in commercial production, and the age of electricity had begun.

11.2   Flow in Space–Time

Current is not incompressible: electrical charge can accumulate as at a capacitor. But, there is a simple relationship between current and charge density. Given any bounded, simply connected region, $R \subset \mathbf{R}^3$, the integral of the flux through ∂R over the time interval $[t_0, t_1]$ is equal to the net change in the total charge inside R from time t_0 to t_1. Mathematically, this is expressed as

$$\int_{t_0}^{t_1} \left(\int_{\partial R} J_1\,dy\,dz + J_2\,dz\,dx + J_3\,dx\,dy\right) dt$$
$$= -\int_R \rho(x, y, z, t_1)\,dx\,dy\,dz + \int_R \rho(x, y, z, t_0)\,dx\,dy\,dz. \tag{11.12}$$

In the first integral, which is an iterated integral, the orientation is assumed to be positive. We want to rewrite this as the integral of a 3-form in x, y, z, t space. The problem before us is to determine which order of dx, dy, dt or dy, dz, dt, or dx, dz, dt constitutes positive orientation.

Lemma 11.1 *The constant 3-forms*

$$dy\,dz\,dt, \quad dz\,dx\,dt, \quad \text{and} \quad dx\,dy\,dt$$

each express negative *orientation in x, y, z, t space.*

Proof: We know that $dx\,dy\,dz$ expresses positive orientation. We need to find orientation preserving transformations taking $dx\,dy\,dz$ to $-dy\,dz\,dt$, $-dz\,dx\,dt$, and $-dx\,dy\,dt$, respectively. Let (x', y', z', t') be a reference frame that, relative to our original frame (x, y, z, t), is moving with constant speed v in the direction of the positive x axis. These two frames are related by the linear transformation:

$$\begin{aligned} x' &= x - vt, \\ y' &= y, \\ z' &= z, \\ t' &= t. \end{aligned}$$

What appears stationary to the first observer is moving to the left with speed v to the second. Our 3-forms are related by

$$\begin{aligned} dx'dy'dz' &= dx\,dy\,dz - v\,dt\,dy\,dz \\ &= dx\,dy\,dz - v\,dy\,dz\,dt. \end{aligned} \tag{11.13}$$

Since we have only changed the viewpoint of the observer, we have not changed the orientation.

Let R be a 3-dimensional region in x, y, z, t space given by

$$R = \{(x, y, z, t) \mid x = v, \ 0 \le y, z, t \le 1\}.$$

This can be viewed as a square in the $x = v$ plane that appears at time 0, lasts for one unit of time, and then disappears. Under our transformation, R corresponds to the region

$$R' = \{(x', y', z', t') \mid x' + vt' = v, \ 0 \le x' \le v, \ 0 \le y', z', t' \le 1\},$$

which appears above $x = v$ at time 0, sweeps to the left along the x axis until it reaches $x = 0$ at time 1, and then disappears. We have the equality

$$\int_{R'} dx'dy'dz' = \int_{R} dx\,dy\,dz - v\,dy\,dz\,dt. \tag{11.14}$$

The projection of R' into x', y', z' space is the rectangular block with $0 \le x' \le v$, $0 \le y' \le 1$, $0 \le z' \le 1$:

$$\int_{R'} dx'dy'dz' = \int_0^1 \int_0^1 \int_0^v dx'dy'dz' = v. \tag{11.15}$$

In the region R, x is constant and therefore $dx\,dy\,dz = 0$:

$$\int_R dx\,dy\,dz - v\,dy\,dz\,dt \;=\; \int_R -v\,dy\,dz\,dt$$

$$=\; v\int_R -dy\,dz\,dt. \qquad (11.16)$$

Combining Equations (11.14), (11.15), and (11.16), we see that

$$v \;=\; v\int_R -dy\,dz\,dt,$$

$$-1 \;=\; \int_R dy\,dz\,dt, \qquad (11.17)$$

confirming that $dy\,dz\,dt$ expresses negative orientation. The proof is similar for $dz\,dx\,dt$ and $dx\,dy\,dt$, with the second frame of reference taken to move in the y and z directions, respectively.

Q.E.D.

Equation (11.12) can be rewritten as

$$\int_{\partial R \times [t_0, t_1]} -J_1\,dy\,dz\,dt - J_2\,dz\,dx\,dt - J_3\,dx\,dy\,dt$$

$$=\; -\int_{R \times \{t_1\}} \rho\,dx\,dy\,dz + \int_{R \times \{t_0\}} \rho\,dx\,dy\,dz. \qquad (11.18)$$

Each side of this equation is an integral over a 3-dimensional manifold sitting in 4-dimensional x, y, z, t space. Since it is difficult to visualize such an object, we shall temporarily eliminate one space variable and see what we are integrating over in x, y, t space. If R is a bounded, simply connected 2-dimensional region in the x, y plane, then ∂R is a closed loop. We represent time as a vertical direction normal to the x, y plane so that $\partial R \times [t_0, t_1]$ is seen to be a vertical cylinder whose horizontal cross sections are ∂R (Figure 11.2). This is the surface over which we are integrating on the left side of Equation (11.18). The orientation is such that the positive normal is directed out of the cylinder.

The integrals on the right of Equation (11.18) are evaluated over the ends of our cylinder. The first is over $R \times \{t_1\}$ with negative (downward) orientation; the second is over $R \times \{t_0\}$ with positive (upward) orientation (Figure 11.3). Note that the direction at both ends is inward. Altogether, these three integrals involve integrating over all sides of the boundary of the solid cylinder $R \times [t_0, t_1]$.

This is precisely what is happening in 4-dimensional space–time. The union of our 3-dimensional objects,

$$\partial R \times [t_0, t_1] \bigcup R \times \{t_0\} \bigcup R \times \{t_1\},$$

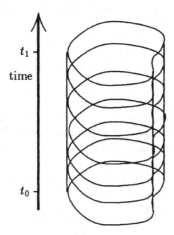

FIGURE 11.2. The space–time cylinder $\partial R \times [t_0, t_1]$.

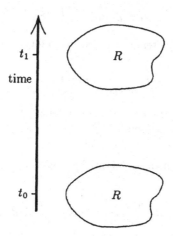

FIGURE 11.3. The surfaces $R \times \{t_1\}$ and $R \times \{t_0\}$.

taken with outward orientations, represents the 3-dimensional boundary of the 4-dimensional cylinder $R \times [t_0, t_1]$:

$$\partial (R \times [t_0, t_1]) = \partial R \times [t_0, t_1] \bigcup R \times \{t_0\} \bigcup R \times \{t_1\}. \qquad (11.19)$$

On either of the faces $R \times \{t_0\}$ or $R \times \{t_1\}$, t is constant and thus the 3-form

$$J_1 \, dy \, dz \, dt + J_2 \, dz \, dx \, dt + J_3 \, dx \, dy \, dt$$

is zero. This means that we do not change the value of the integral on the left of Equation (11.18) if we replace $\partial R \times [t_0, t_1]$ by $\partial(R \times [t_0, t_1])$. Similarly, on $\partial R \times [t_0, t_1]$, x, y, and z are related and thus $dx \, dy \, dz = 0$. The right side of Equation (11.18) is also left unchanged if we integrate over $\partial(R \times [t_0, t_1])$. We have shown that

$$\int_{\partial(R \times [t_0, t_1])} -J_1 \, dy \, dz \, dt - J_2 \, dz \, dx \, dt - J_3 \, dx \, dy \, dt$$

$$= -\int_{\partial(R \times [t_0, t_1])} \rho \, dx \, dy \, dz,$$

or equivalently,

$$\int_{\partial(R \times [t_0, t_1])} \rho \, dx \, dy \, dz - J_1 \, dy \, dz \, dt - J_2 \, dz \, dx \, dt - J_3 \, dx \, dy \, dt = 0. \quad (11.20)$$

Let $\mathbf{J}$ denote this 3-form, called the *total current*:

$$\mathbf{J} = \rho \, dx \, dy \, dz - J_1 \, dy \, dz \, dt - J_2 \, dz \, dx \, dt - J_3 \, dx \, dy \, dt. \qquad (11.21)$$

The fundamental theorem of calculus for 3-forms in $\mathbf{R}^4$ implies that

$$\int_{\partial(R \times [t_0, t_1])} \mathbf{J} = \int_{R \times [t_0, t_1]} d\mathbf{J}. \qquad (11.22)$$

Since the left side of this equation is zero for any bounded, simply connected region, R, $\mathbf{J}$ must be closed:

$$d\mathbf{J} = 0. \qquad (11.23)$$

From this, we obtain the differential equation

$$\begin{aligned}
0 &= d\mathbf{J} \\
&= \frac{\partial \rho}{\partial t} dt \, dx \, dy \, dz - \frac{\partial J_1}{\partial x} dx \, dy \, dz \, dt - \frac{\partial J_2}{\partial y} dy \, dz \, dx \, dt - \frac{\partial J_3}{\partial z} dz \, dx \, dy \, dt \\
&= -\left(\frac{\partial \rho}{\partial t} + \frac{\partial J_1}{\partial x} + \frac{\partial J_2}{\partial y} + \frac{\partial J_3}{\partial z} \right) dx \, dy \, dz \, dt, \qquad (11.24)
\end{aligned}$$

and thus,

$$\frac{\partial \rho}{\partial t} + \frac{\partial J_1}{\partial x} + \frac{\partial J_2}{\partial y} + \frac{\partial J_3}{\partial z} = 0,$$

$$\nabla \cdot \vec{J} = -\frac{\partial \rho}{\partial t}. \tag{11.25}$$

This is known as the *equation of continuity*.

The Antidifferential of J

An incompressible flow in x, y, z space is a closed 2-form. It describes a steady flow in which time is not a variable. For a time-dependent flow, such as electrical current, the natural representation is **J**, our closed 3-form in x, y, z, t space. Note that density is an important part of the description of a time-dependent flow. Such a flow has four coordinates: the rate of flow in each of the three spatial directions and the density which can be viewed as the coordinate of flow in the time direction. A continuous time-dependent flow described in this manner will always be closed.

Since **J** is closed, it must be exact in any simply connected region in which it is defined. The first of Maxwell's accomplishments was to find an antidifferential of **J**. Because **J** is a 3-form, it is the differential of a 2-form in x, y, z, t space.

We begin by finding an antidifferential in the special case in which the current, $\vec{J}$, is constant. If $\vec{J}$ is constant, then so is the charge density, ρ, and we can use Gauss's and Ampère's laws:

$$\epsilon \nabla \cdot \vec{E} = \rho, \tag{11.26}$$
$$\nabla \times \vec{B} = \mu \vec{J}. \tag{11.27}$$

If we redefine **B** to be the 1-form

$$\mathbf{B} = B_1 dx + B_2 dy + B_3 dz$$

and **E** to be the 2-form

$$\mathbf{E} = E_1 dy\, dz + E_2 dz\, dx + E_3 dx\, dy,$$

then we have

$$d\left(\epsilon\mathbf{E} - \frac{1}{\mu}\mathbf{B}\,dt\right) = \epsilon\left(\frac{\partial E_1}{\partial x} + \frac{\partial E_2}{\partial y} + \frac{\partial E_3}{\partial z}\right)dx\,dy\,dz$$
$$+ \left(\epsilon\frac{\partial E_1}{\partial t} - \frac{1}{\mu}\frac{\partial B_3}{\partial y} + \frac{1}{\mu}\frac{\partial B_2}{\partial z}\right)dy\,dz\,dt$$
$$+ \left(\epsilon\frac{\partial E_2}{\partial t} - \frac{1}{\mu}\frac{\partial B_1}{\partial z} + \frac{1}{\mu}\frac{\partial B_3}{\partial x}\right)dz\,dx\,dt$$
$$+ \left(\epsilon\frac{\partial E_3}{\partial t} - \frac{1}{\mu}\frac{\partial B_2}{\partial x} + \frac{1}{\mu}\frac{\partial B_1}{\partial y}\right)dx\,dy\,dt.\ (11.28)$$

Since current and density are assumed to be constant with respect to time, so also is the electrical field $\mathbf{E}$:

$$\frac{\partial E_1}{\partial t} = \frac{\partial E_2}{\partial t} = \frac{\partial E_3}{\partial t} = 0.$$

Equations (11.26) and (11.27) imply that

$$d\left(\epsilon\mathbf{E} - \frac{1}{\mu}\mathbf{B}\,dt\right) = \rho\,dx\,dy\,dz - J_1\,dy\,dz\,dt - J_2\,dz\,dx\,dt - J_3\,dx\,dy\,dt$$
$$= \mathbf{J}. \tag{11.29}$$

It was Maxwell's genius to realize that this equation remained valid even when the current was not constant, in which case Ampère's law becomes

$$\epsilon\frac{\partial\vec{E}}{\partial t} - \frac{1}{\mu}\nabla\times\vec{B} = -\vec{J},$$

or equivalently,

$$\nabla\times\vec{B} - \epsilon\mu\frac{\partial\vec{E}}{\partial t} = \mu\vec{J}. \tag{11.30}$$

Maxwell's Equations

This completes the derivations of Maxwell's equations. Defining $\vec{E}$ to be the electric field, $\vec{B}$ the magnetic field, $\vec{J}$ the current, ρ the charge density, ϵ the dielectric constant, and μ the magnetic permeability, we have shown

$$\nabla\cdot\vec{B} = 0, \tag{11.31}$$

$$\nabla\times\vec{E} + \frac{\partial\vec{B}}{\partial t} = 0, \tag{11.32}$$

$$\epsilon\nabla\cdot\vec{E} = \rho, \tag{11.33}$$

$$\nabla\times\vec{B} - \epsilon\mu\frac{\partial\vec{E}}{\partial t} = \mu\vec{J}. \tag{11.34}$$

In the language of differential forms, the first two equations can be written as

$$d(\mathbf{B} + \mathbf{E}\,dt) = 0, \qquad (11.35)$$

where $\mathbf{B}$ is a 2-form and $\mathbf{E}$ is a 1-form. The last two equations can be written as

$$d\left(\mathbf{E} - \frac{1}{\epsilon\mu}\mathbf{B}\,dt\right) = \frac{1}{\epsilon}\mathbf{J}, \qquad (11.36)$$

where $\mathbf{B}$ is viewed as a 1-form and $\mathbf{E}$ as a 2-form and

$$\mathbf{J} = \rho\,dx\,dy\,dz - J_1\,dy\,dz\,dt - J_2\,dz\,dx\,dt - J_3\,dx\,dy\,dt.$$

11.3   Electromagnetic Potential

The electromagnetic field is a 2-form in space–time with two representations:

$$\mathbf{B} + \mathbf{E}\,dt \quad \text{or} \quad \mathbf{E} - \frac{1}{\epsilon\mu}\mathbf{B}\,dt.$$

Since $\epsilon\mu$ is a constant depending only on the medium through which these forces act, we can pass easily between these two representations. To know one of them is to know both. Equation (11.36) implies that knowledge of the electromagnetic field uniquely determines the charge density and current. Equation (11.35) implies that the electromagnetic field, in the form $\mathbf{B} + \mathbf{E}\,dt$, is exact and therefore is the differential of a 1-form, the *electromagnetic potential*,

$$\Phi = A_1\,dx + A_2\,dy + A_3\,dz + A_4\,dt. \qquad (11.37)$$

Knowing Φ, we can find the electromagnetic field by taking its differential, and thence find $\mathbf{J}$. Knowing $\mathbf{J}$, can we reconstruct Φ? We first find the partial differential equations satisfied by A_1, A_2, A_3, and A_4 in terms of ρ and $\vec{J}$. We assume that our coefficients are twice continuously differentiable so that the second mixed partials may be taken in either order. In electrostatics, we defined ϕ so that $\vec{E} = -\nabla\phi$. Similarly, here we set

$$
\begin{aligned}
\mathbf{B} + \mathbf{E}\,dt &= -d\Phi \\
&= \left(\frac{\partial A_2}{\partial z} - \frac{\partial A_3}{\partial y}\right) dy\,dz + \left(\frac{\partial A_3}{\partial x} - \frac{\partial A_1}{\partial z}\right) dz\,dx \\
&\quad + \left(\frac{\partial A_1}{\partial y} - \frac{\partial A_2}{\partial x}\right) dx\,dy + \left(\frac{\partial A_1}{\partial t} - \frac{\partial A_4}{\partial x}\right) dx\,dt \\
&\quad + \left(\frac{\partial A_2}{\partial t} - \frac{\partial A_4}{\partial y}\right) dy\,dt + \left(\frac{\partial A_3}{\partial t} - \frac{\partial A_4}{\partial z}\right) dz\,dt, (11.38)
\end{aligned}
$$

$$\frac{1}{\epsilon}\mathbf{J} = d\left(\mathbf{E} - \frac{1}{\epsilon\mu}\mathbf{B}\,dt\right)$$

$$= d\left[\left(\frac{\partial A_1}{\partial t} - \frac{\partial A_4}{\partial x}\right)dy\,dz + \left(\frac{\partial A_2}{\partial t} - \frac{\partial A_4}{\partial y}\right)dz\,dx\right.$$

$$+ \left(\frac{\partial A_3}{\partial t} - \frac{\partial A_4}{\partial z}\right)dx\,dy - \frac{1}{\epsilon\mu}\left(\frac{\partial A_2}{\partial z} - \frac{\partial A_3}{\partial y}\right)dx\,dt$$

$$\left. - \frac{1}{\epsilon\mu}\left(\frac{\partial A_3}{\partial x} - \frac{\partial A_1}{\partial z}\right)dy\,dt - \frac{1}{\epsilon\mu}\left(\frac{\partial A_1}{\partial y} - \frac{\partial A_2}{\partial x}\right)dz\,dt\right]$$

$$= \left[-\frac{\partial^2 A_4}{\partial x^2} - \frac{\partial^2 A_4}{\partial y^2} - \frac{\partial^2 A_4}{\partial z^2}\right.$$

$$\left. + \frac{\partial}{\partial t}\left(\frac{\partial A_1}{\partial x} + \frac{\partial A_2}{\partial y} + \frac{\partial A_3}{\partial z}\right)\right]dx\,dy\,dz$$

$$+ \frac{1}{\epsilon\mu}\left[\epsilon\mu\frac{\partial^2 A_1}{\partial t^2} - \frac{\partial^2 A_1}{\partial y^2} - \frac{\partial^2 A_1}{\partial z^2}\right.$$

$$\left. + \frac{\partial}{\partial x}\left(\frac{\partial A_2}{\partial y} + \frac{\partial A_3}{\partial z} - \epsilon\mu\frac{\partial A_4}{\partial t}\right)\right]dy\,dz\,dt$$

$$+ \frac{1}{\epsilon\mu}\left[\epsilon\mu\frac{\partial^2 A_2}{\partial t^2} - \frac{\partial^2 A_2}{\partial x^2} - \frac{\partial^2 A_2}{\partial z^2}\right.$$

$$\left. + \frac{\partial}{\partial y}\left(\frac{\partial A_1}{\partial x} + \frac{\partial A_3}{\partial z} - \epsilon\mu\frac{\partial A_4}{\partial t}\right)\right]dz\,dx\,dt$$

$$+ \frac{1}{\epsilon\mu}\left[\epsilon\mu\frac{\partial^2 A_3}{\partial t^2} - \frac{\partial^2 A_3}{\partial x^2} - \frac{\partial^2 A_3}{\partial y^2}\right.$$

$$\left. + \frac{\partial}{\partial z}\left(\frac{\partial A_1}{\partial x} + \frac{\partial A_2}{\partial y} - \epsilon\mu\frac{\partial A_4}{\partial t}\right)\right]dx\,dy\,dt. \qquad (11.39)$$

Things do not look particularly hopeful at this point. But, we need to remember that Φ is not uniquely determined by $\mathbf{J}$, and we are free to put in an additional constraint. In $\mathbf{R}^3$, we have a natural correspondence between 1-forms and 2-forms. In the same manner, there is a natural correspondence between 1-forms and 3-forms in $\mathbf{R}^4$. Electromagnetic potential can also be represented as a space–time flow:

$$\Phi = A_4\,dx\,dy\,dz + \frac{1}{\epsilon\mu}A_1\,dy\,dz\,dt + \frac{1}{\epsilon\mu}A_2\,dz\,dx\,dt + \frac{1}{\epsilon\mu}A_3\,dx\,dy\,dt.$$

The reason for inserting $1/\epsilon\mu$ will become evident in a moment. If we insist on this flow being closed, then

$$0 = d\Phi = \left[-\frac{\partial A_4}{\partial t} + \frac{1}{\epsilon\mu}\left(\frac{\partial A_1}{\partial x} + \frac{\partial A_2}{\partial y} + \frac{\partial A_3}{\partial z}\right)\right]dx\,dy\,dz\,dt,$$

or

$$\frac{\partial A_1}{\partial x} + \frac{\partial A_2}{\partial y} + \frac{\partial A_3}{\partial z} - \epsilon\mu \frac{\partial A_4}{\partial t} = 0. \tag{11.40}$$

This equation is precisely what we need to simplify the right side of Equation (11.39):

$$\begin{aligned} \frac{1}{\epsilon} \mathbf{J} =\ & -\left(\frac{\partial^2 A_4}{\partial x^2} + \frac{\partial^2 A_4}{\partial y^2} + \frac{\partial^2 A_4}{\partial z^2} - \epsilon\mu\frac{\partial^2 A_4}{\partial t^2} \right) dx\, dy\, dz \\ & -\frac{1}{\epsilon\mu}\left(\frac{\partial^2 A_1}{\partial x^2} + \frac{\partial^2 A_1}{\partial y^2} + \frac{\partial^2 A_1}{\partial z^2} - \epsilon\mu\frac{\partial^2 A_1}{\partial t^2} \right) dy\, dz\, dt \\ & -\frac{1}{\epsilon\mu}\left(\frac{\partial^2 A_2}{\partial x^2} + \frac{\partial^2 A_2}{\partial y^2} + \frac{\partial^2 A_2}{\partial z^2} - \epsilon\mu\frac{\partial^2 A_2}{\partial t^2} \right) dz\, dx\, dt \\ & -\frac{1}{\epsilon\mu}\left(\frac{\partial^2 A_3}{\partial x^2} + \frac{\partial^2 A_3}{\partial y^2} + \frac{\partial^2 A_3}{\partial z^2} - \epsilon\mu\frac{\partial^2 A_3}{\partial t^2} \right) dx\, dy\, dt. \end{aligned} \tag{11.41}$$

Here we introduce a new operator, the *d'Alembertian*:

$$\Box^2 f = \frac{\partial^2 f}{\partial x^2} + \frac{\partial^2 f}{\partial y^2} + \frac{\partial^2 f}{\partial z^2} - \epsilon\mu\frac{\partial^2 f}{\partial t^2}.$$

Equation (11.41) then becomes

$$\Box^2\left(A_4 dx\, dy\, dz + \frac{1}{\epsilon\mu}A_1 dy\, dz\, dt + \frac{1}{\epsilon\mu}A_2 dz\, dx\, dt + \frac{1}{\epsilon\mu}A_3 dx\, dy\, dt \right) + \frac{1}{\epsilon}\mathbf{J} = 0. \tag{11.42}$$

The system of partial differential equations satisfied by the A_i has turned into an equation in the d'Alembertian of the 3-form Φ.

Theorem 11.1 *The electromagnetic potential, Φ, can be expressed as a closed 3-form in x, y, z, t space, satisfying the partial differential equation*

$$\epsilon\,\Box^2\Phi + \mathbf{J} = 0. \tag{11.43}$$

Electromagnetic Waves

Jean Le Rond d'Alembert (1717–1783) now comes into the study of electricity and magnetism because of his work in 1746 on the vibration of strings with fixed endpoints. Let c^2 be the length times the tension divided by the mass of the string, and let $h(x, t)$ denote the height of the point x on the string at time t (Figure 11.4). D'Alembert showed that h satisfies the partial differential equation

$$\frac{\partial^2 h}{\partial x^2} - \frac{1}{c^2}\frac{\partial^2 h}{\partial t^2} = 0. \tag{11.44}$$

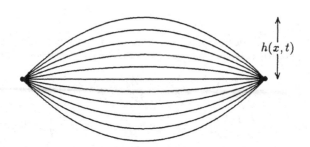

$h(x,t)$

FIGURE 11.4. A vibrating string.

It was Euler, in 1749, who demonstrated that the solution to this equation is a sum of periodic functions, each vibrating with a frequency of $nc/2l$, where l is the length of the string and n can be any positive integer. The existence of these multiple frequencies explains the appearance of overtones.

The fact that the d'Alembertian arises in the context of electricity and magnetism immediately marks electromagnetic potential as a wave phenomenon. If we consider the potential field, Φ, at a point where there is no charge or current, then

$$\Box^2 \Phi = 0;$$

each component of Φ satisfies

$$\frac{\partial^2 A_i}{\partial x^2} + \frac{\partial^2 A_i}{\partial y^2} + \frac{\partial^2 A_i}{\partial z^2} - \frac{1}{c^2}\frac{\partial^2 A_i}{\partial t^2} = 0, \qquad (11.45)$$

where we have set

$$c^2 = \frac{1}{\epsilon\mu}. \qquad (11.46)$$

The only difference between this and d'Alembert's equation is that whereas a string is a 1-dimensional vibrating object, each component of the electromagnetic potential is a 3-dimensional vibrating object.

A typical solution of Equation (11.45) is

$$\sin[k_1(u_1 x + u_2 y + u_3 z - ct) + k_2],$$

where k_1 and k_2 are arbitrary constants and $\vec{u} = (u_1, u_2, u_3)$ is any unit vector. This is a wave traveling in time through space. If we hold the argument constant so that we are watching the progress of a particular point on the wave, we see that

$$\vec{u}\cdot(x, y, z) = \text{constant} + ct.$$

The wave moves in the direction $\vec{u}$ at the speed c.

Maxwell determined the values of ϵ and μ in air and so obtained a value for c:

$$c \simeq 310\,740\,000 \quad \text{m/s}.$$

To this, he compared the speed of light in air as determined by Jean L. Foucault (1819–1868) in 1860:

$$\text{speed of light} \simeq 298\,000\,000 \quad \text{m/s},$$

and concluded that the difference was experimental error. Light is an electromagnetic phenomenon:

$$c = \frac{1}{\sqrt{\epsilon\mu}} = \text{speed of light.} \tag{11.47}$$

Postscript

It was more than 20 years before experimenters were able to observe the electromagnetic waves predicted by Maxwell. This was accomplished in 1887 by Heinrich Rudolf Hertz (1857–1894). In 1895, Guglielmo Marconi (1874–1937) and Alexander S. Popov (1859–1906) independently discovered how to use electromagnetic waves to send messages over large distances. The radio had been invented.

James Clerk Maxwell stands at a watershed in the history of the use of mathematics to interpret our physical universe. Newton had demonstrated the power of mathematical models in understanding and predicting observable reality. Maxwell's equations are the culmination of the great project, begun in the *Principia*, to render the basic processes of nature into the language of calculus. Not a few scientists believed that Maxwell had solved the last of the great problems in physics. But, Maxwell's contribution goes beyond his equations. His tools began to take on a life of their own. They pointed to a new aspect of reality that he and his contemporaries had neither experienced nor anticipated, the existence of electromagnetic waves. This is the direction that Einstein was to pursue so fruitfully, to let the mathematics suggest new and unexpected realities.

11.4   Exercises

Exercises 1 through 4 provide a more careful derivation of Faraday's law. We assume that S is a simply connected, bounded, twice continuously differentiable surface and E and B are continuously differentiable forms.

1. Given that

$$\oint_{\partial S} \mathbf{E} = -\frac{\partial}{\partial t} \int_S \mathbf{B},$$

show that

$$\int_{\partial S \times [t_0, t_1]} \mathbf{E} \, dt = - \int_{S \times \{t_1\}} \mathbf{B} + \int_{S \times \{t_0\}} \mathbf{B}.$$

2. Explain why the second equation in Exercise 1 is equivalent to

$$\int_{\partial (S \times [t_0, t_1])} \mathbf{B} + \mathbf{E} \, dt = 0.$$

3. Using the previous exercise and $\nabla \cdot \vec{B} = 0$, show that

$$\int_{S \times [t_0, t_1]} d(\mathbf{B} + \mathbf{E} \, dt) = 0.$$

4. Why does the equation of Exercise 3 imply that

$$d(\mathbf{B} + \mathbf{E} \, dt) = 0?$$

5. Find the electromagnetic potential at (x, y) that results from a *dipole* consisting of charge m at $(1, 0)$ and a charge $-m$ at $(-1, 0)$. What is the electrostatic field generated by this dipole?

6. A point charge of mass m is located at $(0, 0, 2)$. Find the flux of the field $\mathbf{E}$ due to this charge through the disc $x^2 + y^2 \leq 1$ lying in the plane $z = 0$.

7. The following "proof" that magnetic fields do not exist was published by George Arfken in the *American Journal of Physics*, Vol. 27 (1959), page 526, as a warning to undergraduate students of the dangers of blind application of equations. Find the fallacy.

$$\nabla \cdot \vec{B} = 0, \tag{11.48}$$

from one of Maxwell's equations.

$$\int \nabla \cdot \vec{B} \, dV = \int \vec{B} \cdot \vec{n} \, d\sigma = 0, \tag{11.49}$$

by Gauss's theorem.

$$\vec{B} = \nabla \times \vec{A} \tag{11.50}$$

because an incompressible field is the curl of some vector field.

$$\int \vec{B} \cdot \vec{n} \, d\sigma = \int \nabla \times \vec{A} \cdot \vec{n} \, d\sigma = 0 \tag{11.51}$$

by substitution of Equation (11.50) into Equation (11.49).

$$\int \nabla \times \vec{A} \cdot \vec{n} \, d\sigma = \oint \vec{A} \cdot d\vec{r} = 0 \tag{11.52}$$

by Stokes' theorem.

$$\vec{A} = \nabla \phi \qquad (11.53)$$

because an irrotational field is the gradient of some potential field.

$$\vec{B} = \nabla \times \vec{A} = \nabla \times \nabla \phi = \vec{0} \qquad (11.54)$$

because the curl of a gradient always vanishes. Therefore, magnetic fields do not exist.

8. Using Equation (10.32) from Section 10.8 and Maxwell's equations, prove that

$$\frac{\partial}{\partial t}\left(\frac{1}{c^2}E^2 + B^2\right) = 2\nabla \cdot (\vec{B} \times \vec{E}) - 2\mu\,\vec{J} \cdot \vec{E}, \qquad (11.55)$$

where $E^2 = \vec{E} \cdot \vec{E}$ and $B^2 = \vec{B} \cdot \vec{B}$.

9. Verify that

$$f(x, y, z, t) = \sin[k_1(u_1 x + u_2 y + u_3 z - ct) + k_2]$$

satisfies the partial differential equation

$$\Box^2 f = 0.$$

10. Show that if g is a twice differentiable function, $\vec{u} = (u_1, u_2, u_3)$ is a unit vector, and if $f(x, y, z, t) = g(u_1 x + u_2 y + u_3 z - ct)$, then

$$\Box^2 f = 0.$$

11. Let ϕ be a scalar field depending only on x, z, and t and satisfying d'Alembert's equation: $\Box^2 \phi = 0$. Show that $\vec{B} = (0, -c^{-2}\,\partial\phi/\partial t, 0)$ and $\vec{E} = (\partial\phi/\partial z, 0, -\partial\phi/\partial x)$ satisfy Maxwell's equations in the absence of any charges or currents.

12. Let

$$\Phi = \sin\left(\frac{\sqrt{2}}{2}(x + z) - ct\right)\,dz\,dx\,dt.$$

Verify that Φ is closed and satisfies Equation (11.43) in the absence of any charges or currents. Find the magnetic and electrical fields associated with this potential.

13. Given the electromagnetic potential

$$\Phi = x^2\cos(z - ct)\,dx\,dy\,dz - cx^2\cos(z - ct)\,dx\,dy\,dt,$$

verify that Φ is closed and then find $\vec{B}$, $\vec{E}$, ρ, and $\vec{J}$.

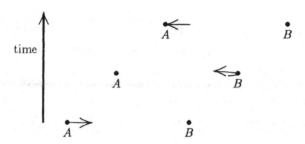

FIGURE 11.5. Measuring the speed of light in the direction of velocity.

11.5   Special Relativity

The Michelson–Morley Experiment

A prominent characteristic of Newton's laws of mechanics is that they are independent of the frame of reference of the observer. As we have seen, this was one of the reasons for developing vector algebra. If the coordinate frame is not important, then there should be a mathematical language for expressing these laws that is independent of any fixed viewpoint. As Newton pointed out, his laws are also invariant under a uniform translation of the reference system. They do not enable us to distinguish between a system at rest and the same system moving at a constant velocity.

By the nineteenth century, light was recognized to be a wave phenomenon. To explain how it could be propagated through a vacuum, it was postulated that all of space was pervaded by an undetectable ether and that it was this ether that carried the light waves. Because light always travels in straight lines, this ether should provide an absolute notion of rest against which all velocities could be measured.

In 1881, Albert A. Michelson (1852–1931) performed an experiment that he hoped would establish the velocity of the earth relative to the ether. If two points, A and B, are traveling at a common velocity, v, in the direction of the line from A to B, and if a beam of light, whose true velocity in the ether is c, is sent from A to B and then reflected back to A (Figure 11.5), then the velocity as observed at A is $c - v$ on the way to B and $c + v$ on the return. If l is the distance between A and B, then the time it takes for light to travel from A to B and back to A is

$$t_1 = \frac{l}{c - v} + \frac{l}{c + v} = \frac{2cl}{c^2 - v^2} = \frac{2l}{c(1 - v^2/c^2)}. \qquad (11.56)$$

We now choose a second point C, the same distance from A, but in

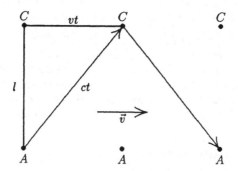

FIGURE 11.6. Measuring the speed of light perpendicular to the velocity.

a direction perpendicular to $\overrightarrow{AB}$ (Figure 11.6). Since A and C are both moving, light traveling from A to C does not reach C until it has moved to the position C'. If t is the elapsed time, then the sides of the right triangle ACC' are $|AC| = l$, $|CC'| = vt$, and $|AC'| = ct$. The Pythagorean theorem tells us that

$$l^2 + v^2t^2 \;=\; c^2t^2,$$

$$t \;=\; \frac{l}{c\sqrt{1 - v^2/c^2}}.$$

The same thing happens on the trip back to A, so that the total elapsed time is

$$t_2 = \frac{2l}{c\sqrt{1 - v^2/c^2}}. \tag{11.57}$$

It follows that the observed speed of light should depend on the direction in which this speed is measured, with a maximum discrepancy of

$$\frac{t_2}{t_1} = \sqrt{1 - \frac{v^2}{c^2}}. \tag{11.58}$$

Michelson constructed his experiment so that the speed of light was simultaneously measured in two orthogonal directions. The entire apparatus was set up on a stone wheel floating on a bed of mercury so that it could be continuously rotated. The result appeared inconclusive. The values of t_1 and t_2 always agreed to within the accuracy of the experiment.

In 1887, Michelson and Edward Williams Morley (1838–1923) performed the experiment a second time with refined equipment that could detect an absolute velocity, v, of as little as 1 m/s, a leisurely stroll. Again, the values of t_1 and t_2 were the same. What was to prove to be the correct explanation for Michelson and Morley's lack of success was suggested by George Francis Fitzgerald (1851–1901) in 1893 and by Hendrick Antoon Lorentz (1853–1928) in 1895: *In the direction of absolute velocity, distances contract by a factor of* $\sqrt{1 - v^2/c^2}$. To quote Lorentz:

> Surprising as this hypothesis may appear at first sight, yet we shall have to admit that it is by no means far-fetched, as soon as we assume the molecular forces are also transmitted through the ether, like the electric and magnetic forces of which we are able at the present time to make this assertion definitely.

Lorentz Transformations

Lorentz's insight that such a contraction does occur in electricity and magnetism arose from his investigation of Maxwell's equations and the electromagnetic potential. As we have seen, the coefficients of Φ are scalar fields satisfying the partial differential equation

$$\frac{\partial^2 f}{\partial x^2} + \frac{\partial^2 f}{\partial y^2} + \frac{\partial^2 f}{\partial z^2} - \frac{1}{c^2}\frac{\partial^2 f}{\partial t^2} = 0. \tag{11.59}$$

A linear transformation, $\vec{L}$, corresponds to a rotation if and only if it has a positive determinant and its transpose is its inverse. If we consider any rotation of the space coordinates:

$$(x', y', z') = \vec{L}(x, y, z),$$

then f satisfies the same partial differential equation with respect to these new variables:

$$\frac{\partial^2 f}{\partial x'^2} + \frac{\partial^2 f}{\partial y'^2} + \frac{\partial^2 f}{\partial z'^2} - \frac{1}{c^2}\frac{\partial^2 f}{\partial t^2} = 0.$$

This is not surprising since electromagnetic phenomena should be independent of the frame of reference.

If we introduce a uniform velocity into our system, moving it at speed v in the direction of the positive x axis, then this corresponds to the linear transformation

$$\begin{aligned}
x' &= x - vt, \\
y' &= y, \\
z' &= z, \\
t' &= t,
\end{aligned} \tag{11.60}$$

where (x, y, z, t) are the coordinates as seen by an observer at rest and (x', y', z', t') are the coordinates as viewed within the moving system. The problem is that Equation (11.59) is not invariant under this transformation.

Lorentz sought a linear transformation to replace Equation (11.60) that would leave the differential equations of electricity and magnetism invariant. It would have to satisfy the following criteria.

1. The determinant is positive.

2. The variables y and z remain unchanged,

$$y' = y, \quad z' = z,$$

and x' and t' are independent of y and z.

3. A particle moving at speed v in the positive x direction, as observed by the resting observer, appears at rest to the moving observer. This implies that there is a constant β such that

$$x' = \beta(x - vt).$$

4. The d'Alembertian of any scalar field f with respect to (x, y, z, t) is equal to the d'Alembertian of f with respect to (x', y', z', t').

It is a straightforward exercise in the use of the chain rule to verify that there is only one transformation satisfying these conditions, namely,

$$\begin{aligned}
x' &= \beta(x - vt), \\
y' &= y, \\
z' &= z, \\
t' &= \beta(t - \tfrac{v}{c^2}x),
\end{aligned} \tag{11.61}$$

where

$$\beta = \frac{1}{\sqrt{1 - v^2/c^2}}. \tag{11.62}$$

To see that this transformation produces the desired contraction of distances, let (x_1, t_1) and (x_2, t_2) be two pairs of position and time variables in the stationary system, (x_1', t_1') and (x_2', t_2') the corresponding pairs in the moving system. From our transformation, Equation (11.61), we see that

$$x_2' - x_1' = \beta(x_2 - x_1) - \beta v(t_2 - t_1), \tag{11.63}$$

$$t_2' - t_1' = \beta(t_2 - t_1) - \frac{\beta v}{c^2}(x_2 - x_1). \tag{11.64}$$

Distance in our moving system only makes sense if both positions are measured simultaneously: $t_2' = t_1'$. Under this restriction, Equation (11.64) becomes

$$t_2 - t_1 = \frac{v}{c^2}(x_2 - x_1), \tag{11.65}$$

Substitution into Equation (11.63) yields

$$\begin{aligned}
x_2' - x_1' &= \beta\left(1 - \frac{v^2}{c^2}\right)(x_2 - x_1) \\
&= \sqrt{1 - \frac{v^2}{c^2}}\,(x_2 - x_1), \tag{11.66}
\end{aligned}$$

precisely the expected contraction.

Einstein was the first to remark that this transformation implied that very strange things were happening with respect to time. Two events that were simultaneous in the moving system ($t_2' = t_1'$) would, in general, *not* be simultaneous as seen from the stationary system [Equation (11.65)]. A contraction is also occurring in the time dimension. If we hold the position constant, $x_2' = x_1'$, then

$$x_2 - x_1 = v(t_2 - t_1),$$

and Equation (11.64) becomes

$$
\begin{aligned}
t_2' - t_1' &= \beta(t_2 - t_1) - \frac{\beta v^2}{c^2}(t_2 - t_1) \\
&= \beta\left(1 - \frac{v^2}{c^2}\right)(t_2 - t_1) \\
&= \sqrt{1 - \frac{v^2}{c^2}}\,(t_2 - t_1).
\end{aligned}
\tag{11.67}
$$

It can be shown that spatial rotations and the transformation of Equation (11.61), as well as the transformations built out of compositions of these, are the *only* linear transformations with a positive determinant that leave Equation (11.59) unchanged. Collectively, these are known as the *Lorentz transformations*. These transformations do more than leave Equation (11.59) invariant, they leave all of the relationships of Maxwell's equations unchanged.

Invariance of Maxwell's Equations

Maxwell's equations can be seen as a cyclical series of operations. Beginning with our electron flow described by the 3-form **J**, we find the 3-form

$$\Phi = \phi_4\, dx\, dy\, dz + \phi_1\, dy\, dz\, dt + \phi_2\, dz\, dx\, dt + \phi_3\, dx\, dy\, dt$$

satisfying

$$d\Phi = 0 \quad \text{and} \quad \epsilon\,\square^2\Phi + \mathbf{J} = 0.$$

The next step is to transform Φ into the 1-form:

$$\frac{\phi_1}{c^2}\, dx + \frac{\phi_2}{c^2}\, dy + \frac{\phi_3}{c^2}\, dz + \phi_4\, dt.$$

The third step is to take the differential of the negative of this 1-form, giving us the electromagnetic field in the representation

$$\mathbf{B} + \mathbf{E}\, dt.$$

For the fourth step, we transform this 2-form into

$$\mathbf{E} - c^2\mathbf{B}\, dt,$$

where we have used the fact that $c^2 = 1/\epsilon\mu$. The fifth step, which completes the cycle, is to take the differential of this last 2-form, returning us to the electron flow, $\mathbf{J}$.

As we have seen, a change of variable by a Lorentz transformation does not change the shape of the partial differential equation

$$\epsilon\,\Box^2\Phi + \mathbf{J} = 0.$$

Only the 3-form $\mathbf{J}$ is transformed. Furthermore, a closed differential form remains closed under any continuous change of variables. Let us therefore start with our potential field Φ and simultaneously follow both it and the potential field obtained by the Lorentz transformation of Equation (11.61):

$$
\begin{array}{ccc}
\begin{aligned}
&\phi_4\, dx\, dy\, dz + \phi_1\, dy\, dz\, dt \\
&\quad + \phi_2\, dz\, dx\, dt + \phi_3\, dx\, dy\, dt
\end{aligned}
&=&
\begin{aligned}
&\beta(\phi_4 + vc^{-2}\phi_1)\, dx'\, dy'\, dz' \\
&\quad + \beta(\phi_1 + v\phi_4)\, dy'\, dz'\, dt' \\
&\quad + \phi_2\, dz'\, dx'\, dt' + \phi_3\, dx'\, dy'\, dt'
\end{aligned}
\\[2mm]
\downarrow & & \downarrow \\[2mm]
\begin{aligned}
&\frac{\phi_1}{c^2}\, dx + \frac{\phi_2}{c^2}\, dy + \frac{\phi_3}{c^2}\, dz + \phi_4\, dt
\end{aligned}
&=&
\begin{aligned}
&\frac{\beta}{c^2}(\phi_1 + v\phi_4)\, dx' + \frac{\phi_2}{c^2}\, dy' \\
&\quad + \frac{\phi_3}{c^2}\, dz' + \beta(\phi_4 + \frac{v}{c^2}\phi_1)\, dt'
\end{aligned}
\\[2mm]
\downarrow & & \downarrow \\[2mm]
\begin{aligned}
&B_1\, dy\, dz + B_2\, dz\, dx \\
&\quad + B_3\, dx\, dy + E_1\, dx\, dt \\
&\quad + E_2\, dy\, dt + E_3\, dz\, dt
\end{aligned}
&=&
\begin{aligned}
&B_1\, dy'\, dz' + \beta(B_2 + \frac{v}{c^2}E_3)\, dz'\, dx' \\
&\quad + \beta(B_3 - \frac{v}{c^2}E_2)\, dx'\, dy' \\
&\quad + E_1\, dx'\, dt' \\
&\quad + \beta(E_2 - vB_3)\, dy'\, dt' \\
&\quad + \beta(E_3 + vB_2)\, dz'\, dt'
\end{aligned}
\\[2mm]
\downarrow & & \downarrow \\[2mm]
\begin{aligned}
&E_1\, dy\, dz + E_2\, dz\, dx \\
&\quad + E_3\, dx\, dy - c^2 B_1\, dx\, dt \\
&\quad - c^2 B_2\, dy\, dt - c^2 B_3\, dz\, dt
\end{aligned}
&=&
\begin{aligned}
&E_1\, dy'\, dz' + \beta(E_2 - vB_3)\, dz'\, dx, \\
&\quad + \beta(E_3 + vB_2)\, dx'\, dy' \\
&\quad - c^2 B_1\, dx'\, dt' \\
&\quad - \beta(c^2 B_2 + vE_3)\, dy'\, dt' \\
&\quad - \beta(c^2 B_3 - vE_2)\, dz'\, dt'
\end{aligned}
\\[2mm]
\downarrow & & \downarrow \\[2mm]
\begin{aligned}
&\rho\, dx\, dy\, dz - J_1\, dy\, dz\, dt \\
&\quad - J_2\, dz\, dx\, dt - J_3\, dx\, dy\, dt
\end{aligned}
&=&
\begin{aligned}
&\beta(\rho - \frac{v}{c^2}J_1)\, dx'\, dy'\, dz' \\
&\quad - \beta(J_1 - v\rho)\, dy'\, dz'\, dt' \\
&\quad - J_2\, dz'\, dx'\, dt' - J_3\, dx'\, dy'\, dt'.
\end{aligned}
\end{array}
$$

At each step, the relationship between successive forms is precisely the same whether we are working in our original coordinate system or the moving system. It can be verified that this also holds for any rotation. Lorentz's insight is summed up in the following theorem.

Theorem 11.2 *A linear transformation of positive determinant leaves the relationships of Maxwell's equations invariant if and only if it is a Lorentz transformation.*

Force and Momentum

Albert Einstein's theory of special relativity, published in *Annalen der Physik* in 1905 under the title "Zur Elektrodynamik bewegter Körper" ("On the Electrodynamics of Moving Bodies"), begins with the assumption that *all* physical laws of nature must be invariant under the Lorentz transformations. Einstein applied this assumption to the equation with which we began this book,

$$F = ma,$$

and asked what it would mean for this law to be Lorentz invariant. He looked at how this law should be interpreted in the world of electromagnetism.

We begin with the easiest side, the right side. Recall that Newton did not say that force equals mass time acceleration. He said, in effect, that force equals the rate at which momentum changes:

$$\vec{F} = \frac{\partial}{\partial t}(m\vec{v}). \tag{11.68}$$

If we assume that the mass is constant, then

$$\frac{\partial}{\partial t}(m\vec{v}) = m\frac{\partial \vec{v}}{\partial t} = m\vec{a}.$$

Let us not make that assumption. We ask for the electromagnetic analog of momentum. If we have a single particle of charge m traveling with constant velocity $\vec{v} = (v_1, v_2, v_3)$, then $m\vec{v}$ describes the current. We have seen that the correct way to view the current $m\vec{v}$ is as the 3-form:

$$m\,dx\,dy\,dz - mv_1\,dy\,dz\,dt - mv_2\,dz\,dx\,dt - mv_3\,dx\,dy\,dt,$$

so that m is the component of momentum in the direction of time. The right side of Equation (11.68) should be

$$\frac{\partial}{\partial t}(m, m\vec{v}).$$

We now want to describe force as a Lorentz invariant 3-form. The electromagnetic field is a 2-form. It is not the force. If we assume that our particle is not moving, $\vec{v} = \vec{0}$, then the force exerted on it is $m\vec{E}$. If we view **J** as the appropriate Lorentz invariant 1-form (see Exercise 7, Section 11.6),

$$\mathbf{J} = m\,dt - \frac{mv_1}{c^2}\,dx - \frac{mv_2}{c^2}\,dy - \frac{mv_3}{c^2}\,dz, \tag{11.69}$$

then $\mathbf{J}(\mathbf{E} - c^2\mathbf{B}\,dt)$ is also Lorentz invariant, yielding

$$
\begin{aligned}
\mathbf{J}(\mathbf{E} - c^2\mathbf{B}\,dt) \;=\; & -\left(m\,dt - \frac{mv_1}{c^2}\,dx - \frac{mv_2}{c^2}\,dy - \frac{mv_3}{c^2}\,dz\right) \\
& \times (E_1\,dy\,dz + E_2\,dz\,dx + E_3\,dx\,dy \\
& \quad\ - c^2 B_1\,dx\,dt - c^2 B_2\,dy\,dt - c^2 B_3\,dz\,dt) \\
=\; & \frac{1}{c^2}(mv_1 E_1 + mv_2 E_2 + mv_3 E_3)\,dx\,dy\,dz \\
& - (mE_1 + mv_2 B_3 - mv_3 B_2)\,dy\,dz\,dt \\
& - (mE_2 + mv_3 B_1 - mv_1 B_3)\,dz\,dx\,dt \\
& - (mE_3 + mv_1 B_2 - mv_2 B_1)\,dx\,dy\,dt.
\end{aligned}
\tag{11.70}
$$

When $\vec{v} = \vec{0}$, the three space coordinates of Equation (11.70) are precisely $m\vec{E}$. This equation provides an expression for the force when $\vec{v} = \vec{0}$ and it is Lorentz invariant, thus giving us the force under any velocity. The space components of the force acting on a moving particle,

$$
\text{Force} \;=\; m\vec{E} \;+\; m\vec{v} \times \vec{B},
\tag{11.71}
$$

had been known before Einstein. What is significant for our purposes is the time component,

$$
\frac{m}{c^2}\vec{v} \cdot \vec{E}.
$$

The Lorentz invariant form of $\vec{F} = m\vec{a}$ for a point charge m moving with velocity $\vec{v}$ is thus

$$
\left(\frac{m}{c^2}\vec{v} \cdot \vec{E},\, m\vec{E} + m\vec{v} \times \vec{B}\right) = \frac{\partial}{\partial t}(m,\, m\vec{v}).
\tag{11.72}
$$

Equating the first coordinates, we see that

$$
m\vec{v} \cdot \vec{E} = c^2\,\frac{\partial m}{\partial t}.
\tag{11.73}
$$

Now, $m\vec{v} \cdot \vec{E}$ is the rate at which work is performed by the field $\vec{E}$ acting on our point charge. It represents the rate at which the *internal energy, E,* of our charge is changing:

$$
m\vec{v} \cdot \vec{E} = \frac{\partial E}{\partial t}.
\tag{11.74}
$$

Making this substitution and integrating both sides of Equation (11.73) with respect to time leads to

$$
\Delta E = c^2\,\Delta m,
\tag{11.75}
$$

where ΔE and Δm represent the change in E and m, respectively. If we assume that zero mass corresponds to zero energy, then

$$
E = mc^2.
\tag{11.76}
$$

11.6 Exercises

1. Why is it that we cannot perform the following experiment? We fix two points, A and B, measure the speed of light as it travels from A to B, and then compare that with the speed of light measured as it travels from B to A. If the segment from A to B is in the direction of the velocity of the system, then these two speeds will be different.

2. Verify that spatial rotations leave the d'Alembertian unchanged.

3. Verify that Equation (11.61) is the only linear transformation satisfying the four conditions given on page 355.

4. What is the inverse of the Lorentz transformation given by Equation (11.61)?

5. Verify that spatial rotations leave the relationships of Maxwell's equations unchanged.

6. Find the magnetic and electric fields generated by a single particle of charge m that passes through the origin at time $t = 0$, moving at a constant speed v in the positive x direction. Hint: use a Lorentz transformation and the fact that a stationary particle of charge m generates the electromagnetic field:

$$\mathbf{B} + \mathbf{E}\,dt = m\gamma r^{-3}(x\,dx + y\,dy + z\,dz)\,dt.$$

7. Show that the 3-form

$$\rho\,dx\,dy\,dz - j_1\,dy\,dz\,dt - j_2\,dz\,dx\,dt - j_3\,dx\,dy\,dt$$

corresponds to the 1-form

$$\rho\,dt - \frac{j_1}{c^2}dx - \frac{j_2}{c^2}dy - \frac{j_3}{c^2}dz$$

by applying the linear transformation of Equation (11.61) to each of them and showing that the corresponding forms in x', y', z', t' space are in precisely the same relationship.

Appendix A

An Opportunity Missed

The following is taken from the Josiah Willard Gibbs Lecture, "Missed Opportunities,"[1] given by Freeman Dyson under the auspices of the American Mathematical Society on January 17, 1972.

"The first clear sign of a breakdown in communication between physics and mathematics was the extraordinary lack of interest among mathematicians in James Clerk Maxwell's discovery of the laws of electromagnetism. Maxwell discovered his equations, which describe the behavior of electric and magnetic fields under the most general conditions, in the year 1861, and published a clear and definitive statement of them in 1865. This was the great event of nineteenth century physics, achieving for electricity and magnetism what Newton had achieved for gravitation two hundred years earlier. Maxwell's equations contained, among other things, the explanation of light as an electromagnetic phenomenon, and the basic principles of electric power transmission and radio technology. These aspects of the theory were of primary interest to physicists and engineers. But in addition to their physical applications, Maxwell's equations had abstract mathematical qualities which were profoundly new and important. Maxwell's theory was formulated in terms of a new style of mathematical concept, a tensor field extending throughout space and time and obeying coupled partial differential equations of peculiar symmetry.

"After Newton's laws of gravitational dynamics had been promulgated in 1687, the mathematicians of the eighteenth century seized hold of these laws and generalized them into the powerful mathematical theory of analytical mechanics. Through the work of Euler, Lagrange and Hamilton, the equations of Newton were analyzed and understood in depth. Out of this deep exploration of Newtonian physics, new branches of pure mathematics ultimately emerged. Lagrange distilled from the extremal properties of dynamical integrals the general principles of the calculus of variations. Fifty years later the work of Euler on geodesic motions led Gauss to the creation of differential geometry. Another fifty years later, the generalization of the Hamilton–Jacobi formulation of dynamics led Lie to the invention of Lie

---

[1]Published in the *Bulletin of the American Mathematical Society*, Volume 78, Number 5, September 1972, pages 635–652; reprinted by permission of the American Mathematical Society and Freeman Dyson.

groups. And finally, the last gift of Newtonian physics to pure mathematics was the work of Poincaré on the qualitative behavior of orbits which led to the birth of modern topology.

"But the mathematicians of the nineteenth century failed miserably to grasp the equally great opportunity offered to them in 1865 by Maxwell. If they had taken Maxwell's equations to heart as Euler took Newton's, they would have discovered, among other things, Einstein's theory of special relativity, the theory of topological groups and their linear representations, and probably large pieces of the theory of hyperbolic differential equations and functional analysis. A great part of twentieth century physics and mathematics could have been created in the nineteenth century, simply by exploring to the end the mathematical concepts to which Maxwell's equations naturally lead.

"There is plenty of documentary evidence showing how Maxwell's theory appeared to the mathematicians of his time. I shall quote two short extracts to illustrate contrasting ways in which the theory was brought to their attention. Both are taken from Presidential Addresses to meetings of the British Association, then as now the chief organization in Britain dedicated to promoting the unity of science. First comes Maxwell himself, speaking in 1870, his announced topic being the relation between mathematics and physics.

" 'According to a theory of electricity which is making great progress in Germany, two electrical particles act on one another directly at a distance, but with a force which, according to Weber, depends on their relative velocity, and according to a theory hinted at by Gauss, and developed by Riemann, Lorenz, and Neumann, acts not instantaneously, but after a time depending on the distance. The power with which this theory, in the hands of these eminent men, explains every kind of electrical phenomena must be studied in order to be appreciated. Another theory of electricity which I prefer denies action at a distance and attributes electric action to tensions and pressures in an all-pervading medium, these stresses being the same in kind with those familiar to engineers, and the medium being identical with that in which light is supposed to be propagated.'

"It is difficult to read Maxwell's address without being infuriated by his excessive modesty, which led him to refer to his epoch-making discovery of nine years earlier as only 'Another theory of electricity which I prefer.' How different is his style from that of Newton, who wrote at the beginning of the third book of his *Principia*, 'It remains that, from the same principles, I now demonstrate the frame of the System of the World.' Since Maxwell himself seemed so half-hearted, it is not surprising that he did not inspire the mathematicians to throw aside their fashionable covariants and quantics and study his equations.

"My second quotation is from the Oxford mathematician, Henry Smith, speaking three years later to the same audience. A few months before he spoke in 1873, Maxwell's great book on electricity and magnetism had appeared.

" 'In the course of the present year a treatise on electricity has been published by Professor Maxwell, giving a complete account of the mathematical theory of that science, ... No mathematician can turn over the pages of these volumes without very speedily convincing himself that they contain the first outlines (and something more than the first outlines) of a theory which has already added largely to the methods and resources of pure mathematics, and which may one day render to that abstract science services no less than those which it owes to astronomy. For electricity now, like astronomy of old, has placed before the mathematician an entirely new set of questions, requiring the creation of entirely new methods for their solution, ... It must be considered fortunate for the mathematicians that such a vast field of research in the application of mathematics to physical enquiries should be thrown open to them, at the very time when the scientific interest in the old mathematical astronomy has for the moment flagged, ...'

"These words show that at least one mathematician did understand the historic nature of the challenge presented to mathematics by Maxwell's work. Smith's perception is the more remarkable, considering that he was a very pure mathematician whose best-known work is in analytic number theory. Unfortunately he was then 46 years old, too old to be a pioneer in a new field. No doubt he contented himself with 'turning over the pages of these volumes' and then returned with relief to his familiar ternary forms. The young men, who might have been stimulated by his words to create new fields of mathematics, were not listening. Hermann Minkowski and Jacques Hadamard, who were to achieve in the twentieth century the fulfilment of some of Smith's prophecies, were boys of nine and eight at the time he spoke. Élie Cartan was three; Hermann Weyl, Jean Leray and Harish-Chandra were not yet born.

"Hermann Minkowski had something to say thirty-five years later about the opportunity which the mathematicians had missed. He was talking to the German equivalent of the British Association, three years after Einstein's discovery of special relativity. He pointed out that the mathematical basis of Einstein's discovery lies in an incompatibility between two groups of transformations of space and time coordinates. On the one hand, the equations of Newtonian mechanics are invariant under a group G_∞ which physicists now call the *Galilei group*. The group G_∞ is six-dimensional; it is generated by three rotations of the space coordinates, and three uniform velocity transformations of the form

$$x \to x - vt, \quad t \to t.$$

On the other hand, the Maxwell equations in the absence of matter are invariant under a goup G_c which physicists call the *Lorentz group*. The group G_c is also six-dimensional; it is generated by three rotations as before, together with three Lorentz transformations of the form

$$x \rightarrow \beta(x - vt), \quad t \rightarrow \beta(t - (vx/c^2)),$$

$$\beta = (1 - (v/c)^2)^{-1/2},$$

where c is the velocity of light. From a purely mathematical point of view, G_c has a simpler structure than G_∞. In fact G_c is a real noncompact form of the semisimple Lie algebra $A_1 \times A_1$, whereas G_∞ is not semisimple. I quote now from Minkowski's 1908 lecture.

"'Group G_c in the limit when $c = \infty$, that is the group G_∞, becomes no other than that complete group which is appropriate to Newtonian mechanics. This being so, and since G_c is mathematically more intelligible than G_∞, it looks as though the thought might have struck some mathematician, fancy-free, that after all, as a matter of fact, natural phenomena do not possess an invariance with the group G_∞, but rather with the group G_c, c being finite and determinate, but in ordinary units of measure, extremely great. Such a premonition would have been an extraordinary triumph for pure mathematics. Well, mathematics, though it now can display only staircase-wit, has the satisfaction of being wise after the event, and is able, thanks to its happy antecedents, with its senses sharpened by an unhampered outlook to far horizons, to grasp forthwith the far-reaching consequences of such a metamorphosis of our concept of nature.'

"Why were the mathematicians of the later nineteenth century blind to these possibilities which Smith had so clearly foreshadowed? There are many reasons. If Maxwell had written in a style as lucid and as confident as that of Newton, the mathematicians would have been more inclined to take him seriously. Another reason for the mathematicians' indifference was the fact that Maxwell's equations were not generally accepted even by physicists for twenty years after their discovery. Until Hertz demonstrated the existence of radio waves in 1887, the majority of physicists considered Maxwell's theory to be a speculative hypothesis.[2] Mathematicians who had themselves lost touch with physics were not able to make an independent assessment of the theory's merits. Lastly, and perhaps most importantly, the mathematicians ignored Maxwell because mathematics in the later nineteenth century had developed in quite other directions. Mathematicians were busy with the theory of functions of complex variables, with analytic number theory, with algebraic forms and invariants. The flowering of these subjects had given the mathematicians definite tastes and aesthetic standards, into which the new physics of Maxwell would not easily fit."

---

[2]The physicist Michael Pupin, in his autobiography, *From immigrant to inventor*, Charles Scribner's Sons, 1924, gives a vivid account of the difficulties encountered by a student trying to learn the Maxwell theory in the 1880's. Pupin traveled from America to England searching in vain for somebody who had mastered the theory; he finally succeeded in learning it from Helmholtz in Berlin.

Appendix B

Bibliography

Selected Textbooks

Apostol, Tom M., *Calculus*, two volumes, second edition, Blaisdell, Waltham, Massachusetts, 1969.

Bleaney, B. I., & Bleaney, B., *Electricity and Magnetism*, Oxford University Press, London and New York, 1976.

Edwards, Harold M., *Advanced Calculus*, reprinted by Robert E. Krieger, Huntingdon, New York, 1980.

Flanders, Harley, *Differential Forms*, Academic Press, New York, 1962.

French, A. P., *Special Relativity*, W. W. Norton, New York, 1968.

Gibbs, J. Willard, & Wilson, Edwin B., *Vector Analysis*, second edition, reprinted by Dover, New York, 1960.

Kaplan, Wilfred, *Advanced Calculus*, second edition, Addison-Wesley, Reading, Massachusetts, 1973.

Protter, Murray H., & Morrey, Charles B., Jr., *Intermediate Calculus*, second edition, Springer-Verlag, New York, 1985.

Simmonds, James G., *A Brief on Tensor Analysis*, Springer, New York, 1982.

Thomas, George B., Jr., & Finney, Ross L., *Calculus and Analytic Geometry*, fifth edition, Addison-Wesley, Reading, Massachusetts, 1981.

Urwin, Kathleen M., *Advanced Calculus and Vector Field Theory*, Pergamon Press, Oxford, 1966.

Histories

Bell, E. T., *Men of Mathematics*, Simon and Schuster, New York, 1937.

Bourbaki, Nicolas, *Éléments d'histoire des mathématiques*, reprinted by Masson, Paris, 1984.

Boyer, Carl B., *A History of Mathematics*, Princeton University Press, Princeton, New Jersey, 1985.

Cajori, Florian, *A History of Mathematics*, Macmillan, New York, 1926.

Dijksterhuis, E. J., *The Mechanization of the World Picture*, translated by Dikshoorn, Princeton University Press, Princeton, New Jersey, 1986.

Kline, Morris, *Mathematical Thought from Ancient to Modern Times*, Oxford University Press, Oxford and New York, 1972.

Meyer, Herbert W., *History of Electricity and Magnetism*, MIT Press, Cambridge, Massachusetts, 1971.

Original Sources

Heaviside, Oliver, *Electromagnetic Theory*, reprinted by Dover, New York, 1950.

Kepler, Johann, *Gesammelte Werke*, 18 volumes, C. H. Beck, München, 1937–1969.

Lagrange, Joseph-Louis, *Oeuvres de Lagrange*, 14 volumes, Gauthier-Villars, Paris, 1867–1892.

Lorentz, H. A., Einstein, A., Minkowski, H., & Weyl, H., *The Principle of Relativity*, a collection of original memoirs with notes by Sommerfeld, translated by Perrett and Jeffrey, reprinted by Dover, New York, 1952.

Maxwell, James Clerk, *Scientific Papers of James Clerk Maxwell*, Dover, New York, 1952.

Newton, Isaac, *Mathematical Principles of Natural Philosophy*, Motte's translation revised by Cajori, University of California Press, Berkeley, 1962.

Poincaré, Henri, *Les méthodes nouvelles de la mécanique céleste*, reprinted by Dover, New York, 1957.

Appendix C

Clues and Solutions

Section 1.4

1. (b) $\left(-\frac{3}{2}\sqrt{2}, \sqrt{2}\right), \left(\frac{3}{2}\sqrt{2}, -\sqrt{2}\right)$.

2. Clue: the slope of the conjugate diameter is $-9/20$.

14. Clue: see E. J. Dijksterhuis, *The Mechanization of the World Picture*.

Section 1.6

1. (a) $(3,0)$, $\left(\frac{3}{2}, \frac{\sqrt{3}}{2}\right)$, $\left(-\frac{3\sqrt{2}}{2}, \frac{\sqrt{2}}{2}\right)$, $(0,-1)$.

2. (a) $(0,1)$, $\left(-\frac{3\sqrt{3}}{2}, \frac{1}{2}\right)$, $\left(-\frac{3\sqrt{2}}{2}, -\frac{\sqrt{2}}{2}\right)$, $(3,0)$.

3. (a) $(-3,0)$, $\left(-\frac{3}{2}, -\frac{\sqrt{3}}{2}\right)$, $\left(\frac{3\sqrt{2}}{2}, -\frac{\sqrt{2}}{2}\right)$, $(0,1)$.

6. $r(t) = t$.

7. $\vec{u}_r = (t-1, \sqrt{2t-t^2})$, $\vec{u}_\theta = (-\sqrt{2t-t^2}, t-1)$.

8. $\frac{dr}{dt} = 1$, $\frac{d\theta}{dt} = -(2t-t^2)^{-1/2}$.

9. $\vec{v} = \vec{u}_r - t(2t-t^2)^{-1/2}\vec{u}_\theta$.

10. $\vec{a} = (t-2)^{-1}\vec{u}_r + (t^2-3t)(2t-t^2)^{-3/2}\vec{u}_\theta$.

16. Clue: show that if r and θ are related by $\frac{r^2\cos^2\theta}{a^2} + \frac{r^2\sin^2\theta}{b^2} = 1$ and if $r^2\frac{d\theta}{dt} = k$, then $\vec{a} = -\frac{rk^2}{a^2b^2}\vec{u}_r$.

17. Clue: show that if $r^2\frac{d\theta}{dt} = k$, then $\vec{a} = -\frac{k^2(1+c^2)}{r^3}\vec{u}_r$.

Section 2.5

1. (a) $(2, -3, 1) + (6, 2, -3) = (8, -1, -2)$, $(2, -3, 1) \cdot (6, 2, -3) = 3$,
   $(2, -3, 1) \times (6, 2, -3) = (7, 12, 22)$.

5. $\vec{r}_s = \left(\frac{-63}{50}, \frac{21}{10}, \frac{-42}{25}\right)$, $\vec{r}_{s\perp} = \left(\frac{13}{50}, \frac{-1}{10}, \frac{-8}{25}\right)$, $\vec{s}_r = \left(\frac{7}{3}, \frac{-14}{3}, \frac{14}{3}\right)$, $\vec{s}_{r\perp} = \left(\frac{2}{3}, \frac{-1}{3}, \frac{-2}{3}\right)$.

7. $2x + 2y + z = 0$.

8. $2x + 2y + z = 6$.

9. (a) $(5, 1, -1)$ is not on this plane; (c) $(1, 1, -2) = 2\vec{r} + \vec{s}$.

10. $(1, -4, 7)$.

11. $(4, 1, 0)$ and $(1, 2, 1)$ (there are many other possible pairs).

12. $x = -1 + 4u + v$, $y = -1 + u + 2v$, $z = v$.

13. $1/\sqrt{66}$.

14. $x - 4y + 7z = 23$.

15. Clue: add $(1, -2, 0)$ to each point so that the problem becomes to find
   the distance from $(1, -4, -3)$ to the line from the origin to $(3, -6, 2)$.
   Find the magnitude of the component of $(1, -4, -2)$ in the direction
   perpendicular to $(3, -6, 2)$.

18. $(4/3, 1/6, -5/6)$.

Section 3.2

5. (a) $(2t, -4, -2t)$, $(2, 0, -2)$; (b) $(\sinh t, \cosh t, 1)$, $(\cosh t, \sinh t, 0)$;
   (c) $(\cos t - t \sin t, \sin t + t \cos t, 0)$, $(-2 \sin t - t \cos t, 2 \cos t - t \sin t, 0)$.

7. (a) $(t^2 + 4)/\sqrt{t^4 + 10t^2 + 16}$, never $\perp$, $\|$ when $t = 0$; (b) $(\sinh(2t) + t)/\sqrt{\cosh^2(2t) + (t^2 + 1)\cosh(2t) + t^2}$, $\perp$ at $t = 0$, never $\|$.

8. (a) $t/\sqrt{t^2 + 2}$, $\perp$ at $t = 0$, never $\|$; (b) $\sinh(2t)/\sqrt{\cosh^2(2t) + \cosh(2t)}$, $\perp$ at $t = 0$, never $\|$.

9. (a) $\int_0^2 \sqrt{8t^2 + 16}\, dt \simeq 10.17$; (b) $\int_0^2 \sqrt{\cosh(2t) + 1}\, dt \simeq 5.13$.

10. (a) $(8, 0, 8)$; (b) $(-\sinh t, \cosh t, -1)$.

11. (a) $x + z = 0$; (b) $x \sinh t - y \cosh t + z = t$.

12. (a) $\frac{1}{2}(t^2 + 2)^{-3/2}$; (b) $1/(1 + \cosh(2t))$.

15. $\vec{v} = \sin\theta\,\vec{u}_r + 2r^{-1}\vec{u}_\theta$, $\vec{a} = (2r\cos\theta - 4)r^{-3}\vec{u}_r = -4r^{-2}\vec{u}_r$.

Section 3.4

1. Clue: write $\vec{u}_r = (\cos\theta, \sin\theta, 0)$, $\vec{u}_\theta = (-\sin\theta, \cos\theta, 0)$.

2. Clue: use Equation (3.39) to prove that $\vec{v}$ is perpendicular to $\vec{r}$ if and only if $\vec{r}$ is parallel to $\vec{\varepsilon}$. Use Equation (3.36) to show that if $\varepsilon \neq 0$ then r is maximized (apogee) or minimized (perigee) when $\theta = 0$ or π.

4. $v = 7727$ m/s, period $= 5448$ s $= 90$ min, 48 s.

5. $\varepsilon = .35675$, $K = 6.03 \times 10^{10}$, perigee $= 6.7 \times 10^6$ m, apogee $= 1.4 \times 10^7$ m, period $= 10560$ s $= 176$ min.

6. $\varepsilon = .5878$, $K = 5.2 \times 10^{10}$, perigee $= 4.3 \times 10^6$ m (note that this would not be a practical orbit), apogee $= 1.65 \times 10^7$ m, period $= 10560$ s $= 176$ min.

7. Clue: use Equations (3.45) and (3.50).

8. $\varepsilon = .029$, $v_1 = 7900$ m/s, $v_2 = 7450$ m/s.

9. $(12 - 8)/(12 + 8) = 0.2$.

10. $98°$.

11. $v = 7800$ m/s, time $= 1440$ s $= 24$ min.

12. (a) $\varepsilon = -.16$, $K = 4.8 \times 10^{10}$, $v = 7448$m/s, time $= 26$ min,
　　 (b) $\varepsilon = -.29$, $K = 4.6 \times 10^{10}$, $v = 7277$m/s, time $= 32$ min,
　　 (c) $\varepsilon = -.39$, $K = 4.42 \times 10^{10}$, $v = 7246$m/s, time $= 37$ min,
　　 (d) $\varepsilon = -.52$, $K = 4.16 \times 10^{10}$, $v = 7380$m/s, time $= 46$ min.

13 Clue: let (r, θ) be the position of the rocket (in polar coordinates) when the burn ends. For the specific problem at hand, $r = 6.56 \times 10^6$ m and $\theta = -49°$. We have

$$\frac{dr}{dt} = \frac{GM\varepsilon\sin\theta}{K}, \quad r\frac{d\theta}{dt} = \frac{K}{r} = \frac{GM(1 + \varepsilon\cos\theta)}{K}.$$

After substitution and simplification,

$$v = \sqrt{\left(\frac{dr}{dt}\right)^2 + \left(r\frac{d\theta}{dt}\right)^2} = GM\sqrt{\frac{\varepsilon^2 - 1}{K^2} + \frac{2}{GMr}}.$$

Since r is constant, v is minimized when $(\varepsilon^2 - 1)/K^2$ is minimized. Show that this happens when the apogee is

$$\frac{r}{2}(1 + \cos\theta + \sin\theta).$$

Section 4.3

1. (a) -19; (b) 19; (c) -9; (d) -8; (e) -11.

5. Clue: let $(x, y, z) = \vec{a} + \vec{b}t, \alpha \le t \le \beta$, be one parametrization, $(x, y, z) = \vec{c} + \vec{d}t, \gamma \le t \le \delta$, be another. Both parametrizations start and end at the same points:

$$\vec{r} = \vec{a} + \vec{b}\alpha = \vec{c} + \vec{d}\gamma, \quad \vec{s} = \vec{a} + \vec{b}\beta = \vec{c} + \vec{d}\delta.$$

Show that the respective pullbacks of $\int_{\vec{r}}^{\vec{s}} k_1\, dx + k_2\, dy + k_3\, dz$ are $\int_\alpha^\beta \vec{k} \cdot \vec{b}\, dt$ and $\int_\gamma^\delta \vec{k} \cdot \vec{d}\, dt$ and show that these integrals are equal.

6. Clue: the path is from $(0, 1, 0)$ to $(-2, 0, 3)$.

7. (a) 6; (b) -6; (c) -5; (d) $9/2$; (e) $-7/2$.

9. Clue: show that the force is directed toward the origin and has magnitude GMr^{-2}.

13. (a) $c\pi/2$; (b) $-3c\pi/4$; (c) $-3c\pi/4$.

14. (a) 10; (b) -10; (c) 1; (d) 9; (e) 0.

15. (a) 9; (b) -9; (c) -5; (d) $39/2$; (e) 4.

16. Net rate $= 0$, negative from $(2, -2)$ to $(2, 1)$, positive from $(2, 1)$ to $(2, 4)$.

17. Net rate $= 28$, negative from $(4, -2)$ to $(13/4, 1)$, positive from $(13/4, 1)$ to $(2, 6)$.

Section 4.5

1. (a) 6; (b) -9; (c) 9; (d) 7.5.

2. (a) $x = 2u - 2v, y = 1 - v, z = -3 + 8u + 9v$; (c) $x = 2 - 4u - v, y = 7 - 4u - 3v, z = -1 + u + 2v$; (e) $x = -1 + u + 3v, y = 2 + u + 3v, z = 3 + u + 3v$.

3. (a) $-82du\,dv$; (c) $-5du\,dv$; (e) 0.

4. (a) -41; (c) $-5/2$; (e) 0.

5. (a) 28; (c) 0; (e) 1; (g) 0.

8. $dz = \frac{2}{3}dx + \frac{13}{3}dy$, $6\,dx\,dy$.

9. Area$(S) = 3/2$, $\int_S 6\,dx\,dy = 9$.

10. -24.

Section 4.8

2. (a) $x = -2 + 3u + 5v + 6w$, $y = u + 5v - 2w$, $z = 3 - 5u - 3v - 2w$;
(c) $x = 5 - 2u - 4v + 2w$, $y = -1 + 2u + v + w$, $z = 2 - 2u - 2w$.

3. (a) $144\,du\,dv\,dw$; (c) 0.

4. (a) 24; (c) 0.

5. Clue: the mapping from u, v, w space to x, y, z space is

$$(x, y, z) = \vec{a} + (\vec{b} - \vec{a})u + (\vec{c} - \vec{a})v + (r(\vec{b} - \vec{a}) + s(\vec{c} - \vec{a}))w.$$

6. (a) $-23dx_1dx_2 + 15dx_1dx_3 - 8dx_1dx_4 - 3dx_2dx_3 + 20dx_2dx_4 - 12dx_3dx_4$;
(c) $-30dx_1dx_2dx_3 + 13dx_1dx_2dx_4 - 15dx_1dx_3dx_4 - 21dx_2dx_3dx_4$.

12. -22.

13. Clue: for a pyramid of height 1 with base of area A, the area of the cross section at height t is t^2A and the total volume is $\int_0^1 t^2A\,dt$. For a k simplex whose kth dimension has length 1 and whose base has hypervolume H, explain why the cross section at t is $t^{k-1}H$.

17. $25/6$.

18. $5/12$.

Section 5.3

1. (a) $3/4$; (c) 1; (e) $5/6$.

3. (a) $1/2$; (c) $1/2$; (e) $1/2$.

4. (a) 2; (c) 2.

5. (a) 8/3; (c) 3.

6. (a) $-\pi$; (b) π; (c) $-\pi$; (d) π; (e) $-\pi$; (f) π.

9. 8π.

Section 5.6

1. (a) 16/3; (c) 649/48; (e) 11/6.

2.

$$\int_0^1 \int_{x^2}^1 x\sqrt{1-y^2}\, dy\, dx = \int_0^1 \int_0^{\sqrt{y}} x\sqrt{1-y^2}\, dx\, dy = \frac{1}{6}.$$

5. 1/24.

6. Area $= 4\sqrt{15} - \cosh^{-1} 4 = 13.426\ldots$, mass $= 10\sqrt{15} = 38.729\ldots$, $\overline{x} = (124\sqrt{15} - \cosh^{-1} 4)/40\sqrt{15} = 3.086\ldots$, $\overline{y} = 0$.

12. The boundary line is the line connecting $(0, 1, -1)$ to $(1, 10/23, -6/23)$.

13. 1/2.

14. (a) 14/3; (c) 1/48.

15. (a) $\int_0^1 \int_0^2 \int_x^1 f\, dy\, dz\, dx$, $\int_0^2 \int_0^1 \int_x^1 f\, dy\, dx\, dz$.

16. 2048/3.

18. The integral is

$$\int_0^1 \int_0^{1-x_1} \int_0^{1-x_1-x_2} \cdots \int_0^{1-x_1-x_2-\cdots-x_{n-1}} dx_n\, dx_{n-1}\, \cdots\, dx_1.$$

The evaluation of this integral can be accomplished by induction: After the first integration we have:

$$\int_0^1 \int_0^{1-x_1} \cdots \int_0^{1-x_1-x_2-\cdots-x_{n-2}} (1-x_1-x_2-\cdots-x_{n-1})\, dx_{n-1}\, \cdots\, dx_1.$$

Assume that after j integrations we have:

$$\frac{1}{j!} \int_0^1 \int_0^{1-x_1} \cdots \int_0^{1-x_1-x_2-\cdots-x_{n-1-j}} (1-x_1-x_2-\cdots-x_{n-j})^j\, dx_{n-j}\, \cdots\, dx_1.$$

Show that this implies that after the next integration we have:

$$\frac{1}{(j+1)!} \int_0^1 \int_0^{1-x_1} \cdots$$
$$\cdots \int_0^{1-x_1-x_2-\cdots-x_{n-1-(j+1)}} (1 - x_1 - x_2 - \cdots - x_{n-(j+1)})^{j+1} \, dx_{n-(j+1)} \cdots dx_1.$$

Therefore, after $n - 1$ integrations, we have

$$\text{Volume} = \frac{1}{(n-1)!} \int_0^1 (1 - x_1)^{n-1} \, dx_1 = \frac{1}{n!}.$$

Section 6.4

2. a,c,d,e, and f.

3. (a) $\begin{pmatrix} 1 & c \\ 0 & 1 \end{pmatrix}$, det $= 1$; (c) $\begin{pmatrix} a & 0 \\ 0 & b \end{pmatrix}$, det $= ab$;

  (d) $\begin{pmatrix} \cos\phi & -\sin\phi \\ \sin\phi & \cos\phi \end{pmatrix}$, det $= 1$; (e) $\begin{pmatrix} \frac{a^2}{a^2+b^2} & \frac{ab}{a^2+b^2} \\ \frac{ab}{a^2+b^2} & \frac{b^2}{a^2+b^2} \end{pmatrix}$, det $= 0$;

  (f) $\begin{pmatrix} \frac{a^2-b^2}{a^2+b^2} & \frac{2ab}{a^2+b^2} \\ \frac{2ab}{a^2+b^2} & \frac{b^2-a^2}{a^2+b^2} \end{pmatrix}$, det $= -1$.

4. Clue: first deal with the case $a_{11} = a_{12} = 0$. Now, assume that a_{11} or a_{12} is not 0. If $a_{11} = 0$, then $a_{12} \neq 0$. A 90° rotation takes you to a transformation where $a'_{11} \neq 0$. Show that this transformation can be obtained by a blowup followed by a shear from a transformation where $a''_{11} = 1$ and $a''_{12} = 0$. Now, work on the second coordinates.

5. $\mathbf{L} = \begin{pmatrix} 1 & 0 & 2 \\ 0 & 3 & -1 \\ -2 & 1 & 1 \end{pmatrix}$, $\mathbf{M} = \begin{pmatrix} 0 & 5 & 1 \\ -1 & 1 & 3 \\ 3 & -1 & 1 \end{pmatrix}$,

  $\mathbf{LM} = \begin{pmatrix} 6 & 3 & 3 \\ -6 & 4 & 8 \\ 2 & -10 & 2 \end{pmatrix}$, $\mathbf{ML} = \begin{pmatrix} -2 & 16 & -4 \\ -7 & 6 & 0 \\ 1 & -2 & 8 \end{pmatrix}$.

8. (a) 13; (b) 0; (c) 1; (d) 38; (e) 0; (f) 1.

9. (a) -30; (b) 0; (c) 63.

Section 6.6

1. Clue: prove that $\mathbf{BA}(\mathbf{BA} - \mathbf{I}) = \mathbf{0}$ and then use this to prove that $\mathbf{BA} = \mathbf{I}$.

2. δ^{n-1}.

4. (a) $\begin{pmatrix} -2 & -3 \\ -1 & 2 \end{pmatrix}$; (c) $\begin{pmatrix} -1 & 2 & -1 \\ 2 & -4 & 2 \\ -1 & 2 & -1 \end{pmatrix}$;

(e) $\begin{pmatrix} -1 & -1 & 2 & -1 \\ 2 & -1 & -1 & -1 \\ -1 & 2 & -1 & -1 \\ -1 & -1 & -1 & 2 \end{pmatrix}$.

5. (a) $\begin{pmatrix} 2/7 & 3/7 \\ 1/7 & -2/7 \end{pmatrix}$; (c) no inverse;

(e) $\begin{pmatrix} 1/3 & 1/3 & -2/3 & 1/3 \\ -2/3 & 1/3 & 1/3 & 1/3 \\ 1/3 & -2/3 & 1/3 & 1/3 \\ 1/3 & 1/3 & 1/3 & -2/3 \end{pmatrix}$.

6. (a) $x = 3$, $y = -9$, $z = -5$.

Section 7.2

8. See Figure C.1.

10. $f(0,0) = 1$.

11. Clue: compare $f(\delta, 0)$ with $f(0, \delta)$.

Section 7.5

1. (a) 0; (b) $15/\sqrt{2}$; (c) $49/5$; (d) $62 + 51\sqrt{3}/2$.

2. (a) 2; (b) 8; (c) $18/5$.

3. Clue: you *must* use the limit definition of the directional derivative.

4. Clue: if $f(x, y, z)$ is differentiable at $(0, 0, 0)$, then the previous exercise implies that the derivative at $(0, 0, 0)$ must send every vector to 0: $L_c(x, y, z) = 0$. The derivative exists if and only if

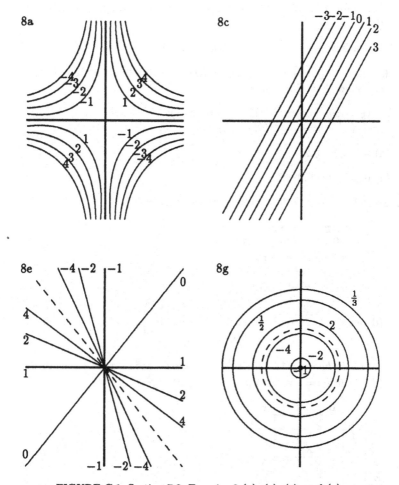

FIGURE C.1. Section 7.2, Exercise 8 (a), (c), (e), and (g).

$$f(x,y,z) - f(0,0,0) \;=\; L_c(x,y,z) \;+\; \sqrt{x^2 + y^2 + z^2}\, E(\vec{0};(x,y,z)),$$

where

$$\lim_{(x,y,z)\to(0,0,0)} E\left[\vec{0};(x,y,z)\right] = 0.$$

Solve for E and decide if it approaches 0 as (x,y,z) approaches $(0,0,0)$.

6. It *is* continuous. It is *not* differentiable at $(0,0)$.

7. Clue: let $r = \sqrt{x^2 + y^2}$, then (a) $\partial f/\partial x = 2x\sin(r^{-2}) + r^{-2}\cos(r^{-2})$; (c) Show that $E = r\sin(r^{-2})$.

14. (a) $(y\cos xy, \, x\cos xy - z\sin yz, \, -y\sin yz)$; (c) $-(x^2 + y^2 + z^2)^{-3/2}(x,y,z)$.

15. (a) 0.1; (c) -0.2.

16. (a) 6.28%; (c) 0.084%.

17. Error in volume $= 3\%$, error in surface area $= 2\%$.

21. (b) $\begin{pmatrix} \sin y & x\cos y & 0 \\ 0 & -z\sin y & \cos y \end{pmatrix}$;

(c) $\begin{pmatrix} e^{yz} & xze^{yz} & xye^{yz} \\ yze^{xz} & e^{xz} & xye^{xz} \\ yze^{xy} & xze^{xy} & e^{xy} \end{pmatrix}$;

(d) let $r = \sqrt{x^2 + y^2}$: $\begin{pmatrix} \frac{-2xy}{r^4} & \frac{x^2-y^2}{r^4} & 0 \\ \frac{x^2-y^2}{r^4} & \frac{2xy}{r^4} & 0 \\ 0 & 0 & 1 \end{pmatrix}$.

22. (c) $du\, dv\, dw = e^{xy+yz+xz}(2x^2y^2z^2 - x^2yz - xy^2z - xyz^2 + 1)dx\, dy\, dz$; (d) $du\, dv\, dw = -r^{-4}dx\, dy\, dz$, $(r = \sqrt{x^2 + y^2})$.

27.

	A	B	C	D	E
A	X				
B	X	X	X		X
C			X		
D	X	X	X	X	X
E			X		X

Section 7.8

1. (a) $F'(t) = -(2t^3 + e^{2t})(t^4 + e^{2t})^{-3/2}$, (c) $F'(t) = \tan t(1 + \cos 2t)$.

3. (a) $\partial f/\partial u = 2v^2 e^{2u} + uv e^{u+v} + v e^{u+v}$, $\partial f/\partial v = 2v e^{2u} + u e^{u+v} + uv e^{u+v}$;
(c) $\partial f/\partial u = 1, \partial f/\partial v = 1$.

5. (a) $\partial^2 f/\partial u^2 = 4v^2 e^{2u} + uv e^{u+v} + 2v e^{u+v}$; (c) $\partial^2 f/\partial u^2 = 0$.

6. $f'(|\vec{x}|)\, x_j/|\vec{x}|$.

9. $\frac{\partial f}{\partial r} = \cos\theta \frac{\partial f}{\partial x} + \sin\theta \frac{\partial f}{\partial y}$, $\frac{\partial f}{\partial \theta} = -r\sin\theta \frac{\partial f}{\partial x} + r\cos\theta \frac{\partial f}{\partial y}$, $\frac{\partial^2 f}{\partial r^2} = \cos^2\theta \frac{\partial^2 f}{\partial x^2} + \sin 2\theta \frac{\partial^2 f}{\partial x \partial y} + \sin^2\theta \frac{\partial^2 f}{\partial y^2}$, $\frac{\partial^2 f}{\partial r \partial \theta} = \frac{r}{2}\sin 2\theta \left(\frac{\partial^2 f}{\partial y^2} - \frac{\partial^2 f}{\partial x^2} \right) + r\cos 2\theta \frac{\partial^2 f}{\partial x \partial y} - \sin\theta \frac{\partial f}{\partial x} + \cos\theta \frac{\partial f}{\partial y}$, $\frac{\partial^2 f}{\partial \theta^2} = r^2 \sin^2\theta \frac{\partial^2 f}{\partial x^2} + r^2 \cos^2\theta \frac{\partial^2 f}{\partial y^2} - r^2 \sin 2\theta \frac{\partial^2 f}{\partial x \partial y} - r\cos\theta \frac{\partial f}{\partial x} - r\sin\theta \frac{\partial f}{\partial y}$.

12. (a) $(2,1)$; (b) $(-1,2)$; (c) $(-3,-9,-4)$; (d) gradient is $\vec{0}$, all directional derivatives are 0.

17. (a) $6x - 2y - z = 12$; (b) $2y + z = 2$.

18. Clue: show that the equation of the tangent plane at (x_0, y_0, z_0) is
$$x_0 x + y_0 y - z_0 z = 0.$$

20. Tangent plane: $\sqrt{2}x - y = 1$, osculating plane: $y = 1$.

Section 8.3

1. 35.

3. 7/24.

8. ρ.

9. $\rho^2 \cos\phi$.

12. Clue: use the fact that for $n \geq 2$

$$\int_{-\pi/2}^{\pi/2} \cos^n \phi\, d\phi = \frac{n}{n-1} \int_{-\pi/2}^{\pi/2} \cos^{n-2} \phi\, d\phi.$$

Section 8.6

1. (a) $d\theta\, d\phi$; (b) $dr\, d\theta$; (c) $dr\, d\theta$; (d) $d\phi\, d\theta$; (e) $d\theta\, dr$; (f) $d\theta\, dr$.

2. (a) $7\pi/6$; (c) $5\pi/2$; (d) $-7\pi/6$; (f) $-5\pi/2$.

3. (a) $-\pi$; (c) $-\pi$; (d) π; (f) π.

5. Clue: use the parametrization: $x = 2\cos\theta$, $y = 2\sin\theta$, $z = z$.

6. Clue: show that $d\sigma = 2\,d\theta\,dz$.

9. $(2/1215)(484\sqrt{22} - 338\sqrt{13} + 32) = 1.7835\ldots$.

10. 8.

Section 9.3

4. If y is held constant then $\partial x/\partial u = \log 3$, if v is held constant, then $\partial x/\partial u = 2\log 3 - 3$.

6. Clue: show that $\partial y/\partial x = (z-x)/(y-z)$, $\partial y/\partial z = (x+y-2z)/(y-z)$. Find $\frac{\partial}{\partial x}\left[(x+y-2z)/(y-z)\right]$, treating y (but not z) as a function of x.

7. Clue: $\frac{\partial}{\partial x}\left(\frac{\partial f}{\partial x}\right) = \frac{\partial^2 f}{\partial x^2} + \frac{\partial^2 f}{\partial z\partial x}\frac{\partial z}{\partial x}$, $\frac{\partial}{\partial x}\left(\frac{\partial f}{\partial z}\right) = \frac{\partial^2 f}{\partial x\partial z} + \frac{\partial^2 f}{\partial z^2}\frac{\partial z}{\partial x}$.

17. Clue: solve for x and y in terms of u and v.

Section 9.6

1. & 2. (a) Saddle point at $(-3/2, 1/2)$; (c) local maximum at $(-3/10, 3/10)$; (e) local minimum at $(-1/2, 3/2)$; (g) saddle points at $(n\pi, (m + \frac{1}{2})\pi)$, local maxima at $((n + \frac{1}{2})\pi, m\pi)$ when n amd m are both even or both odd, local minima at $((n + \frac{1}{2})\pi, m\pi)$ when n and m have opposite parity; (i) saddle point at $(0,0)$, local maximum at each point on the curve $x^2 - y^2 = 1$; (k) saddle points everywhere on the y axis $(x = z = 0)$ and on the z axis $(x = y = 0)$.

6. $1 + x + y + \frac{1}{2}x^2 + xy + \frac{1}{2}y^2$.

12. Clue: let $\vec{c}_1, \ldots, \vec{c}_n$ be the column vectors of $\mathbf{A} - \mathbf{B}$. Explain why we must have constants: $\alpha_1, \alpha_2, \ldots, \alpha_n$, not all zero, such that $\alpha_1\vec{c}_1 + \alpha_2\vec{c}_2 + \ldots + \alpha_n\vec{c}_n = \vec{0}$. Show that $\vec{A} - \vec{B}$ sends $(\alpha_1, \ldots, \alpha_n)$ to $\vec{0}$.

16. 0, $(5 + \sqrt{13})/2$, $(5 - \sqrt{13})/2$.

17. Saddle point at $(-8/3, 7/3, 1/4)$.

Section 9.8

1. (a) Extrema at $\left(\pm\sqrt{1/6},\pm\sqrt{5/6}\right)$ and at $(\pm1,0)$, abs. max. at $(1,0)$, abs. min. at $(-1,0)$; (c) abs. max. at $(7/\sqrt{29},5/\sqrt{29})$, abs. min. at $(-7/\sqrt{29},-5/\sqrt{29})$; (e) abs. max. at $(1/2,-1/2,-4)$ and at $(-1/2,1/2,-4)$, abs. min. at $\left(\sqrt{(3+\sqrt5)/2},\sqrt{(3-\sqrt5)/2},1\right)$ and at $\left(\sqrt{(3-\sqrt5)/2},\sqrt{(3+\sqrt5)/2},1\right)$.

2. $\left(-\sqrt{1/3},\sqrt{1/3}\right),\left(\sqrt{1/3},-\sqrt{1/3}\right)$.

6. $\frac13\sqrt{42}$.

7. Clue: maximize xyz.

8. Clue: the equation $\lambda^4-\lambda^2-7\lambda+2=0$ has real roots at $\lambda=2$ and $\lambda=\frac13\left\{[\,(\sqrt{4725}+65)/2\,)^{1/3}-(\,(\sqrt{4725}-65)/2\,]^{1/3}-2\right\}=0.27568\ldots$.

9. $\frac23\sqrt6$.

12. 1.

Section 10.3

7. Clue: let C_1 and C_2 be any two curves encircling the origin. If they do not cross, then insert two connecting curves, C_3 and C_4 (see Figure C.2). Let C_1^+ and C_2^+ denote the top halves of C_1 and C_2, respectively, C_1^-, C_2^-, the bottom halves. Let Γ_1 be the closed curve consisting of C_2^+ followed by C_3 followed by C_1^+ in the negative direction followed by C_4, and let Γ_2 be the closed curve consisting of C_2^- followed by C_4 in the negative direction followed by C_1^- in the negative direction followed by C_3 in the negative direction. Show that

$$\oint_{C_2}\omega-\oint_{C_1}\omega=\oint_{\Gamma_1}\omega+\oint_{\Gamma_2}\omega=0+0.$$

11. & 12. (a) Closed and exact, $=d(x^2y+xy^2)$; (b) not closed; (e) closed and exact, $=d(xy+xz+yz)$; (f) closed and exact, $=d(xy^2+yz^2)$; (h) closed and exact, $=d\left(\frac{-1}{2}(x^2+y^2+z^2)^{-1}\right)$; (i) closed and exact, $=d\left(\frac12\log(x^2+y^2+z^2)\right)$.

14. Clue: there are three equations. The first is $\partial f/\partial y=\partial g/\partial x$.

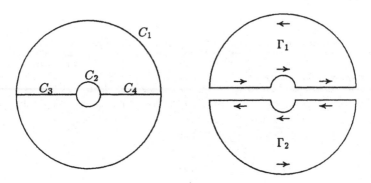

FIGURE C.2. Section 10.3, Exercise 7.

15. Clue: there are six equations.

16. 7.

19. $x^2 + yzt - zt + xy$.

Section 10.5

1. -4.

3. -8π.

5. $-4/3$.

7. 72.

8. Clue: let C be parametrized by $x(t)$, $y(t)$, $a \le t \le b$. Pull both integrals back to integrals in t space.

11. 5/2.

13. Clue: use the parametrization $x = 6r \cos \theta \cos \phi$, $y = 3r \sin \theta \cos \phi$, $z = 2r \sin \phi$, $0 \le r \le 1$, $0 \le \theta \le 2\pi$, $-\pi/2 \le \phi \le \pi/2$.

14. Clue: use the parametrization $x = (4 + \rho \cos \phi) \cos \theta$, $y = (4 + \rho \cos \phi) \sin \theta$, $z = \rho \sin \phi$, $0 \le \rho \le 1$, $0 \le \theta \le 2\pi$, $0 \le \phi \le 2\pi$.

18. (a) $\pi/24$.

Section 10.8

1. 3π, take $(0,0,1)$ as the positive direction.

3. 60, take $(-1,-2,-3)$ as the positive direction.

5. $\frac{\partial}{\partial u}\left(f\frac{\partial x}{\partial v} + g\frac{\partial y}{\partial v} + h\frac{\partial z}{\partial v}\right) - \frac{\partial}{\partial v}\left(f\frac{\partial x}{\partial u} + g\frac{\partial y}{\partial u} + h\frac{\partial z}{\partial u}\right)$

$\quad = \left(\frac{\partial h}{\partial y} - \frac{\partial g}{\partial z}\right)\frac{\partial(y,z)}{\partial(u,v)} + \left(\frac{\partial f}{\partial z} - \frac{\partial h}{\partial x}\right)\frac{\partial(z,x)}{\partial(u,v)} + \left(\frac{\partial g}{\partial x} - \frac{\partial f}{\partial y}\right)\frac{\partial(x,y)}{\partial(u,v)}.$

7. 2π.

12. -4π.

Section 11.4

6. $-(2/5)\pi m\gamma(5 - 2\sqrt{5}) = -(m)/(2\epsilon)(5 - 2\sqrt{5}).$

13. $\vec{B} = (0, -2xc^{-1}\cos(z - ct), 0)$, $\vec{E} = (-2x\cos(z - ct), 0, 0)$,
$\quad \rho = -2\epsilon\cos(z - ct)$, $\vec{J} = (0, 0, -2c\epsilon\cos(z - ct))$.

Section 11.6

2. Clue: use the fact that a linear transformation from $\mathbf{R}^3$ to $\mathbf{R}^3$ is a rotation if and only if it has a positive determinant and its inverse equals its transpose.

6. The charge density in the moving system is $m\beta^{-1}$,
$\beta = \left(1 - (v^2/c^2)\right)^{-1/2}$. The magnetic and electric fields in the stationary system are
$\mathbf{B} = m\gamma v c^{-2}\left(\beta^2(x - vt)^2 + y^2 + z^2\right)^{-3/2}(-z\,dz\,dx + y\,dx\,dy),$
$\mathbf{E} = m\gamma\left(\beta^2(x - vt)^2 + y^2 + z^2\right)^{-3/2}\left((x - vt)\,dx + y\,dy + z\,dz\right).$

Index

absolute maximum, 239, 262
absolute minimum, 239, 262
Alembert, Jean Le Rond d'
    (1717–1783), 347
Alfonso X (1221–1284), 1
algebraic topology, 291
Ampère, André Marie (1775–
    1836), 335
Ampère's law, 336
antiderivative, 279
antidifferential, 292
apogee, 69
Apostol, Tom, vii, 365
Appolonius of Perga (262–
    170 B.C.), 9
Arago, Dominique François Jean
    (1786–1853), 335
arc length, 58
    integration with respect to,
    115
Archimedes of Syracuse (287–212
    B.C.), 111
area, 131
Arfken, George, 350

Barrow, Isaac (1630–1677), 1, 279
Bell, Eric Temple, 366
Beltrami, Eugenio (1832–1900),
    78
Bernoulli, Daniel (1700–1782),
    333
Bernoulli, Jacques (1654–1705),
    255
Bernoulli, Jean (1667–1748), 255,
    333
Biot, Jean Baptiste (1774–1862),
    335
Biot–Savart law, 335
Bleaney, B., 365
Bleaney, B. I., 365

blowup, 159
Bourbaki, Nicolas, 366
Boyer, Carl B., 366
Brahe, Tycho (1546–1601), 4

Cajori, Florian, 366
Cartan, Élie (1869–1951), 78, 363
Cauchy, Augustin-Louis (1789–
    1857), 121, 158
Cavendish, Henry (1731–1810),
    333
Cayley, Arthur (1821–1895), 157
center of mass, 117, 134
chain rule, 201
chaos theory, 74
characteristic value, 261
characteristic vector, 261
charge density, 334
Christoffel, Elwin Bruno (1829–
    1900), 78
closed curve, 281, 288
closed form, 288
cofactor, 165
conjugate diameters, 10
continuity, 177, 184
continuously differentiable, 238,
    283
Copernicus, Nicolaus (1473–
    1543), 4
Cotes, Roger (1682–1716), 1
Coulomb, Charles Augustin
    (1736–1806), 333
Coulomb's law, 333
Cramer, Gabriel (1704–1752),
    157
Cramer's rule, 157, 167, 236
cross product, 39, 105, 321
curl, 317, 321
curvature, 62

d'Alembertian, 347

Davy, Humphry (1778–1829), 335
del, 188
derivative, 53, 141, 192
determinant, 147, 150, 154, 157
   three-dimensional, 148
diameter, 10
dielectric constant, 334
differentiability, 141, 194, 280
differential, 79
differential form, 29, 77
   constant 1-form, 79, 80
   constant 2-form, 89
   constant k-form, 101
   rules for multiplication, 95, 157
   0-form, 103
Dijksterhuis, E. J., 366, 367
directional derivative, 181, 188
Dirichlet, Peter Gustav Lejeune-(1805–1859), 327
Dirichlet problem, 327
discriminant, 157
divergence, 306, 322
divergence theorem, 284
dot product, 34, 105, 321
Du Bois-Reymond, Paul (1831–1889), 121
Dyson, Freeman, viii, 361

eccentricity, 70
ecliptic, 3
Edwards, Harold, vii, 107, 112, 113, 239, 285, 365
eigenvalue, 261
eigenvector, 261
Einstein, Albert (1879–1955), vii, 78, 79, 228, 333, 349, 356, 358, 362, 363, 366
electric permittivity, 334
electromagnetic field, 338
electromagnetic potential, 345, 346, 347
electromagnetic waves, 349
electrostatic potential, 334
ellipse, 10

epicycle, 3
equant, 3
equation of continuity, 343
error estimation, 190
escape velocity, 70
Euler, Leonhard (1707–1783), 285, 348, 361, 362
exact form, 286
exterior algebra, 96
exterior product, 95

Faraday, Michael (1791–1867), 336, 338
Faraday's law, 338
Fermat, Pierre (1601–1665), 279
Finney, Ross L., 79, 365
Fitzgerald, George Francis (1851–1901), 354
Flanders, Harley, 365
flow, 227
fluid flow, 84, 91
flux, 329
foci, 12
Foucault, Jean L. (1819–1868), 349
Fourier, Joseph (1768–1830), 111, 326
Fourier series, 111, 326
French, A. P., 365
fundamental theorem of calculus, 283, 322
   for scalar function of one variable, 279
   for 1-forms, 280

Galilei group, 363
Gauss, Carl Friedrich (1777–1855), 78, 157, 162, 281, 329, 361
Gaussian elimination, 162
Gauss's law, 334
Gauss's theorem, 284, 303, 306, 323
geosynchronous orbit, 72

Gibbs, Josiah Willard (1839–
        1903), 29, 361, 365
gradient, 188, 205, 321
Grassman, Hermann Günther
        (1809–1877), 29
gravitational potential, 327
Green, George (1793–1841), 281,
        284, 327
Green's theorem, 283, 297
Gregory, James (1638–1675), 255,
        279

Hadamard, Jacques (1865–1963),
        363
Hamilton, William (1805–1865),
        29, 79, 188, 361
Harish-Chandra (1923–1983),
        363
harmonic function, 326
heat flow, 111, 228
Heaviside, Oliver (1850–1925),
        29, 46, 366
Helmholtz, Hermann Ludwig von
        (1821–1894), 364
Heron's formula, 52
Hertz, Heinrich Rudolf (1857–
        1894), 349, 364
Hesse, Ludwig Otto (1811–1874),
        194, 254
Hessian, 194, 254
Hilbert, David (1862–1943), 327
Hipparchus (d. circa 125 B.C.), 3
hypervolume, 134

implicit differentiation, 233
incompressibility, 77, 306
integrability, 112
inversion number, 109
invertibility, 239
invertible transformation, 147
irrotational, 317
isotherm, 178

Jacobi, Carl Gustav Jacob (1804–
        1851), 158, 361

Jacobian, 158, 191, 204
Jacobian matrix, 158, 191, 203

Kaplan, Wilfred, 365
Kelvin, William Thomson (1824–
        1907), 281, 284, 285, 327
Kepler, Johann (1571–1630), 2, 4,
        5, 18, 366
Kepler's laws, 4, 18, 67, 72
Kline, Morris, 157, 366

Lagrange, Joseph-Louis (1736–
        1813), 157, 158, 213,
        254, 256, 326, 361, 366
Lagrange multiplier, 270
Laplace, Pierre-Simon (1749–
        1827), 285, 326, 329
Laplace's equation, 285
Laplacian, 311, 322, 326
Lamé, Gabriel (1795–1870), 78
law of gravity, 67
Lebesgue, Henri (1875–1941), 112
Leibniz, Gottfried Wilhelm
        (1646–1716), 1, 78, 157,
        255
Leray, Jean, 363
level curve, 174
level surface, 178
Levi-Civita, Tullio (1873–1941),
        78
Lie, Sophus (1842–1899), 361
limit, 171
line, 32
linear combination, 33
linear transformation, 139
Lipschitz, Rudolf (1832–1903), 78
local coordinates, 21
local extremum, 249
local maximum, 248, 262
local minimum, 248, 262
loop integral, 293
Lorentz, Hendrick Antoon (1853–
        1928), 354, 357, 366
Lorentz group, 363
Lorentz transformation, 356

Lorenz, Ludwig (1829–1891), 362

MacLaurin, Colin (1698–1746), 157, 255
MacLaurin's series, 255
magnetic permeability, 336
main diagonal, 162
manifold, 283
    differentiable, 283
Marconi, Guglielmo (1874–1937), 349
mass, 132, 133
matrix inverse, 145, 166
matrix minor, 165
matrix of cofactors, 166
matrix product, 144
Maxwell, James Clerk (1831–1879), 29, 333, 346, 349, 361, 362, 363, 364, 366
Maxwell's equations, 228, 285, 332, 344, 354, 356
    statement of, 344
mean distance, 69
mean value theorem, 186
mesh, 112
Meyer, Herbert W., 366
Michelson, Albert A. (1852–1931), 352, 353
Minkowski, Hermann (1864–1909), 363, 364, 366
Möbius strip, 281
monkey saddle, 252
Morley, Edward Williams (1838–1923), 353
Morrey, Charles B., Jr., 365
multiple integral, 120
mutually orthogonal, 43

nabla, 188
Napoléon I (1769–1821), 326
negative orientation, 89, 99
Neumann, Carl G. (1832–1925), 327
Neumann, Franz Ernst (1798–1895), 362

Newton, Isaac (1642–1727), vii, 1, 5, 7, 9, 12, 14, 18, 120, 255, 279, 331, 333, 349, 352, 358, 361, 362, 364, 366
Newton's laws of motion, 5
norm, 112

Oersted, Hans Christian (1777–1851), 334
oriented tetrahedron, 99
oriented triangle, 89
osculating plane, 59, 210
Ostrogradski, Michel (1801–1861), 281, 284

parallel, 31
parallelogram rule, 6
partial derivative, 183
    mixed, 185, 200
partition, 112
perigee, 69
plane, 33
Poincaré, Henri (1854–1912), 78, 290, 327, 362, 366
Poincaré's lemma, 290, 301, 323
Poisson, Siméon-Denis (1781–1840), 329
Poisson's equation, 331, 334
Popov, Alexander S. (1859–1906), 349
positive orientation, 89, 99
potential field, 294, 318
principal latus rectum, 15
principal normal, 58
projection, 159
Protter, Murray H., 365
Ptolemy (d. 168 A.D.), 3
pullback, 82, 93
Pupin, Michael (1858–1935), 364

Q.E.D., 8
quaternions, 29, 188

radius of curvature, 62
reflection, 159

Ricci-Curbastro, Gregorio (1853–1925), 78
Riemann, Georg Friedrich Bernhard (1826–1866), 78, 111, 362
Riemann integrability, 112
Riemann integral, 112
Riemann sum, 112
Riemannian geometry, 78
right-hand rule, 41
rotation, 159

saddle point, 252, 262
Savart, Félix (1791–1841), 335
scalar field, 77, 321
scalar function, 20
scalar product, 34
scalar triple product, 47
semimajor axis, 10
semiminor axis, 10
shear, 77, 159
Simmonds, James G., 365
simplex, 101
simply connected, 289
simply loop connected, 318
Smith, Henry (1826–1883), 362, 364
solenoidal, 306
span, 33, 34
speed, 58
speed of light, 349
standard k simplex, 101
stationary point, 248
Stirling, James (1692–1770), 256
Stokes, George Gabriel (1819–1903), 285
Stokes' theorem, 285, 317, 322
surface area, 224

tangent, 58
tangent plane, 206, 210
Taylor, Brook (1685–1731), 255
Taylor formula, 256
Taylor series, 255
tensor analysis, 77, 79

Thomae, Karl J. (1840–1921), 121
Thomas, George B., Jr., 79, 365
Thomson, William, see Kelvin, William Thomson
total derivative, 192
translation, 159
transpose, 161
triangle inequality, 37
twice continuously differentiable, 283

unit tangent, 58
unit vector, 21
Urwin, Kathleen M., 365

vector, 6
vector algebra, 29
vector analysis, 77
vector field, 77, 321
vector function, 20
vector product, 39
volume, 133, 149, 158, 212

Weber, Wilhelm Edward (1804–1891), 362
wedge product, 95, 126
Weyl, Hermann (1885–1955), 363, 366
Wilson, Edwin B. (1879–1964), 29, 365

Undergraduate Texts in Mathematics

Anglin: Mathematics: A Concise History and Philosophy.
Readings in Mathematics.
Anglin/Lambek: The Heritage of Thales.
Readings in Mathematics.
Apostol: Introduction to Analytic Number Theory. Second edition.
Armstrong: Basic Topology.
Armstrong: Groups and Symmetry.
Axler: Linear Algebra Done Right.
Bak/Newman: Complex Analysis.
Banchoff/Wermer: Linear Algebra Through Geometry. Second edition.
Berberian: A First Course in Real Analysis.
Brémaud: An Introduction to Probabilistic Modeling.
Bressoud: Factorization and Primality Testing.
Bressoud: Second Year Calculus.
Readings in Mathematics.
Brickman: Mathematical Introduction to Linear Programming and Game Theory.
Browder: Mathematical Analysis: An Introduction.
Cederberg: A Course in Modern Geometries.
Childs: A Concrete Introduction to Higher Algebra. Second edition.
Chung: Elementary Probability Theory with Stochastic Processes. Third edition.
Cox/Little/O'Shea: Ideals, Varieties, and Algorithms.
Croom: Basic Concepts of Algebraic Topology.
Curtis: Linear Algebra: An Introductory Approach. Fourth edition.
Devlin: The Joy of Sets: Fundamentals of Contemporary Set Theory. Second edition.
Dixmier: General Topology.
Driver: Why Math?
Ebbinghaus/Flum/Thomas: Mathematical Logic. Second edition.
Edgar: Measure, Topology, and Fractal Geometry.
Elaydi: Introduction to Difference Equations.
Fischer: Intermediate Real Analysis.
Flanigan/Kazdan: Calculus Two: Linear and Nonlinear Functions. Second edition.
Fleming: Functions of Several Variables. Second edition.
Foulds: Combinatorial Optimization for Undergraduates.
Foulds: Optimization Techniques: An Introduction.
Franklin: Methods of Mathematical Economics.
Hairer/Wanner: Analysis by Its History.
Readings in Mathematics.
Halmos: Finite-Dimensional Vector Spaces. Second edition.
Halmos: Naive Set Theory.
Hämmerlin/Hoffmann: Numerical Mathematics.
Readings in Mathematics.
Iooss/Joseph: Elementary Stability and Bifurcation Theory. Second edition.
Isaac: The Pleasures of Probability.
Readings in Mathematics.
James: Topological and Uniform Spaces.
Jänich: Linear Algebra.
Jänich: Topology.

Undergraduate Texts in Mathematics

(continued)

Kemeny/Snell: Finite Markov Chains.
Kinsey: Topology of Surfaces.
Klambauer: Aspects of Calculus.
Lang: A First Course in Calculus. Fifth edition.
Lang: Calculus of Several Variables. Third edition.
Lang: Introduction to Linear Algebra. Second edition.
Lang: Linear Algebra. Third edition.
Lang: Undergraduate Algebra. Second edition.
Lang: Undergraduate Analysis.
Lax/Burstein/Lax: Calculus with Applications and Computing. Volume 1.
LeCuyer: College Mathematics with APL.
Lidl/Pilz: Applied Abstract Algebra.
Macki-Strauss: Introduction to Optimal Control Theory.
Malitz: Introduction to Mathematical Logic.
Marsden/Weinstein: Calculus I, II, III. Second edition.
Martin: The Foundations of Geometry and the Non-Euclidean Plane.
Martin: Transformation Geometry: An Introduction to Symmetry.
Millman/Parker: Geometry: A Metric Approach with Models. Second edition.
Moschovakis: Notes on Set Theory.
Owen: A First Course in the Mathematical Foundations of Thermodynamics.
Palka: An Introduction to Complex Function Theory.
Pedrick: A First Course in Analysis.
Peressini/Sullivan/Uhl: The Mathematics of Nonlinear Programming.
Prenowitz/Jantosciak: Join Geometries.
Priestley: Calculus: An Historical Approach.
Protter/Morrey: A First Course in Real Analysis. Second edition.
Protter/Morrey: Intermediate Calculus. Second edition.
Ross: Elementary Analysis: The Theory of Calculus.
Samuel: Projective Geometry.
Readings in Mathematics.
Scharlau/Opolka: From Fermat to Minkowski.
Sigler: Algebra.
Silverman/Tate: Rational Points on Elliptic Curves.
Simmonds: A Brief on Tensor Analysis. Second edition.
Singer/Thorpe: Lecture Notes on Elementary Topology and Geometry.
Smith: Linear Algebra. Second edition.
Smith: Primer of Modern Analysis. Second edition.
Stanton/White: Constructive Combinatorics.
Stillwell: Elements of Algebra: Geometry, Numbers, Equations.
Stillwell: Mathematics and Its History.
Strayer: Linear Programming and Its Applications.
Thorpe: Elementary Topics in Differential Geometry.
Troutman: Variational Calculus and Optimal Control. Second edition.
Valenza: Linear Algebra: An Introduction to Abstract Mathematics.
Whyburn/Duda: Dynamic Topology.
Wilson: Much Ado About Calculus.